# Polymer Blends

*VOLUME 2*

# Contributors

T. Alfrey, Jr.
J. C. Andries
K. E. Atkins
K. C. Baranwal
C. B. Bucknall
C. F. Hammer
J. V. Koleske
Gerard Kraus
E. N. Kresge
E. T. McDonel
Seymour Newman
D. R. Paul
A. P. Plochocki
W. J. Schrenk
L. H. Sperling
D. A. Thomas

# POLYMER BLENDS

*Edited by*
## D. R. PAUL
*The University of Texas, Department of Chemical Engineering*
*Austin, Texas*

## SEYMOUR NEWMAN
*Ford Motor Company, Plastics, Paint, and Vinyl Division*
*Detroit, Michigan*

## VOLUME 2

ACADEMIC PRESS, INC.
**Harcourt Brace Jovanovich, Publishers**
San Diego   New York   Berkeley   Boston
London   Sydney   Tokyo   Toronto

ACADEMIC PRESS, INC.
1250 Sixth Avenue, San Diego, California 92101


*United Kingdom Edition published by*
ACADEMIC PRESS, INC. (LONDON) LTD.
24/28 Oval Road, London NW1 7DX


Library of Congress Cataloging in Publication Data

Main entry under title:

Polymer blends.

    Bibliography: p.
    1. Polymers and polymerization. I. Paul,
Donald R. II. Newman, Seymour, Date
QD381.P613      668        77-19020
ISBN 0-12-546802-4 (v. 2)


PRINTED IN THE UNITED STATES OF AMERICA

88 89 90 91 92          10 9 8 7 6 5 4 3

# Contents

## Chapter 11    Interpenetrating Polymer Networks

### D. A. Thomas and L. H. Sperling

## Chapter 12    Interfacial Agents ("Compatibilizers") for Polymer Blends

### D. R. Paul

## Chapter 13    Rubber Modification of Plastics

### Seymour Newman

## Chapter 14  Fracture Phenomena in Polymer Blends

### *C. B. Bucknall*

## Chapter 15  Coextruded Multilayer Polymer Films and Sheets

### *W. J. Schrenk and T. Alfrey, Jr.*

## Chapter 16  Fibers from Polymer Blends

### *D. R. Paul*

## Chapter 17  Polymeric Plasticizers

### *C. F. Hammer*

Contents                vii

## Chapter 22    Blends Containing Poly(ε-caprolactone) and Related Polymers

*J. V. Koleske*

## Chapter 23    Low-Profile Behavior

*K. E. Atkins*

## Appendix    Conversion Factors to SI Units                          415

# List of Contributors

Numbers in parentheses indicate the pages on which the authors' contributions begin.

T. Alfrey, Jr. (129), The Dow Chemical Company, Midland, Michigan 48640

J. C. Andries (263), The B. F. Goodrich Company, Brecksville, Ohio 44141

K. E. Atkins (391), Chemicals and Plastics, Research and Development Department, Union Carbide Corporation, South Charleston, West Virginia

K. C. Baranwal (263), The B. F. Goodrich Company, Brecksville, Ohio 44141

C. B. Bucknall (91), Cranfield Institute of Technology, Cranfield, Bedford MK43 OAL, England

C. F. Hammer (219), Plastic Products and Resins Department, E. I. du Pont de Nemours & Company, Wilmington, Delaware 19898

J. V. Koleske (369), Chemicals and Plastics, Research and Development Department, Union Carbide Corporation, South Charleston, West Virginia

Gerard Kraus (243), Research and Development Department, Phillips Petroleum Company, Bartlesville, Oklahoma 74004

E. N. Kresge (293), Elastomers Technology Division, Exxon Chemical Company, Linden, New Jersey 07036

E. T. McDonel (263), The B. F. Goodrich Company, Brecksville, Ohio 44141

Seymour Newman (63), Plastics, Paint, and Vinyl Division, Ford Motor Company, Detroit, Michigan 48239

D. R. Paul (35, 167), Department of Chemical Engineering, The University of Texas, Austin, Texas 78712

*A. P. Plochocki* (319), Industrial Chemistry Institute, Plastics Division, Warsaw 86, Poland

*W. J. Schrenk* (129), The Dow Chemical Company, Midland, Michigan 48640

*L. H. Sperling* (1), Materials Research Center, Coxe Laboratory, Lehigh University, Bethlehem, Pennsylvania 18015

*D. A. Thomas* (1), Materials Research Center, Coxe Laboratory, Lehigh University, Bethlehem, Pennsylvania 18015

# Preface

At different times in the history of polymer science, specific subjects have come to "center stage" for intense investigation because they represented new and important intellectual challenges as well as technological opportunities. Dilute solution behavior, chain statistics, rubber elasticity, tacticity, single crystal formation, and viscoelastic behavior have all had their day of peak interest and then taken their place for continuing investigation by the community of polymer scientists. These periods of concentrated effort have served to carve out major new areas of macromolecular science to add to and build on the efforts of previous workers.

Polymer blends have now come to the fore as such a major endeavor. Their current and potential technological importance is remarkable and their ubiquitous presence in consumer products is testimony to their commercial importance.

Furthermore, pursuit of our understanding of the physical and mechanical properties of blends has uncovered new principles, refined earlier fundamental concepts, and revealed further opportunities for research and practical problem solving. In this last respect, polymer blends or "polyblends" offer a strong analogy to the previously established role of copolymerization as a means of combining the useful properties of different molecular species, but blends allow this to be done through physical rather than chemical means.

Our purpose, therefore, in organizing this work has been to underscore the importance of mixed polymer systems as a major new branch of macromolecular science as well as to provide academic and industrial research scientists and technologists with a broad background in current principles and practice. A wide range of subjects must be covered to meet these objectives, and no individual could be expected to have the knowledge and experience required to write an authoritative treatment for any significant fraction of these subjects. Consequently, the book was written by a group of authors—selected by the editors because of their particular expertise and contributions. However, a strong effort was made to have the outcome represent a cohesive treatment rather than a collection of separate contributions. The final outcome is the result of numerous decisions, and clearly different decisions could have been made. Considerable thought and consultation were devoted to a selection of topics that covered the main principles within the constraints of time and space.

Once the authors were selected, the final chapter outlines evolved through continuing consultation among the editors and authors. Each chapter has been reviewed and altered as necessary. In some cases, editorial footnotes have been added for clarification. There is extensive cross-referencing among chapters to emphasize the relations among chapters and to reduce duplication of content. A loose system of common nomenclature was developed, but exceptions were permitted to allow each subject to be developed in terms most commonly used in the literature and best understood by the reader. A mixture of unit systems appears here because various disciplines of science and technology are in different stages of conversion to the metric and SI systems at this time; hence, we have included in the Appendix selected conversion factors to help minimize the problems this creates.

We have elected to define a polymer blend as any combination of two or more polymers resulting from common processing steps. In keeping with this broad definition, we have elected to include, for example, the chapter on multilaminate films by Schrenk and Alfrey and the treatment of bicomponent fibers by Paul, both in Volume 2. However, since block and graft copolymers have been treated extensively in other recent publications, we have included only two chapters specifically devoted to these systems, i.e., Block Copolymers in Blends with Other Polymers by Kraus (Volume 2) and Interfacial Agents (''Compatibilizers'') for Polymer Blends by Paul (Volume 2). However, additional references to such systems are made in other chapters.

For thermodynamic reasons, most polymer pairs are immiscible. Nevertheless, the degree of compatibility may vary widely, and since this aspect of polymer blends is of such underlying importance to morphology and properties, we have set aside a considerable portion of the book for this purpose. Krause's chapter, Polymer–Polymer Compatibility (Volume 1), provides an authoritative, comprehensive summary and interpretation of available information on a great number of systems reported in the literature. A review of more recent statistical mechanical theories for mixtures with particular attention given to lower critical solution temperatures is presented by Sanchez in Volume 1, whereas an interpretation of experimental work on phase separation and boundaries using NMR and other techniques is found in the contribution by Kwei and Wang, also in Volume 1.

Compatible polymers represent such unique cases in the realm of blends that the editors have sought to single out some of these for special discussion. Aside from relevant discussions by Krause and Kwei and Wang, the reader is referred to the following chapters: Blends Containing Poly($\epsilon$-caprolactone) and Related Polymers, by Koleske (Volume 2);

Solid State Transition Behavior of Blends, by MacKnight, Karasz, and Fried (Volume 1); Polymeric Plasticizers, by Hammer (Volume 2); and Transport Phenomena in Polymer Blends, by Hopfenberg and Paul (Volume 1).

A second cornerstone along with thermodynamics in the building of the often complex structure of incompatible disperse phases is the interaction of melt flow with interfacial behavior and the viscoelastic properties of multicomponent systems. We have sought to lay the groundwork for this fundamental aspect of polyblends in Wu's chapter, Interfacial Energy, Structure, and Adhesion between Polymers (Volume 1), and Van Oene's chapter, Rheology of Polymer Blends and Dispersions (Volume 1). More technologically oriented extensions of this subject will be found in Polyolefin Blends: Rheology, Melt Mixing, and Applications, by Plochocki (Volume 2), and Rubbery Thermoplastic Blends, by Kresge (Volume 2).

Overall, mixed polymers provide an incredible range of morphological states from coarse to fine. One need only glance at Interpenetrating Networks (Thomas and Sperling, Volume 2) and Rubbery Thermoplastic Blends (Kresge) to sense the range of possibilities. However, aside from unusual possibilities of disperse phase, size, shape, and geometric arrangement such as described in Fibers from Polymer Blends (Paul, Volume 2) even more subtle complexities are possible with crystalline polymers wherein intricate arrangements of the crystalline and amorphous phases are possible as is shown by Stein in Optical Behavior of Blends (Volume 1). Here special investigatory techniques have been brought to bear on the disposition of molecules in the submicroscopic range. This chapter also presents the basic principles for understanding the optical properties of blends important in some applications.

Properties, along with composition and morphology, represent the third major area of interest. Historically, the modification of glassy polymers with a disperse rubber phase was one of the first major synthetic efforts directed to useful polymer blends. It is of particular importance because the mechanical properties of these synergistic mixtures are not related simply to the sum of the properties of the components. We have attempted to trace the growth of the concept of rubber modification in Newman's chapter, Rubber Modification of Plastics (Volume 2). A more detailed study of toughening theory for a restricted range of systems is presented in Fracture Phenomena in Polymer Blends, by Bucknall (Volume 2). These chapters deal largely with ultimate properties. Small strain behavior is dealt with in Mechanical Properties (Small Deformations) of Multiphase Polymer Blends, by Dickie (Volume 1).

Another major area of commercial importance of blending is the rubber industry, which capitalizes on the combination of properties each component contributes to the blend. The chapter by McDonel, Baranwal, and

Andries in Volume 2 illustrates the advantages and compromises in the very large area of application, Elastomer Blends in Tires. Somewhat related are the contrasting combinations of rubbery polymers and plastics treated by Hammer in Polymeric Plasticizers and Kresge in Rubbery Thermoplastic Blends.

The connections between polymer blends and the field of reinforced composites and its techniques of analysis are clearly evident in the chapters by Dickie on mechanical properties and by Hopfenberg and Paul on sorption and permeation. This is also seen in the chapters on films by Schrenk and Alfrey and fibers by Paul and to a lesser extent in the chapters by Thomas and Sperling and Newman.

A remarkable balance of diverse properties are achievable with blends as is evidenced by the chapters mentioned in the previous paragraphs. One unusual case of great commercial importance heretofore largely unnoticed in fundamental research is represented in Low Profile Behavior (Atkins, Volume 2), where polyblends of thermoplastics in cross-linked styrenated polyesters are used to control the volume of molded objects. The list of useful performance characteristics that can be controlled by polyblending is as long as the list of properties themselves.

An introductory chapter, written after all of the other chapters were completed and reviewed, has been included to fill some gaps and to provide a certain degree of perspective.

We have selected the above topics in order to achieve our goal of a comprehensive treatment of the science and technology of polymer blends. However, the growth of the field has made it necessary to publish this treatise in two volumes. We have attempted to divide the chapters equally between the two and to concentrate the fundamental and general topics in Volume 1 and the more specific and commercially oriented subjects in Volume 2.

In Volume 2, the topics dealt with are more specific and have a greater orientation toward commercial interests. It has been our purpose to make each chapter self-contained without making it necessary for the reader to be referred continually to other chapters. Nevertheless, some overlap is unavoidable in a cooperative undertaking. This does, however, permit the presentation of different points of view and thereby contributes to a balanced presentation. On the other hand, we have also sought to tie together the various chapters, and this has been accomplished in part by the addition of cross-references.

Many people have contributed to the development and completion of these two volumes through their advice and encouragement. However, we wish to mention specifically the help of F. P. Baldwin, J. W. Barlow, R. E. Bernstein, C. A. Cruz, S. Davison, P. H. Hobson, and J. H. Saunders.

# Contents of Volume 1

# Polymer Blends

*VOLUME 2*

# Chapter 11

# Interpenetrating Polymer Networks

## D. A. Thomas and L. H. Sperling

*Materials Research Center*
*Coxe Laboratory*
*Lehigh University*
*Bethlehem, Pennsylvania*

1

## I. INTRODUCTION

### A. Definition of an Interpenetrating Polymer Network

The term *interpenetrating polymer network* (IPN) denotes an entire class of materials, rather than a single molecular topology. In its broadest definition, an IPN is any material containing two polymers, each in network form. A practical restriction requires that the two polymers have been synthesized or cross-linked in the presence of each other. Both simultaneous and sequential types of syntheses have been explored; both yield distinguishable materials [1].

One type of sequential IPN begins with the synthesis of cross-linked Polymer I. Monomer II, plus its own cross-linker and initiator, are swollen into Polymer I, and polymerized in situ [2]. *Simultaneous interpenetrating networks* (SINs) begin with a solution of both monomers and cross-linkers, which are then polymerized by noninterfering modes, such as addition and condensation reactions [3–5]. The two rates of polymerization, and their approach to gelation may be the same, or significantly different. Again, different products ensue. A third mode of IPN synthesis takes two latexes of linear polymers, mixes and coagulates them, and cross-links both components simultaneously. The product is called an *interpenetrating elastomeric network* (IEN) [6]. As is seen below, there are, in fact, many different ways that an IPN can be prepared; each yields a distinctive topology.

The term IPN implies an interpenetration of the two polymer networks of some kind, and was coined before the full consequences of phase separation were realized [7]. Molecular interpenetration only occurs in the case of total mutual solubility; however, most IPNs phase separate to a greater or lesser extent. Given that the synthetic mode yields two networks, the extent of continuity of each network needs to be examined. If both networks are continuous throughout the sample, and the material is phase separated, the phases must interpenetrate in some way [8], and, thus, some IPN compositions are thought to contain two continuous phases.

Thus, molecular interpenetration may be restricted or shared with supermolecular levels of interpenetration. True molecular interpenetration is thought to take place only at the phase boundaries in some cases. If the two polymers are chemically identical, the product is called a *Millar IPN* [7]. In this case, true compatibility is achieved, and the network chains are believed actually to interpenetrate at the molecular level (but see Section IV.A.5 for a possible reinterpretation).

When only one of the polymers is cross-linked, the product is called a semi-IPN [9]. If the polymerizations are sequential in time, four semi-IPNs may be distinguished. If Polymer I is cross-linked and Polymer II is linear, the product is called a semi-IPN of the first kind, or semi-1 for short. If

Polymer I is linear, and Polymer II cross-linked, a semi-2 results. The remaining two compositions are materialized by inverting the order of polymerization. Several recent reviews [10–13] have been written on the subject of IPNs.

### B.   Relationships among IPNs, Blends, and Grafts

A polymer blend may be defined as a combination of two polymers without any chemical bonding between them (Volume 1, Chapters 1 and 2). A graft copolymer refers to a product prepared by the polymerization of Monomer II in the intimate presence of Polymer I, with greater or lesser extent of actual graft copolymer formation. Recent electron microscope and kinetic evidence suggests that grafting in many systems is less extensive than previously believed, but still important. The graft copolymer may behave as a nonaqueous surface active agent, binding the two phases together at their interface (see Chapter 12). Several topologies of interest are illustrated schematically in Fig. 1. Since most IPNs involve polymerization of one polymer in the immediate presence of the other, they are also generally graft copolymers. They constitute a special class, however, since one or both polymers contain cross-links. The interesting properties of

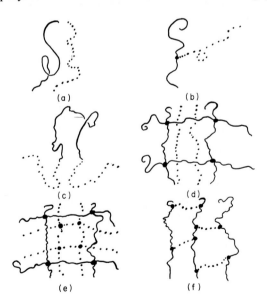

Fig. 1   Schematic diagram of some simple two-polymer combinations: (a) a polymer blend; (b) a graft copolymer; (c) a block copolymer; (d) a semi-IPN; (e) an IPN; (f) an AB cross-linked copolymer. The solid line represents Polymer I; dotted line represents Polymer II. Enlarged intersections represent cross-link sites.

IPNs emerge when the deliberately introduced cross-links outnumber the accidentally introduced grafts. When this condition prevails, the cross-links dominate and control the morphology and hence influence most of the physical and mechanical behavior. However, the grafts are still present and usually contribute favorably to the behavior of the IPN as a material [14a,b].

## C. Historical Aspects

The concept of an IPN goes back at least as far as 1941, when Staudinger and Hutchinson [15] first applied for a British patent. Their United States patent of 1951 describes the use of an IPN topology to prepare improved optically smooth plastic surfaces. The first use of the term "interpenetrating polymer network" was by Millar [7] in 1960. Frisch *et al.* [6] and Sperling [16] independently arrived at the IPN topology through different thought processes. Frisch conceived of the IPNs as the macro-molecular analog of the catenanes. The catenanes consist of interlocking ring structures, which are physically bound together like chain links. Sperling evolved IPNs as a method of producing finely divided phase domains without the need for mechanical mixing.

## D. IPN Nomenclature

Some of the terminology used in this chapter is as follows:

1. IPN: interpenetrating polymer network, the general term. Also used to indicate the time-sequential synthesis product.
2. IEN: interpenetrating elastomeric network, originally used by Frisch *et al.* [6] to denote materials made by mixing two latexes, coagulating them, followed by independent cross-linking reactions.
3. SIN: simultaneous interpenetrating network where both polymers are synthesized simultaneously, that is, by addition and condensation reactions.
4. Semi-IPN: compositions, generally of one cross-linked polymer and one linear polymer.
5. Semi-1, semi-2: semi-IPNs where, respectively, Polymer I or Polymer II is the cross-linked component.
6. Interstitial composites: notation of Allen *et al.* [17] for materials which would be described as semi-SINs in the above notation system; both polymers are synthesized simultaneously, but one is linear, and another cross-linked.
7. Gradient IPN: an IPN of nonuniform macroscopic composition, usually by nonequilibrium swelling in Monomer II, and polymerizing rapidly.

A more complete nomenclature of this type was developed by Klempner and Frisch [18]. A more mathematically oriented nomenclature based on group and ring theory is presented below.

## II. ISOMERIC GRAFT COPOLYMERS AND IPNs

An examination of the scientific and patent literature reveals that over 200 topologies of IPNs and related materials have been synthesized. By classical standards, most would be designated simply as "graft copolymers," with little to distinguish one composition from another. In an attempt to correct the inadequacies of the nomenclature system, Sperling derived two types of systematic notation. The first is based on group theory concepts [19–21], and presents a detailed approach to the nomenclature of polymer blends, grafts, and IPNs. Later, a nomenclature system based on the ideas of mathematical rings was developed [22]. The latter is less detailed in nomenclature and offers improved insight into the interrelationships of the many materials. Even with simplifying assumptions, the two techniques show that more than 50,000 distinguishable combinations of the same two polymers are possible. The basics of the two systems are outlined in the following paragraphs.

### A. Group Theory Notation

In terms of group theory concepts, the symbols $P_i$, $C_i$, and $G_{ij}$ represent the formation of the linear $i$th polymer, the cross-linked $i$th polymer, and the graft of the $j$th polymer onto the $i$th polymer, respectively, in the generalized multicomponent system. The symbols $m_i$, $c_i$, and $g_i$ stand for the monomer, cross-linker, and grafting agents, respectively. Parentheses indicate simultaneous processes, and brackets indicate materials synthesized separately, and later mixed and/or reacted. For example,

$$[m_1 P_1][m_2 P_2](C_1 C_2) \qquad (1)$$

represents two linear polymers synthesized separately that were blended together and then both cross-linked [6].

### B. Rings and Binary Notation

It has been assumed in the above that the system has some of the important characteristics of a mathematical group. Now, the subject will be cast in the form of a mathematical ring, as this leads to improved insight into the nature of the combinations and their proposed nomenclature. Two simplifications must be made:

1.  The use of reactive unit symbolism is being discarded.
2.  Materials that are slightly or accidentally grafted are considered to be blends for the present discussion. Only systems that are intentionally or extensively grafted are so considered.

A system with ringlike characteristics has two binary operations, which in ordinary algebra are addition and multiplication. We take all binary polymer combinations not involving bonding between them as the "addition," and its binary operation is designated $o_1$. Thus,

$$P_2 \, o_1 \, P_1 = M_{12} \tag{2}$$

means a blend of Polymers 1 and 2, reading from right to left.[†] (The group notation reads from left to right.) In important ways the symbol $o_1$ replaces the brackets used earlier, which stood for blending.

The combinations involving chemical bonding between polymers is the "multiplication," and its binary operation is designated $o_2$. Thus,

$$P_1 \, o_2 \, P_1 = C_1 \tag{3}$$

indicates a cross-linked Polymer I, and

$$P_2 \, o_2 \, P_1 = G_{12} \tag{4}$$

indicates the grafting of Polymer II onto a Polymer I backbone.

Table I

"Addition" Table for Polymer Blends[a]

| $o_1$ | $P_1$ | $P_2$ | $C_1$ | $C_2$ | $\cdots$ |
|---|---|---|---|---|---|
| $P_1$ | $P_1$ | $M_{12}$ | $S_{11}^2$ | $S_{12}^2$ | |
| $P_2$ | $M_{21}$ | $P_2$ | $S_{21}^2$ | $S_{22}^2$ | |
| $C_1$ | $S_{11}^1$ | $S_{12}^1$ | $I_{11}$ | $I_{12}$ | |
| $C_2$ | $S_{21}^1$ | $S_{22}^1$ | $I_{21}$ | $I_{22}$ | |
| $\vdots$ | | | | | |

[a] In the corresponding addition table for ordinary arithmetic, $o_1$ means "plus." Thus, $3 \, o_1 \, 5 = 8$, $2 \, o_1 \, 3 = 5$, etc., which can be set up in the same form as this polymer blend table. Ordinary addition and multiplication, in fact, form a ring, conforming to all its requirements.

[†] *Note added in proof*:  A left-to-right reconstruction may be undertaken following comments based on the presentation discussed in the note on p. 9.

Table II

"Multiplication" Table for Cross-Linked and Grafted Systems

| $o_2$ | $P_1$ | $P_2$ | $C_1$ | $C_2$ | ... |
|-------|-------|-------|-------|-------|-----|
| $P_1$ | $C_1$ | $G_{12}$ | $C_1G_{11}$ | $C_2G_{12}$ | |
| $P_2$ | $G_{21}$ | $C_2$ | $C_1G_{21}$ | $C_2G_{22}$ | |
| $C_1$ | $G_{11}C_1$ | $G_{12}C_1$ | $C_1$ | $C_2G_{12}C_1$ | |
| $C_2$ | $G_{21}C_2$ | $G_{22}C_2$ | $C_1G_{21}C_2$ | $C_2$ | |
| $\vdots$ | | | | | |

The binary operation tables for $o_1$ and $o_2$ are shown in Tables I and II. The tables are infinite in size in the general case, with very complex structures possible. Two specific points should be noted:

1. In the blend table, the commutative law of addition holds, and $M_{12} = M_{21}$, where M stands for mixture or blend.
2. Coefficients are omitted. In general, $xP_1 \; o_1 \; yP_1 = (x + y)P_1$, etc.

However, simple combinations of the two tables are more illustrative. For example,

$$P_2 \; o_2 \,(P_2 o_1 \, P_1) = P_2 o_2 \, M_{12} = C_2 o_1 \, G_{12} \qquad (5)$$

which makes use of the distributive property of multiplication over addition, and shows some of the characteristics of the notation. It should be noted that the accidentally or slightly grafted polymer combinations fit the blend table better than the cross-link-graft table. Thus, most of the IPNs synthesized to date may be considered analogous to a chemically induced blend of two cross-linked polymers (not a mechanical blend, certainly):

$$C_2 \, o_1 \, C_1 = I_{12} \qquad (6)$$

where I stands for an IPN.

Beside the symbols defined above, it is also convenient to include the semi-IPN, represented by S, with one polymer cross-linked and one linear. In general, the notation $S_{ik}^h$ is a semi-IPN of polymers $i$ and $k$, and $h$ is 1 or 2, depending on whether the first or the second polymer so introduced contains the cross-links. It is obvious that much more complicated symbols could be evolved. For simplicity, however, combinations of existing symbols are employed, as in the right-hand side of Eq. (5).

In Tables I and II, the elements in the left-hand column are employed first, and are "operated on" by the elements in the top row. This specifies the time order of the operations. Upon examination of the tables, a striking symmetry becomes apparent. Elements on either side of the diagonal from the upper left to the lower right are clearly related. For example, the qualitative relationship between $G_{12}$ and $G_{21}$ is obvious, but now we observe that they are quantitatively related in Table II, lying in symmetric positions across the diagonal. In fact, a function $\gamma$ may be defined, which will cause the element to be moved to its corresponding position across the diagonal, and physically adopt its new structure. In general, there are two $\gamma$ functions: $\gamma_1$ for blends (Table I), and $\gamma_2$ for grafts and cross-links (Table II). The transformation of element $x$ into element $y$ by the function $\gamma$ is shown in Fig. 2. For example,

$$\gamma_1 I_{12} = \gamma_1 (C_2 \circ_1 C_1) = C_1 \circ_1 C_2 = I_{21} \tag{7}$$

Thus, $\gamma_1$ transforms $I_{12}$ to $I_{21}$ within the blends, Table I. Since the more complex combinations lack "common names," these are shown in Tables I and II as combinations of symbols.

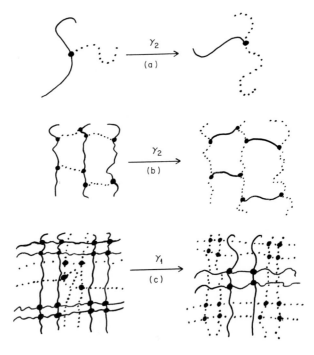

Fig. 2   Illustrations on the application of $\gamma$, the function that calls for inverse materials within the cross-link graft table: (a) graft copolymers; (b) AB cross-linked polymers; (c) IPNs. The IPN case schematically shows the cellular structure of the second polymerized network.

Also, more complex notations such as

$$\left(\begin{array}{c} P_2 \\ o_2 \\ P_1 \end{array}\right) \quad o_2 \quad \left(\begin{array}{c} P_1 \\ o_2 \\ P_2 \end{array}\right) \tag{8}$$

are possible, representing the AB cross-linked copolymers. Equation (8) is read both vertically and horizontally, with time sequences moving from right to left. Quantities within the parenthesis are taken simultaneously. Equation (8) permits a fair (but still not entirely complete) representation of an AB-cross-linked copolymer [23], as distinguished from the simple graft copolymer analog, which does not constitute a network.[†]

It should be emphasized that the above schemes do not fit all the requirements of a formal group or ring, but have sufficient similarities to warrant a comparison.

## III. SURVEY OF SYNTHETIC METHODS

As explained above, the term IPN has gradually assumed a broader interpretation. Although the exact topologies possible run to many tens of thousands, only a relatively few laboratory techniques have been explored in depth.

Sperling and Friedman [16] explored sequential IPNs, where Polymer Network I, poly(ethyl acrylate), was simultaneously photopolymerized and cross-linked with tetraethylene glycol dimethacrylate (TEGDM). Then portions of styrene monomer, cross-linker, and activator were swollen in, and allowed to equilibrate before a second photopolymerization was initiated (Fig. 3).

Most of the sequential IPNs synthesized by Sperling *et al.* can be described through the group theory notation as

$$(m_1 c_1)(P_1 C_1)(m_2 c_2)(P_2 C_2) \tag{9}$$

where the contents within parentheses indicate simultaneous processes. In terms of the ringlike notation, the sequential IPNs can be described as

$$C_2 \, o_1 \, C_1 \tag{10}$$

Some variations on this theme have included latex IPNs [24], where each

[†] *Note added in proof:* In a newer paper presented by E. M. Corwin and L. H. Sperling at the American Chemical Society meeting in Anaheim, California, March 1978, an improved nomenclature scheme was presented. See also P. R. Scarito, E. M. Corwin, and L. H. Sperling, *Polymer Prepr.* **19**(1), 127 (1978).

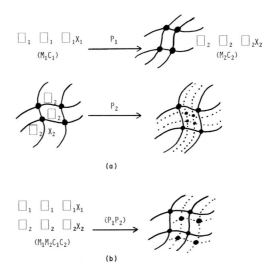

Fig. 3   Schematic of the synthetic steps to make (a) IPNs and (b) SINs. Group theory notation is used. (From Touhsaent *et al.* [3].)

latex particle consists ideally of two molecules. In one set of experiments, however, linear styrene–butadiene rubber (SBR) was later cross-linked, then used for IPN formation with polystyrene (PS) [9, 25].

If the cross-linked Polymer I is swollen in Monomer mix II for a limited period of time, and Polymerization II is carried out before equilibrium can be achieved, the product will contain a composition gradient. Gradient IPNs were prepared by Akovali *et al.* [26], Predecki [27], and Sperling and Thomas [28].

An alternative synthetic technique involves the simultaneous polymerization of both polymers. The synthesis of SINs requires two independent and noninterfering reactions that can be carried out in the same reactor under the same conditions of heat, light, etc. Kim *et al.* [5] investigated simultaneous syntheses of polyurethanes (condensation polymerization) and polystyrene or poly(methyl methacrylate) (addition polymerization). Touhsaent *et al.* [3, 4] investigated the epoxy–poly($n$-butyl acrylate) system, which also involves condensation and addition polymerization reactions run simultaneously. The group theory designation of the SIN formation reads

$$(m_1 c_1 m_2 c_2)(P_1 P_2) \tag{11}$$

while the equivalent ringlike notation reads

$$(P_2 o_2 P_2) o_1 (P_1 o_2 P_1) \tag{12}$$

Of course, both polymers need not be cross-linked. Many types of semi-IPNs exist. Donatelli *et al.* [9] synthesized both semi-1 and semi-2 compositions prepared from SBR elastomer and polystyrene, employing sequential polymerizations (see Fig. 1). Allen *et al.* [17] mixed urethane prepolymers with a triol curing agent, methyl methacrylate monomer, and activator. The ringlike designation for this synthesis reads

$$P_2 \circ_1 P_1 \circ_2 P_1 \qquad (13)$$

Allen designated his materials "interstitial composites." A series of semi-SINs composed of polyurethanes and polyacrylates was recently reported by Yoon *et al.* [29].

The AB cross-linked copolymers represent another interesting mode of joining two polymers. In this case, two polymers are required to make one network, as illustrated in Fig. 1f. Bamford *et al.* [23] have explored this method in some detail. Gardner and Baldwin [30] reported on the case of an AB cross-linked copolymer, where Polymer II was deliberately cross-linked. The graft copolymers prepared from epoxy compositions and CTBN elastomers [31] are recognized as a product having a topology much like that studied by Baldwin and Gardner [32]. In the latter case, the epoxy resin ordinarily cures to a densely cross-linked material. The carboxyl groups at the ends of the butadiene–nitrile rubber react with the epoxy to form the AB cross-linked analog. Another important case involves the castable polyesters [33]. In this last example, an unsaturated polyester is dissolved in styrene monomer, and on polymerization, the polystyrene grafts to and cross-links the polyester.

## IV. MORPHOLOGY

Most IPNs and related materials investigated to date show phase separation. The phases, however, vary in amount, size, shape, sharpness of their interfaces, and degree of continuity. These aspects together constitute the morphology of the material, and the multitude of possible variations controls many of the material properties.

Some aspects of morphology can be observed directly by transmission electron microscopy of stained and ultramicrotomed thin sections [34–36]. Unfortunately, the usual osmium tetroxide staining method is most successful only for polymers containing carbon–carbon double bonds, and many materials are not easily studied. Other aspects of morphology, such as phase continuity and interface characteristics, are best determined by combining information from chemical and mechanical tests with electron microscopy.

## A. Factors Affecting Morphology

The factors that control the morphology of IPNs are now reasonably clear; they include chemical compatibility of the polymers, interfacial tension, cross-link densities of the networks, polymerization method, and the IPN composition. While these factors may be interrelated, they can often be varied independently. Their effects are summarized in this section.

### *1. Compatibility*

A degree of compatibility between polymers is necessary for IPNs, because monomers or prepolymers must form solutions or swollen networks during synthesis.[†] Phase separation generally ensues in the course of polymerization, but the resulting phase domain size is smaller for higher-compatibility systems [38]. Huelck *et al.* [2] varied compatibility systematically in sequential IPNs, with poly(ethyl acrylate) (PEA) as Polymer I and copolymers of methyl methacrylate (MMA) and styrene (S) as Polymer II (Fig. 4). For

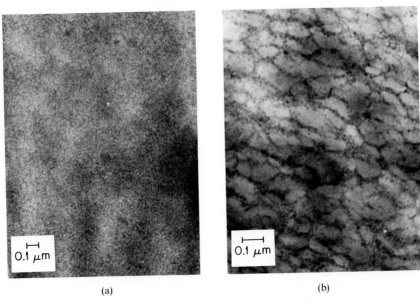

(a)                          (b)

Fig. 4   Electron micrographs of IPNs of (a) 75/25 poly(ethyl acrylate)/poly(methyl methacrylate), and (b) 50/50 poly(ethyl acrylate)/polystyrene. A small amount of butadiene was copolymerized with the poly(ethyl acrylate) to aid in osmium tetroxide staining. (From Huelck *et al.* [2]. Reprinted with permission from *Macromolecules*. Copyright by the American Chemical Society.)

[†] This restriction may be relaxed for IPNs based on latex polymers [6, 24, 37].

PEA–PMMA, in which the components are isomeric and nearly compatible, dispersed phase domains less than 100 Å in size (fine structure) are found (Fig. 4a). This occurs because the high compatibility precludes phase separation until high conversion to PMMA. (An alternate explanation for the fine structure is given later.) With the much less compatible system PEA–

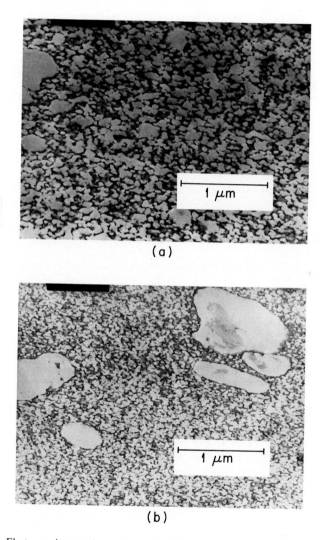

(a)

(b)

Fig. 5  Electron micrographs of IPNs of 20/80 styrene–butadiene rubber/polystyrene. The rubber is cross-linked with (a) 0.1%, and (b) 0.2% dicumyl peroxide. (From Donatelli *et al.* [39].)

PS (Fig. 4b), an additional cellular structure of about 1000 Å in size is found. Here, phase separation is thought to take place earlier in cellular regions rich in PS and S monomer. A second phase separation follows, leading to smaller dispersed PEA and PS domains in the cell walls.

Styrene–butadiene rubber–polystyrene (SBR–PS) IPNs are relatively incompatible, even though they are both nonpolar polymers and their solubility parameters differ by only about 1.0. They show distinct phase separation and a cellular domain structure (Fig. 5) [39].

## 2. Cross-Linking

Increased cross-link density in Polymer network I in an IPN clearly decreases the domain size of Polymer II [39]. This is illustrated by comparison of Figs. 5a and b. This effect is reasonable because ·the tighter initial network must restrict the size of the regions in which Polymer II can phase separate. The effect has also been rationalized by a semiempirical thermodynamic model [40], described later. Variation of cross-link density in the PS has little effect on the IPN morphology, indicating that the first network is controlling. In the extreme case of linear PS (forming a semi-1), the morphologies were somewhat less uniform, but cross-link density in the SBR had the same effect. With no cross-linking in the SBR, on the other hand, the SBR remained continuous but both phases were much coarser and irregular.

Allen *et al.* [17] observed the same effect of the first network in polyurethane (PU)–PMMA materials, made by "interstitial polymerization" according to Eq. (13). The morphology shows roughly spherical domains of PMMA in a matrix of PU (Fig. 6). The PU network formed first from a solution of reactants including the MMA, but reaction conditions permitted varying tightness of the network. Although the PMMA was uncross-linked, its domain size varied from about 650 Å for tight PU networks to about 1800 Å for loose networks [41].

The AB cross-linked copolymers of Bamford *et al.* [23] are not strictly IPNs, since B (Polymer II, polychloroprene) chains are part of the same network as A [Polymer I, poly(vinyl trichloroacetate)] chains. However, like graft copolymers and IPNs, they do phase separate into distinct domains [42]. The Polymer II chains are seen to form the discontinuous phase, perhaps for two reasons: (1) similar to the sequential IPNs and nonphase inverted graft copolymers, Polymer II develops the less continuous phase morphology; and (2) Polymer II is also the minor component in this case.

Meier [43, 44] considered the morphologies of diblock copolymers, and found that for dispersed spheres

$$R = k\alpha CM^{1/2} \tag{14}$$

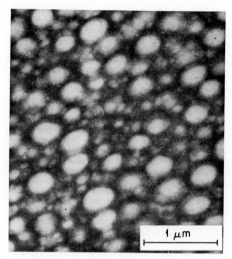

Fig. 6   Electron micrograph of an "interstitial polymer" of 80/20 polyurethane/poly(methyl methacrylate). Poly(butadiene diols) in the polyurethane permit staining with osmium tetroxide. (From Allen *et al.* [41].)

where $R$ represents the domain radius, $k$ is a constant characteristic of the morphology (1.33 for spheres), $C$ represents an experimental constant relating the unperturbed root-mean-square end-to-end distance of a chain to the square root of its molecular weight, and $\alpha$ is the ratio of the perturbed to unperturbed chain dimensions. From morphological dimensions, Eastmond and Smith [42] found

$$R = 0.06 M_n^{1/2} \tag{15}$$

for the domain sizes.

In both the block copolymer case and the AB cross-linked copolymer case, the phase domain dimensions depend upon the molecular weight of the polymer in the domain. In the case of IPNs presented below, the phase domain sizes of Polymer II are shown to depend upon the molecular weight between cross-links ($M_c$) of Polymer I.

### 3.   Polymerization Method

For sequential IPNs, the effects of compatibility and cross-linking have already been described. When the polymers are reversed in sequence, however, the new morphology is again controlled principally by the first network [2].

D. A. Thomas and L. H. Sperling

Fig. 7  Electron micrograph of SIN of 75/25 polyurethane/poly(methyl methacrylate). Polyurethane stained with osmium tetroxide. (From Kim *et al.* [5]. Reprinted with permission from *Macromolecules*. Copyright by the American Chemical Society.)

In simultaneous interpenetrating networks (SINs), the networks form during the same time period, although not necessarily at the same rate, and more complex morphologies result. An example is shown in Fig. 7 [5]. The reaction rates are similar here, so the majority component probably forms a network first and becomes the more continuous phase. Touhsaent *et al.* [3,4] controlled the relative reaction rates in epoxy–poly($n$-butyl acrylate) P($n$BA) SINs and found that the finest two-phase dispersion occurred when the polymerization reactions for both networks reached the Flory gelation point simultaneously.

In the already mentioned semi-IPNs of Allen *et al.* [17], the PU reactions occurred first and their cross-link level established the morphology.

Latex polymers offer a different route to IPNs, and the latex size and morphology establish the morphology of the cast or molded bulk product. In the simplest case, mixed and subsequently cross-linked latexes of Frisch *et al.* have both phases continuous in certain composition ranges. The morphology shows coarse phases, with the urethane latex particles still identifiable (Fig. 8). Molecular interpenetration is expected only at the

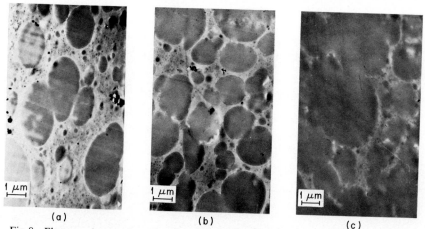

(a)  (b)  (c)

Fig. 8   Electron micrographs of interpenetrating elastomer networks from mixed polyacrylate and polyurethane latexes: (a) 70/30; (b) 50/50; (c) 30/70 polyacrylate/polyurethane. Polyurethane stained with osmium tetroxide. (From Matsuo *et al.* [45].)

interfaces between the two types of latexes. Two-component core–shell IPN latexes have also been produced and cast or molded into bulk materials [46]. Their morphologies are undoubtedly more complex, probably resembling the ABS latex morphology [47].

## 4. Composition

The IPN composition determines the relative amounts of the two phases present after polymerization. Increasing amounts of Polymer II generally lead to increasing domain sizes, but the effect also depends on the polymerization method.

For sequential IPNs based on PEA–PS, Huelck *et al.* [2] found only slight increases in domain sizes as the composition ranged from 75/25 to 25/75. Donatelli *et al.* [9] found the same result for SBR–PS over a range limited to PS-rich materials. Note that for sequential IPNs, an upper limit on the amount of Polymer II is set by the equilibrium swelling of Monomer II plus cross-linker into the first network.

Similar changes were observed by Allen *et al.* [41, 48] as their semi-IPNs contained larger amounts of PMMA. Here, however, the effect was attributed, in part, to the looser PU network that formed from the increasingly dilute urethane prepolymer solutions in MMA.

For PU–PMMA materials with similar rates of polymerization, the SINs of Kim *et al.* [5], show that phase inversion occurred in the composition range of 20–40% PMMA. When either component exceeds 85%, however, the morphology becomes more complicated. This work and that of Touhsaent et al. [3, 4] both indicate that phase domain size and continuity depend sensitively on reaction rates for the two polymers, as well as on composition.

### 5. A Quantitative Expression for Phase Domain Size

Donatelli et al. [40] derived a semiempirical equation for the phase domain size of semi-IPNs of the first kind. The principal variables are the cross-link density of polymer I, the mass fraction of Polymer II, and the interfacial tension. The equation can be written

$$\left(\frac{\nu_1}{C^2 K^2}\right)\left(\frac{\nu_1}{1 - w_2} + \frac{2}{M_1}\right)D_2{}^3 + \left(\frac{w_2}{M_2} - \frac{\nu_1}{2}\right)D_2 = \frac{2\gamma w_2}{RT} \qquad (16)$$

where $\nu_1$ is the cross-link density of Polymer I, $M_1$ the primary molecular weight of Polymer I, $D_2$ the domain size of Polymer II, $M_2$ the molecular weight of Polymer II, $w_2$ the mass fraction of Polymer II, $\gamma$ the interfacial energy, $C$ a constant of the order of $\sqrt{2}$, and $K = (r_0/M^{1/2})$ for Polymer I, where $r_0$ is the unperturbed root-mean-square end-to-end distance.

The case for the full IPN (both polymers cross-linked) can be approximated by taking $M_2 = \infty$. This equation has been applied to three systems to date:

1. SBR–PS [9], where the cross-link density of Polymer I was the principal variable. The results are shown in Table III.

Table III

Experimental and Theoretical PS Domain Sizes in IPNs and Semi-1 Compositions ($\gamma = 3$ dyn/cm)[a]

| Type | Composition SBR/PS (%) | Experimental PS domain size (Å) | Theory (Å) |
|------|------------------------|----------------------------------|------------|
| (Low)[b] semi-1 | 20/80 | 1500 | 1250 |
| (High) semi-1 | 18/82 | 600 | 500 |
| (Low) IPN | 22/78 | 1100 | 1300 |
| (High) IPN | 21/79 | 650 | 480 |

[a] From Donatelli et al. [40]. Reprinted with permission from *Macromolecules*. Copyright by the American Chemical Society.
[b] Low refers to 0.1% cross-linking agent in the SBR; high refers to 0.2%.

2. Castor oil–urethane–PS IPNs, where both composition and cross-link level were varied [49]. The results are summarized in Table IV. In both Tables III and IV, the comparison between theory and experiment is better than warranted from the semiempirical nature of the equation.

Table IV

Polystyrene Domain Sizes in Castor Oil–Urethane IPNs[a]

| IPN[b] sample | NCO/OH ratio | Weight fraction PS | Polystyrene domain size (Å) | |
|---|---|---|---|---|
| | | | Experimental[c] | Theoretical[d] |
| 1 | 0.95 | 0.68 | 250 | 323 |
| 2 | 0.95 | 0.60 | 300 | 325 |
| 3 | 0.95 | 0.47 | 370 | 327 |
| 4 | 0.85 | 0.64 | 300 | 344 |
| 5 | 0.85 | 0.60 | 350 | 345 |
| 6 | 0.85 | 0.50 | 430 | 347 |
| 7 | 0.75 | 0.71 | 350 | 432 |
| 8 | 0.75 | 0.64 | 410 | 443 |
| 9 | 0.75 | 0.50 | 550 | 452 |

[a] From Yenwo *et al.* [49].
[b] The castor oil phase in all the above IPNs was cross-linked with 2,4-TDI.
[c] Estimated from electron microscopy.
[d] Theoretical, from Eq. (16).

3.  PEA–poly[styrene-*co*-(methyl methacrylate)] [2], where the interfacial tension is the main variable [50].

The most interesting result from this last application is that domains of 60–100 Å are predicted for $\gamma = 0$ in Eq. (16). Domains of this size were found in the isomeric IPN 75/25 PEA/PMMA (where $\gamma$ must be very close to zero, see Fig. 4a). This modifies earlier conclusions of semi-compatibility for this system, suggesting that fine structure is always to be expected in sequential IPNs. Physically, this might result from the con-centration of Monomer II in small regions of space (within Polymer I) that accidentally have lower than average cross-linking levels.[†]

## B.  Review of Two-Phase Polymer Morphology

Polymer I generally forms the more continuous phase in IPNs, and this tends to control the two-phase morphology. The morphology consists of interpenetrating phases, on a scale of 100–1000 Å, although Polymer II shows less phase continuity and may even appear to be dispersed in a matrix of Polymer I. The morphology may be further modified by the degree of compatibility of the two polymers, the polymerization method, and the IPN composition.

Several guidelines concerning phase continuity may be set down that apply to IPNs and, more broadly, to many two-polymer materials [8]:

[†] Domains of 50–100Å were recently shown in PS/PS Millar IPNs [50].

1. For simple melt blends, the polymer with the higher concentration or the lower viscosity tends to form the continuous phase (see Volume 1, Chapter 7).

2. For bulk or solution graft copolymerizations, the polymer first synthesized forms the more continuous phase. Polymer II usually forms cellular domains within Polymer I.

3. Stirring of bulk or solution-type graft copolymerizations, especially during the early portion of the polymerization of Polymer II, may cause phase inversion, especially if Polymer I is the minority component.

4. For emulsion polymerizations, Polymer II tends to form the continuous phase, after subsequent molding or film formation. In general, the molding of shell–core particulates into macroscopic structures leads to greater continuity of the shell component.

5. For diblock polymers, the relative proportions of the two polymers determine the phase structure and continuity. Midrange compositions have two continuous phases.

6. Cross-linking of either Polymer I or Polymer II tends to promote phase continuity. Materials with both polymers cross-linked (IPNs) tend to develop two continuous phases.

## V. PHYSICAL AND MECHANICAL BEHAVIOR

### A. General Properties

The properties of IPNs depend on: (1) the properties of the component polymers, (2) the phase morphology, and (3) interactions between the phases.

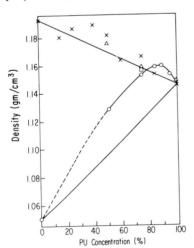

Fig. 9 Density versus polyurethane concentration (straight lines are based on volume additivity: (×) UCMC IPNs; (○) UCSC IPNs; (△) linear blends; (□) pseudo-IPNs. (From Kim *et al.* [51]. Reprinted with permission from *Macromolecules.* Copyright by the American Chemical Society.)

As with other two-component materials, some properties of IPNs are approximately simple averages of the properties of the component polymers; an example is density. However, careful density measurements on polyurethane–polystyrene simultaneous IPNs [51], at a temperature between the glass transition temperatures ($T_g$'s) of the components, showed a density 3 % higher than expected for intermediate IPN compositions (Fig. 9). This was attributed to partial mixing or interpenetration of chains of the rubbery and glassy polymer components, which, Kwei *et al.* [52] reasoned, accounted for slight densification in compatible linear blends.

Optical transparency, on the other hand, departs completely from simple averaging. The IPNs of two amorphous, transparent polymers, such as PEA and PS, are hazy and translucent in thin sheets because the phase domains have different refractive indices and scatter light.[†] Phase separation in less compatible IPNs such as SBR–PS results in white, opaque materials because of increased scattering as the size of the domains approaches the wavelength of light.

We next consider several properties of IPN materials, including special effects conferred by the morphology and interfacial interactions resulting from molecular or phase interpenetration.

## B.  Glass Transitions and Viscoelastic Behavior

When two polymers form a phase-separated mixture, each retains its glass transition. In general, the transitions may be broadened or shifted by mixing, and, of course, in the limit of mutual solubility, only one transition is observed (see Volume 1, Chapter 5).

The loss and storage moduli for IPNs of PEA–PS and PEA–PMMA are shown as a function of temperature and composition in Fig. 10 [2]. With the exchange of PMMA for PS, the two polymers become more compatible because of increased attractive forces. The glass transition behavior of an incompatible polymer pair is illustrated in Fig. 10a, while

---

[†] In general, the quantity of scattered light, and hence the opacity, depends upon two parameters:

1.  The amount of scattered light increases as the square of the difference in refractive index of the two phases. Thus, polymer blends, grafts, and IPNs will be clear if the refractive indexes nearly match.

2.  For particles small compared to the wavelength of the light, scattering increases as the sixth power of the particle diameter. For domains near the wavelength in size, scattering increases as the square of the wavelength. This latter accounts for the blue-white opalescence frequently seen in IPNs and graft copolymers. For objects large compared to the wavelength of light (tables, chairs, etc.) the wavelength dependence on scattering is effectively zero, and the phenomenon is commonly called reflection. (See also Volume 1, Chapter 9.)

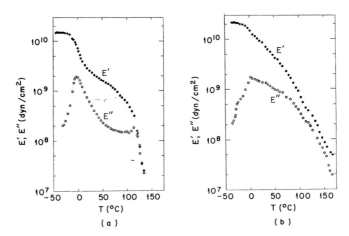

Fig. 10   Storage ($E'$) and loss ($E''$) moduli: (a) 48.8/51.2 PEA/PS; (b) 47.1/52.9 PEA/PMMA IPNs both at 110 Hz. (From Huelck *et al.* [2]. Reprinted with permission from *Macromolecules*. Copyright by the American Chemical Society.)

the modulus changes with temperature in Fig. 10b indicate greater molecular mixing. No IPN has been found that behaves in a manner similar to the equivalent random copolymer, that is, one sharp glass transition. Electron microscopy of these same materials (Fig. 4) confirms the mechanical data, and shows more explicitly the phase domain sizes.

The effect of composition on the modulus for acrylic–urethane IENs is shown in Fig. 11 [53]. Two distinct transitions are observed, with the acrylic copolymer softening at the higher temperature. The slight shifting and broadening of the moduli indicate a modest degree of molecular mixing. However, the large shifts in $E'$ indicate changes in phase continuity. The reader should compare the moduli in Fig. 11 with the morphology of IENs shown in Fig. 8.

Values of storage modulus at 23°C are plotted as a function of composition in Fig. 12. Takayanagi's parallel model for the mechanical behavior of a two-component system [54] corresponds to the case in which the stiffer component is continuous, while his series model corresponds to the case in which the softer component is continuous. Clearly the experimental results agree best with the parallel model, and hence confirm the observations (based on electron microscopy) that a phase inversion takes place at about 30% acrylic component.

The behavior of castor oil–urethane–polystyrene sequential IPNs was investigated by Yenwo *et al.* [49]. These materials have small phase domains of 300–500 Å, caused by the rather high cross-link level of the castor oil–urethane elastomer used as Polymer I. The extent of mixing is also illustrated by its dynamic mechanical behavior (Fig. 13).

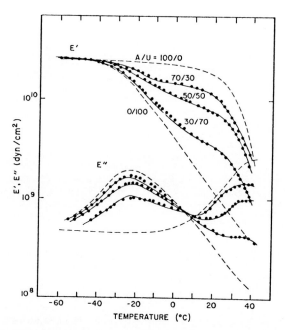

Fig. 11 Temperature dependences of the dynamic storage modulus $E'$ and the loss modulus $E''$ of the interpenetrating elastomeric networks: (●) observed values; (——) calculated; (––––) component homopolymers. (From Matsuo *et al.* [45].)

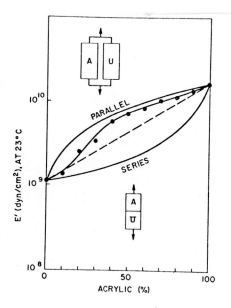

Fig. 12 Plot of the dynamic molulus $E'$ at 23°C against the polyacrylate content. The upper and the lower solid lines are calculated with a parallel and a series model, respectively. (From Matsuo *et al.* [45].)

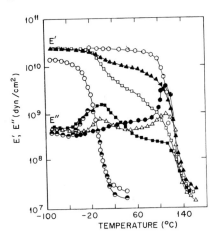

Fig. 13 Dynamic mechanical behavior of some castor oil(CO)–urethane–polystyrene sequential IPNs, with composition as the principal variable. CO/PS IPNs: ($\bigcirc$, $\bullet$) 0/100; ($\blacktriangle$, $\triangle$) 30/70; ($\blacksquare$, $\square$) 44/56; ($\bigcirc$, $\bullet$) 100/0. All at NCO/OH = 0.75. (From Yenwo *et al.* [49].)

The various aspects of dual phase continuity may also be approached through an equation developed by Davies [55]:

$$G^{1/5} = \phi_1 G_1^{1/5} + \phi_2 G_2^{1/5} \tag{17}$$

where $G$ is the shear modulus, and $\phi$ is the volume fraction. Equation (17) was derived on the basis of continuity of both phases. Allen *et al.* [56] tested the Davies equation along with the Kerner [57] and Haskin and Shtrikman [58] equations and others (see Fig. 14). Conformity to the Davies equation

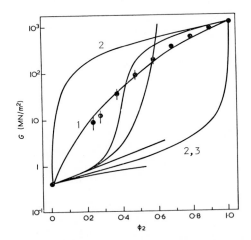

Fig. 14 Log₁₀ (shear modulus) $G$ as a function of volume fraction ($\phi_2$) of PMMA for PMMA–PU interstitial composites. $\bullet$ are experimental points, and numbers on theoretical curves refer as follows: (1) Davies [55]; (2) Kerner [57]; (3) Hashin–Shtrikman [58]; other lines refer to earlier theories. (From Allen *et al.* [56].)

was also noted by Donatelli *et al.* [9] for SBR–PS IPNs and semi-1 compositions, but not for semi-2 compositions.

## C. Ultimate Behavior

A number of IPN compositions exhibit considerable toughness, as measured by stress–strain curves or impact strength. The main features required are similar to those needed by tough graft copolymers: elastomeric phase domains of 500–5000 Å in size, and a $T_g$ of the elastomeric phase below $-40°C$. The phase domains of the sequential IPNs are at the lower limit of the size range, due to the cross-linking of Polymer I. However, the appearance of a cellular-type morphology throughout, and the probable dual phase continuity does contribute to the toughening processes. As shown in Fig. 15 [9], impact values of about 5 ft lb per inch of notch are attainable with 20–25 % elastomer by volume.

Touhsaent *et al.* [3, 4] studied the SIN system epoxy–poly($n$-butyl acrylate). Their objective was to evaluate SINs when both polymer networks reached the Flory gelation point simultaneously. It was reasoned that simultaneous gelation should produce the finest phase structure, which was observed [3]. Measurements of mechanical behavior, however, revealed that the point of simultaneous gelation also resulted in a minimum in tensile strength and other properties [4]. The tensile strength on a computer generated contour map is shown in Fig. 16, with the epoxy prereaction time and $n$-butyl acrylate initiator level as variables. The diagonal from the lower left to the upper right roughly yields a line of simultaneous gelation conditions. Higher tensile strengths are

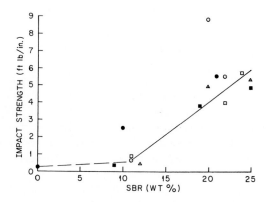

Fig. 15 Impact strength of SBR–PS IPNs as a function of SBR content: (○) Series 3; (△) Series 4; (□) Series 5; (●) Series 12; (■) Series 13. (From Donatelli *et al.* [9]. Reprinted with permission from *Macromolecules.* Copyright by the American Chemical Society.)

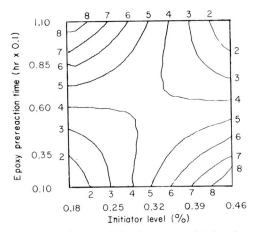

Fig. 16    A contour map of tensile strength for the epoxy–poly(*n*-butyl acrylate) SIN pair. The cross-linker for the poly(*n*-butyl acrylate), DEGDM, was held at the constant level of 0.80%. (From Touhsaent *et al.* [4]. Reprinted with permission of the copyright owner, the American Chemical Society, from *Advances in Chemistry Series 154.*)

| Contour value (psi)[a] | Contour symbol |
|---|---|
| 6.0407397E | 1 |
| 6.4660395E | 2 |
| 6.8913399E | 3 |
| 7.3166403E | 4 |
| 7.7419407E | 5 |
| 8.1672411E | 6 |
| 8.5925415E | 7 |
| 9.4431413E | 8 |

[a] All +0.03.

observed in the upper left and lower right portions of the map, on either side of the simultaneous reaction locus.

All of the materials discussed above had uniform compositions on the macroscopic scale. However, if Polymer Network I is swollen briefly with Monomer Mix II on either one or both sides, and polymerized immediately, a gradient composition results. Sperling and Thomas [28] employed gradient IPNs to obtain a hard exterior, a soft interior, and a composition-gradient intermediate zone for the purposes of developing materials useful for noise and vibration damping.

Akovali *et al.* [26] reported on the general behavior of gradient IPNs. The stress–strain behavior of PMMA–PMA IPNs, random copolymers, and gradient compositions are illustrated in Fig. 17. The gradient material is seen to have the highest yield stress.

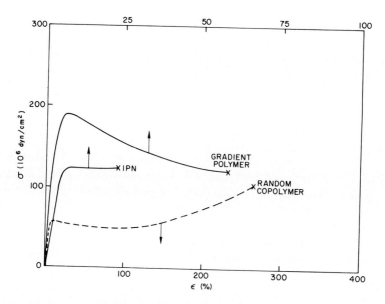

Fig. 17  Stress–strain curves of gradient polymer, interpenetrating networks, and random copolymer of methyl methacrylate and methyl acrylate (ratio 60/40); temperature, 80°C; strain rate, $0.03 \, \text{sec}^{-1}$. (From Akovali *et al.* [26].)

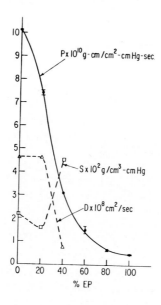

Fig. 18  Water vapor permeation, p, versus IPN network composition of a polyether PU–EP SIN-type IPN; $p_{H_2O}$, 2.01 cm Hg; temperature, 22°C. (From Frisch [13].)

*D. A. Thomas and L. H. Sperling*

## D.  Recent Studies

Results from two recent experiments are described here because of their significance.

1. *Permeability.* The permeability behavior of a polyurethane (PU)–epoxy (EP) SIN-type material is shown in Fig. 18 [13] to be dependent upon composition in a nonlinear fashion, probably due to phase inversion and dual phase continuity (see Volume 1, Chapter 10).

2. *Decross-linking and annealing.* Neubauer *et al.* [59a] studied semi-1 IPNs synthesized from PEA and PS. The cross-linker for the PEA was composed of various ratios of diethylene glycol dimethacrylate (DEGDM), a permanent type of cross-linker, and acrylic acid anhydride (AAA), an hydrolyzable cross-linker (Fig. 19). As the quantity of AAA was increased from zero to 100 %, the modulus change on annealing increased very significantly, suggesting morphological changes.

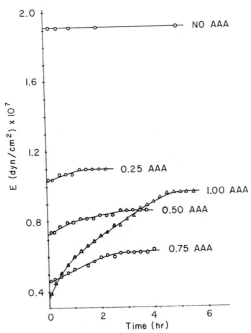

Fig. 19   Annealing of decross-linked PEA–PS semi-1, containing various proportion of AAA and DEGDM, at 110°C: a modulus–time study of annealing effects of 50/50 PEA/PS semi-1 IPNs at 110°C. The ratios of AAA to DEGDM were varied; but their total in the PEA equaled 4 mole %. (From Neubauer *et al.* [59a]. Reprinted with permission of *Coatings in Plastics* (*Preprints*). Copyright by the American Chemical Society.)

The decross-linking of an IPN or semi-1 composition results in a chemically induced blend. Systematic and controlled decross-linking offers a new mode of morphological control through which new or improved physical and mechanical behavior patterns can be achieved.

## VI. APPLICATIONS AND USES

### A. Existing Patents

The patent literature reveals that there are many patents utilizing an IPN or closely related materials. Patents mention production of optically smooth plastic surfaces, tough plastics, pressure-sensitive adhesives, ion-exchange resins, noise- and vibration-damping materials, impact modifiers, contact lenses, etc. An abbreviated list of IPN-related patents is shown in Table V, with emphasis on the network topology [22], and Table VI, emphasizing applications.

Table V

Application of Group and Ringlike Nomenclature to Selected IPN-Related Patent Literature[a]

| Common designation | Group theory designation | Ringlike designation | Ref. |
|---|---|---|---|
| Semi-1 | $(m_1c_1)(P_1C_1)m_2P_2$ | $P_2o_1\,C_1$ | [59b] |
| Graft copolymer | $m_1P_1m_2(P_2G_{12})$ | $P_2o_2\,P_1$ | [59c] |
| IPN | $[m_1P_1][(m_2P_2)(c_1m_3)](C_1P_3G_{23})$ | $\begin{pmatrix} P_1 & o_2 & P_1 \\ o_1 & & o_1 \\ P_3 & o_2 & P_2 \end{pmatrix}$ | [59d] |
| Semi-IPN | $(m_1c_1m_2)(P_1C_1)P_2$ | $P_2o_1\,C_1$ | [59e] |
| Semi-IPN | $[m_1P_1][m_2P_2]C_1$ | $P_1o_2\begin{pmatrix} P_1 \\ o_1 \\ P_2 \end{pmatrix}$ | [59f] |
| IPN | $(m_1c_1)(P_1C_1)(m_2c_2)(P_2C_2)$ | $C_2o_1\,C_1$ | [59g] |
| IPN | $[m_1P_1(m_2c_3)P_2G_{12}][m_3P_3]C_3$ | $P_3o_2\,P_3o_1\begin{pmatrix} P_2 \\ o_2 \\ P_1 \end{pmatrix}o_2\begin{pmatrix} P_1 \\ o_2 \\ P_2 \end{pmatrix}$ | [59h] |

[a] From Sperling *et al.* [22]. Reprinted with permission of *Macromolecules*. Copyright by the American Chemical Society.

Table VI

Actual or Anticipated Uses for IPNs and Related Materials

| Mode of combination | Application | Ref. |
|---|---|---|
| Natural leather–rubber | Improved leather | [60] |
| Anionic–cationic | Piezodialysis membranes | [61] |
| Anionic–cationic | Ion exchange resin | [62] |
| Plastic–rubber | Noise damping | [28] |
| Rubber–plastic semi-IPN | Impact-resistant plastic | [63] |
| Plastic–plastic | Optically smooth surfaces | [15] |
| Rubber–rubber | Pressure sensitive adhesive | [64] |
| Plastic–plastic | Compression molding composition | [65] |
| Plastic–rubber | Tough plastic | [66] |
| Water swellable–nonwater swellable | Soft contact lenses | [67] |
| Rubber–plastic | Impact modifier | [68] |

## B.   Potential Applications

Besides the patent literature, there is a growing number of suggested uses in the scientific literature. Predecki [27] mentions arteriovenous shunts, and in their SIN method Toushaent *et al.* [3] suggest uses as casting syrups. Toughened elastomers, impact-resistant plastics, piezodialysis membranes [61], and wire insulation [69, 70] have been mentioned.

Although polymers made from petroleum sources are emphasized in the above, castor oil-based IPNs [71] and polymer-modified leather [60] both make use of renewable resources.

In general, IPNs and semi-IPNs can be used wherever graft copolymers and polymer blends have been used. In fact, a practical definition of an IPN is a graft copolymer in which the number of deliberately introduced cross-links outnumbers the accidentally introduced grafts. The cross-links allow a new mode of control over two-phase morphology, and hence over properties.

A possible disadvantage, at least for some applications, arises from the thermosetting nature of these materials. However, latex-type formulations employing either the core–shell or IEN modes, or SIN formation permits broader potential uses. It should be noted that for such applications as adhesives, coatings, or toughened elastomers, cross-linked thermoset products are usually required.

### Acknowledgments

The authors wish to acknowledge the support in part of the National Science Foundation through Grant No. DMR73-0255 A01, Polymers Program.

# REFERENCES

1. J. A. Manson and L. H. Sperling, "Polymer Blends and Composites," Chapter 8. Plenum, New York, 1976.
2. V. Huelck, D. A. Thomas, and L. H. Sperling, *Macromolecules* **5**, 340, 348 (1972).
3. R. E. Touhsaent, D. A. Thomas, and L. H. Sperling, *J. Polym. Sci. Part C* **46**, 175 (1974).
4. R. E. Touhsaent, D. A. Thomas, and L. H. Sperling, in "Toughness and Brittleness of Plastics" (R. D. Deanin and A. M. Crugnola, eds.), Adv. Chem. Ser. 154. Amer. Chem. Soc., Washington, D.C., 1976.
5. S. C. Kim, D. Klempner, K. C. Frisch, W. Radigan, and H. L. Frisch, *Macromolecules* **9**, 258 (1976).
6. H. L. Frisch, D. Klempner, and K. C. Frisch, *J. Polym. Sci. Part B* **7**, 775 (1969).
7. J. R. Millar, *J. Chem. Soc.* **263**, 1311 (1960).
8. L. H. Sperling, *Polym. Eng. Sci.* **16**, 87 (1976).
9. A. A. Donatelli, L. H. Sperling, and D. A. Thomas, *Macromolecules* **9**, 671, 676 (1976).
10. L. H. Sperling, *Encycl. Polym. Sci. Technol. Suppl. 1* 288 (1976).
11. H. A. J. Battaerd, *J. Polym. Sci. Part C* **49**, 149 (1975).
12. D. S. Kaplan, *J. Appl. Polym. Sci.* **20**, 2615 (1976).
13. H. L. Frisch, in "Polymer Alloys" (D. Klempner and K. C. Frisch, eds.), p. 97. Plenum, New York, 1977.
14a. L. H. Sperling, *Macromol. Rev.* **12**, 141 (1977).
14b. L. H. Sperling, *J. Polym. Sci. C* (Symp. issue no. 60; S. L. Aggarwal, ed.) **15**, 175 (1977).
15. J. J. P. Staudinger and F. Hutchinson, U.S. Patent 2,539,377 (1951).
16. L. H. Sperling and D. W. Friedman, *J. Polym. Sci. Part A-2* **7**, 425 (1969).
17. G. Allen, M. J. Bowden, D. J. Blundell, F. G. Hutchinson, G. M. Jeffs, and J. Vyvoda, *Polymer* **14**, 597 (1973).
18. D. Klempner, H. L. Frisch, and K. C. Frisch, *J. Polym. Sci. Part A-2* **8**, 921 (1970).
19. L. H. Sperling and K. B. Ferguson, *Macromolecules* **8**, 691 (1975).
20. L. H. Sperling, in "Recent Advances in Polymer Blends, Grafts and Blocks" (L. H. Sperling, ed.). Plenum, New York, 1974.
21. L. H. Sperling, in "Toughness and Brittleness of Plastics" (R. D. Deanin and A. M. Crugnola, eds.), Adv. in Chem. Ser. No. 154. Amer. Chem. Soc., Washington, D.C., 1976.
22. L. H. Sperling, K. B. Ferguson, J. A. Manson, E. M. Corwin, and D. L. Siegfried, *Macromolecules* **9**, 743 (1976).
23. D. H. Bamford, G. C. Eastmond, and D. Whittle, *Polymer* **16**, 377 (1975).
24. J. A. Grates, D. A. Thomas, E. C. Hickey, and L. H. Sperling, *J. Appl. Polym. Sci.* **19**, 1731 (1975).
25. A. J. Curtius, M. J. Covitch, D. A. Thomas, and L. H. Sperling, *Polym. Eng. Sci.* **12**, 101 (1972).
26. G. Akovali, K. Biliyar, and M. Shen, *J. Appl. Polym. Sci.* **20**, 2419 (1976).
27. P. Predecki, *J. Biomed. Mater. Res.* **8**, 487 (1974).
28. L. H. Sperling and D. A. Thomas, U.S. Patent 3,833,404 (1974).
29. H. K. Yoon, D. Klempner, K. C. Frisch, and H. L. Frisch, *Amer. Chem. Soc. Coatings Plast. Chem.* **36**(2), 631 (1976).
30. I. J. Gardner and F. P. Baldwin, presented at the *Meeting Rubber Div. Amer. Chem. Soc., 108th, New Orleans, Louisiana* (1975).
31. A. C. Soldatos and A. S. Burhans, in "Multicomponent Polymer Systems" (N. A. J. Platzer, ed.), Adv. Chem. Ser. 99. Amer. Chem. Soc., Washington, D.C., 1971.

32. F. P. Baldwin and I. J. Gardner, *Amer. Chem. Soc. Organ. Coatings Plast. Chem.* **36**(2), 643 (1976).

33. D. Katz and A. V. Tobolsky, *J. Polym. Sci. Part A-2* **2**, 1587 (1964).

34. K. Kato, *Jpn. Plast.* **2** (April), 6 (1968).

35. M. Matsuo, *Jpn. Plast.* **2** (July), 6 (1968).

36. D. A. Thomas, *J. Polym. Sci. C* (Symp. issue no. 60; S. L. Aggarwal, ed.) **15**, 189 (1977).

37. J. E. Lorenz, D. A. Thomas, and L. H. Sperling, *in* "Emulsion Polymerization" (I. Piirma and J. L. Gardon, eds.), Amer. Chem. Soc. Symp. Ser. No. 24, 1976.

38. S. C. Kim, D. Klempner, K. C. Frisch, H. L. Frisch, and H. Ghiradella, *Polym. Eng. Sci.* **15**, 339 (1975).

39. A. A. Donatelli, D. A. Thomas, and L. H. Sperling, *in* "Recent Advances in Polymer Blends, Grafts, and Blocks" (L. H. Sperling, ed.). Plenum, New York (1974).

40. A. A. Donatelli, L. H. Sperling, and D. A. Thomas, *J. Appl. Polym. Sci.* **21**, 1189 (1977).

41. G. Allen, M. J. Bowden, D. J. Blundell, G. M. Jeffs, J. Vyvoda, and T. White, *Polymer* **14**, 604 (1973).

42. G. C. Eastmond and E. G. Smith, *Polymer* **17**, 367 (1976).

43. D. J. Meier, *J. Polym. Sci. Part C* **26**, 81 (1969).

44. D. J. Meier, *Polym. Prepr.* **11**, 400 (1970).

45. M. Matsuo, T. K. Kwei, D. Klempner, and H. L. Frisch, *Polym. Eng. Sci.* **10**, 327 (1970).

46. L. H. Sperling, T. W. Chiu, C. Hartman, and D. A. Thomas, *Int. J. Polym. Mater.* **1**, 331 (1972).

47. K. Kato, *Polym. Lett.* **4**, 35 (1966).

48. G. Allen, M. J. Bowden, G. Lewis, D. J. Blundell, and G. M. Jeffs, *Polymer* **15**, 13 (1974).

49. G. M. Yenwo, L. H. Sperling, J. Pulido, J. A. Manson, and A. Conde, *Polym. Eng. Sci.* **17**(4), 251–256 (1977).

50. D. L. Siegfried, J. A. Manson, and L. H. Sperling, *J. Polym. Sci. Phys. Ed.* **16**, 583 (1978).

51. S. C. Kim, D. Klempner, K. C. Frisch, and H. L. Frisch, *Macromolecules* **9**, 263 (1976).

52. T. K. Kwei, T. Nishi, and R. F. Roberts, *Macromolecules* **7**, 5,667 (1974).

53. D. Klempner, H. L. Frisch, and K. C. Frisch, *J. Polym. Sci. Part A-2* **8**, 921 (1970).

54. M. Takayanagi, H. Harima, and Y. Iwata, *Mem. Fac. Eng. Kyushu Univ.* **23**, 1 (1963).

55. W. E. A. Davies, *J. Phys. D* **4**, 318 (1971).

56. G. Allen, M. J. Bowden, S. M. Todd, D. J. Blundell, G. M. Jeffs, and W. E. A. Davies, *Polymer* **15**, 28 (1974).

57. E. H. Kerner, *Proc. Phys. Soc.* **69**, 808 (1956).

58. Z. Hashin and S. Shtrikman, *Mech. Phys. Solids* **11**, 127 (1963).

59a. E. A. Neubauer, D. A. Thomas, and L. H. Sperling, *Polymer* **19**, 188 (1978); E. A. Neubauer, N. Devia-Manjarrés, D. A. Thomas, and L. H. Sperling, *Coatings in Plastics Prepr.* **37**(1) 252 (1977).

59b. M. J. Hatch, U.S. Patent 3,041,292 (1962).

59c. T. A. Solak and J. T. Duke, U.S. Patent 3,426,102 (1969).

59d. R. D. Hibelink and G. H. Peters, U.S. Patent 3,657,379 (1972).

59e. F. G. Hutchinson, British Patent 1,239,701 (1971).

59f. J. M. Hawkins, British Patent 1,197,974 (1970).

59g. G. S. Solt, British Patent 728,508 (1955).

59h. W. H. Parrkiss and R. Orr, British Patent 786,102 (1957).

60. S. H. Feairheller, A. H. Korn, E. H. Harris, E. M. Filachione, and M. M. Taylor, U.S. Patent 3,843,320 (1974).

61. L. H. Sperling, V. A. Forlenza, and J. A. Manson, *J. Polym. Sci. Polym. Lett. Ed.* **13**, 713 (1975).

62. G. S. Solt, British Patent 728,508 (1955).

63. B. Volmert, U.S. Patent 3,055,859 (1962).
64. H. A. Clark, U.S. Patent 3,527, 842 (1970).
65. Anonymous (Ciba, Ltd.) British Patent 1,223,338 (1971).
66. K. C. Frisch, H. L. Frisch, and D. Klempner, German Patent 2,153,987 (1972).
67. J. J. Falcetta, G. D. Friends, and G. C. C. Niu, German Offen. Patent 2,518,904 (1975).
68. C. F. Ryan and R. J. Crochowski, U.S. Patent 3,426,101 (1969).
69. G. Odian and B. S. Bernstein, *Nucleonics* **21**, 80 (1963).
70. T. J. Szymczak and J. A. Manson, *Mod. Plast.* **51**(8), 66 (1974).
71. G. N. Yenwo, J. A. Manson, J. Pulido, L. H. Sperling, A. Conde, and N. Devia-Manjarrés, *J. Appl. Polym. Sci.* 21, 1531 (1977).

# Chapter 12

# Interfacial Agents ("Compatibilizers") for Polymer Blends

*D. R. Paul*

*Department of Chemical Engineering*
*The University of Texas*
*Austin, Texas*

## I. INTRODUCTION

Ideally, two or more polymers may be blended together to form a wide variety of random or structured morphologies to obtain products that potentially offer desirable combinations of characteristics. However, it may be difficult or impossible in practice to achieve these potential combinations through simple blending because of some inherent and fundamental problems. Frequently, the two polymers are thermodynamically immiscible, which precludes generating a truly homogeneous product. This may not be a problem per se since often it is desirable to have a two-phase structure. However, the situation at the interface between these two phases very often does lead to problems. The typical case is a high interfacial tension and poor

35

adhesion between the two (see Volume 1, Chapter 6). This interfacial tension contributes, along with high viscosities, to the inherent difficulty of imparting the desired degree of dispersion to random mixtures and to their subsequent lack of stability to gross separation or stratification during later processing or use (see Volume 1, Chapter 7). Poor adhesion leads, in part, to the very weak and brittle (or cheesy) mechanical behavior often observed in dispersed blends [1] and may render some highly structured morphologies (see Chapters 15 and 16) impossible.

In the scientific literature the term *compatibility* is often used in a thermodynamic sense to be synonymous with miscibility; however, in the technological literature it is used to characterize the relative ease of fabrication or the properties of two polymers in a blend [1–3]. Components that resist gross phase segregation and/or give desirable blend properties are frequently said to have a degree of "compatibility" even though in a thermodynamic sense they are not miscible. Confusion results from this dual usage but both practices are widespread. No attempt is made here to change this; however, to reduce confusion this term, when applied in the technological way, will be accompanied by quotation marks in this chapter.

It is widely known that the presence of certain polymeric species, usually block or graft copolymers suitably chosen, can alleviate to some degree the problems mentioned above, and it is generally believed that this is a result of their ability to alter the interfacial situation [3–13]. Such species, as a consequence, are often referred to as "compatibilizers" [3–14], which is analogous to the term "solubilization" used in the colloid field to describe the effect surfactants have on the ability to mix oil and water [15]. The general view is that a properly chosen block or graft copolymer can preferentially locate at the interface between the two phases, as shown in Fig. 1. Ideally this component should be a block of graft with different segments that are chemically identical to those in the respective phases. However, the desired effect may still result if one of the arms of the block or graft were to be miscible with or adhered to one of the phases. This

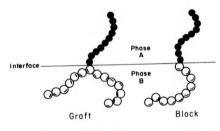

Fig. 1   Ideal location of block-and-graft copolymers at the interface between polymer phases A and B.

type of surface activity should (1) reduce the interfacial energy between the phases, (2) permit a finer dispersion during mixing, (3) provide a measure of stability against gross segregation, and (4) result in improved interfacial adhesion [16].

There are many important applications of this concept particularly in the area of rubber modified plastics to achieve improved toughness (see Chapters 13 and 14). However, the concept has more general applicability than this, and with the increased synthetic capability to tailor graft and block copolymers developed in the last few years [17], it is expected that many new and important applications will emerge.

The purpose of this chapter is to examine and review (1) the basic evidence that supports the envisioned surface active character of block-and-graft copolymers, (2) the fundamental information available to guide the design and use of this concept, and (3) examples of where blocks and grafts have been used to obtain desirable mechanical performance in specific systems. The latter do not include conventional impact modification of plastics, which is dealt with in fuller detail in Chapters 13 and 14. Although this is a widely recognized concept of great utility, it is interesting to note that there is only a modest amount of information available that illustrates its principle in an unambiguous manner.

## II. DIRECT EVIDENCE FOR INTERFACIAL ACTIVITY OF BLOCK COPOLYMERS

It has been recognized for some time that many block-and-graft copolymers segrate into two phases in the solid state [17–21]. The careful control of domain structure with appropriate selection of segment types has led to elastomeric behavior in systems that do not require chemical vulcanization. Most notable examples are the "thermoplastic elastomers" based on styrene and butadiene or isoprene and the segmented polyurethanes for elastomeric fibers and other applications. From a theoretical basis it is also now understood [22] that even in dilute solutions certain individual block copolymer molecules depart from the usual random-coil conformation to form a dumbbell shape [16] owing to the repulsion of the chemically different segments. This intramolecular repulsion is the basis for the surface active behavior of conventional detergents in which individual molecules have nonpolar and polar (or ionic) parts that prefer to associate with oil and water phases, respectively. As a result, important applications have been found for block copolymers as nonionic detergents. A commercially available series, tradename Pluronics, are ABA-type block copolymers [23, 24] in which the A part is composed of ethylene oxide sequences (hydrophilic), whereas the B part is composed of propylene oxide sequences (hydrophobic). A wide range of sequence lengths are available [23]. Interestingly, random copolymers of ethylene oxide and propylene oxide or blends of the homopolymers do not show interfacial activity [23]. A related series known as Tetronics is also available [23, 24]. Similarly, block copolymers based on styrene and isoprene or butadiene have surface-active behavior for two immiscible organic liquids such as dimethylformamide and hexane. Such materials are therefore referred to as oil-in-oil emulsifiers [25–27].

Owen *et al.* [28, 29] have presented detailed studies that clearly demonstrate the surface activity of some ABA-type block copolymers. In one case, the A part was a poly(dimethylsiloxane) chain of fixed length and the B part was a polyether (equal parts by weight of a random ethylene oxide and propylene oxide sequence) whose length was varied [29]. By surface and interfacial tension measurements, they showed that block copolymers with high siloxane contents were preferentially adsorbed at the air–water interface, whereas the copolymers with longer polyether sequences were preferentially absorbed at a silicone fluid–water interface. They also studied the case in which the A part was polystyrene and the B part was poly-(dimethylsiloxane). Certain mixtures of these polymers at the 1% level with polystyrene produced a polymer surface essentially like that of silicone in terms of wetting and physical feel. Gaines and Bender [30] also studied the kinetics of migration of the similar AB copolymer to the polystyrene melt surface.

Legrand and Gaines [31] have shown that block copolymers of silicone and polycarbonate blended with polycarbonate will locate at the poly-carbonate–glass surface when films are cast. Other work has shown that these block copolymers promote adhesion between polycarbonate and glass [32].

The above selected studies (see also Volume 1, Chapter 6) show directly that block copolymers do adsorb preferentially at appropriate liquid–air, liquid–liquid, polymer–glass, and other polymer interfaces [33]. There appear to be no studies by these techniques on the effect of the appropriate block copolymer on a polymer–polymer interface. The extent of the interfacial activity in these cases has been inferred from other observations as described later.

## III.  ROLE OF GRAFT COPOLYMERS
## IN IMPACT-MODIFIED PLASTICS

A severe limitation of many plastics especially glassy ones is their brittle nature which results in poor energy absorbing capability or impact strength.

Very early it was learned for polystyrene that this problem could be combated by polymerizing styrene in the presence of rubber under proper conditions [34, 35] to produce a rubber modified blend. It was also found that simply blending the rubber with polystyrene did not produce the same desired effect. Lunstedt and Bevilacqua [4] showed subsequently that, if a graft of styrene onto rubber was made which in turn was simply blended with polystyrene, that a significant increase in impact strength was produced, as the top curve in Fig. 2 illustrates. The lower curves in Fig. 2 illustrate the range of effects found by merely blending rubber with polystyrene. Curves similar to these are now available for a wide variety of systems [4, 36, 37]. It has since been recognized [16, 38] that during the polymerization of styrene in the presence of rubber that some grafting takes place so that in effect a similar situation to that in the experiments of Lundstedt and Bevilacqua is achieved by these reactor-produced blends.

The detailed mechanism by which the proper combination with rubber acts to improve the impact strength of glassy polymer is very complex and is dealt with extensively in Chapters 13 and 14. Many factors are involved in making the "proper" combination of which regulation of phase morphology is an important example. It is now understood for high-impact polystyrene (HIPS) that occlusion of polystyrene into the rubber phase plays an important role [39, 40]. This morphology is controlled by mixing conditions during polymerization [41–43]. Similarly, in other systems like poly(vinyl chloride) (PVC) in which impact modifiers are made in separate processes and compounded in after polymerization, the proper size and dispersal of the rubbery phase is also important [36, 37, 44]. However, in nearly all cases of impact modification of plastics, grafting reactions seem to be a necessary factor. On the one hand, it is believed that the graft

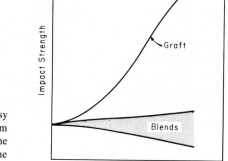

Fig. 2 Effect of adding rubber to glassy polymers on impact strength. The bottom band shows results for simple blending, the upper curve illustrates the situation when some graft copolymer is also present.

copolymer provides improved adhesion between the glassy and rubbery phases, which is crucial to stress transfer between these phases. However, in addition to this, it is quite likely that the interfacial tension between phases may also be lowered by this interfacial activity, which would play an important role in blend morphology. Subsequently, the graft at the interface would contribute to stability against segregation during processing.

In nearly all cases in which graft copolymers are made during the course of impact modification, the grafting is not complete; hence, the blend consists of two homopolymers and some graft copolymer. One of the very severe limitations to fundamental interpretation and design of these systems is the lack of completely adequate techniques to characterize the graft copolymer. In addition, the ability to control these reactions is equally limited.

It is interesting to note that properly chosen block copolymers are now being used as impact modifiers for plastics (see Chapter 18). Their effectiveness for this purpose is also due to their interfacial activity in the blend, which permits fine dispersion, good adhesion, and stability. Techniques are now available for precisely controlling and characterizing the molecular structure of block copolymers, and it will be interesting to see if this will result in a better understanding of impact modification of plastics as well as superior products. The opportunity exists with block copolymers to determine more precisely the exact extent of the interfacial activity and therefore to assess its role in the overall process.

## IV.  FUNDAMENTAL CONSIDERATIONS OF "COMPATIBILIZER" SYSTEMS

The beneficial effects of block or graft copolymer surface activity in blends of two immiscible polymers are qualitatively event, but the magnitude of these benefits is not so clear. It is obvious, however, that the degree of improvement will depend on the structure of the block or graft chosen. An attempt will be made in this section to bring together information that relates to the effective design or selection of surface active additives. No complete theory or set of experimental observations are available for this purpose, but there is some useful information with which tentative guidelines can be formulated. The general requirements are discussed first. This will be followed by a summary of pertinent experimental observations of a morphological nature, and finally an attempt is made to extract useful ideas from recent theories. Major emphasis is placed on molecular weight since this is one of the main variables once the homopolymers are specified. Most of the discussion centers on block copolymers rather than grafts since little fundamental information is available for the latter.

While there is a considerable literature relevant to the present objective, it is not possible to obtain the complete picture desired because most studies have not considered all of the factors of importance. Typically a single response, for example, phase morphology or dynamic mechanical properties [45, 46], has been studied thoroughly, but a complete evaluation of the system is rarely done. It is not possible at present to extrapolate the role of these single responses to see their effect, if any, on the practical and more critical issues of phase stability, adhesion, or ultimate mechanical properties.

## A.  General Requirements

For the block or graft copolymer to be effective it must locate preferentially at the blend interface. One ideal way to do this is shown schematically in Fig. 1. Other conformational modes are possible, such as segments adsorbed onto the surface of one polymer rather than penetrating it. Conformational restraints are important [47, 48], and on this basis a block copolymer is expected to be superior to a graft [5, 49]. A graft with one branch is shown in Fig. 1; however, multiple branches would restrict the opportunities for the backbone to penetrate its homopolymer phase. This, of course, would not preclude adhesion of the backbone to this phase. For the same reasons, diblock copolymers might be more effective than triblocks [5, 49]. Relative proportions and molecular weights of the two types of segments have important effects, as is shown later. The most obvious choice is to select block or graft copolymers whose segments are chemically identical to the homopolymer phases, that is, in Fig. 1 there would be an AB block and an A–*g*–B or B–*g*–A graft. An equivalent alternative exists when the block or graft has a segment miscible with Phase A although it is not chemically the same. A suitable but not equivalent alternative exists when this segment adheres well to Phase A but is not miscible with it.

For the block or graft copolymer to locate at the blend interface it should have the propensity to segregate into two phases. This tendency in block-and-graft copolymers depends on the interactions between the two segments and on their molecular weights. This question is dealt with theoretically in Section IV.C. Furthermore, the block or graft should not be miscible as a whole molecule in one of the homopolymer phases. This tendency also depends on segmental interactions and molecular weights.

Since block-and-graft copolymers are likely to be expensive, it would be of interest to maximize their efficiency so that only small amounts are required. How much is needed depends on many factors but the conformation at the interface and the overall molecular weight are two important ones. The ideal conformation for the block copolymer shown in Fig. 1 (complete penetration of the two phases) is used here to illustrate quantitatively the effect of molecular weight on efficiency. The following calculation is an estimate of the amount of block copolymer of molecular weight $M$ required to saturate all of the interface in a blend. Block copolymer in excess of this is clearly wasted. A blend that contains a volume fraction $\phi_A$ of Polymer A as spherical particles of radius $R$ in a matrix of B has an interfacial area per unit volume of original blend equal to $3\phi_A/R$. If each block copolymer molecule occupies an area $a$ at this interface, then the mass of block copolymer required is

$$\text{Mass of block copolymer/Original volume of blend} = 3\phi_A M/aRN \quad (1)$$

where $N$ is Avogadro's number. A minimum possible value for $a$ is about $50\,\text{Å}^2$ so for $\phi_A = 0.2$ and $R = 1\,\mu\text{m}$ it would take about 20% by weight of a block copolymer with $M = 10^5$ to fill up the interface, whereas this drops to 2% when $M = 10^4$. Conformational restrictions will prohibit filling the interface with block copolymer and may reduce the amount calculated here by an order of magnitude [47, 48]. Because of the large differences in cohesive and chemical bond energies adequate adhesion should result at much lower surface occupation densities [49].

While the above clearly points out the advantages of lower molecular weights, other factors in the overall efficiency point to the need for higher molecular weights. For example, the segments should be long enough to have sufficient cohesive forces to anchor them firmly into the domains they penetrate. Gaylord [3] has proposed that a degree of polymerization of 10–15 should be adequate. However, the molecular weight effects considered in the next two sections appear to be far more stringent than this, and thus even higher values will be required. It is evident that there is likely to be an optimum molecular weight when the additive is examined on a cost–benefit basis.

## B.  Experimental Morphological Observations

The morphology of ternary blends of Polymers A and B with AB block copolymers has been studied extensively. Most studies have involved systems based on styrene and butadiene or isoprene because of their ease of preparation and commercial importance. Molau [16, 50, 51] was apparently among the first to pursue this area in any detail, while later work was done by Riess [5–7, 49], Inoue [52–54], and Skoulios [55]. The motivation for this work, in part, was to model high-impact polystyrene. Early studies involved solutions of these blends in a common solvent. This work recognized conceptually the interfacial activity of the block copolymer and demonstrated its ability to emulsify the solutions of incompatible polymers to produce oil-in-oil emulsions. The effect of the block copolymer on domain size and the profound increase in the stability of the dispersion were clearly shown. Molau [50, 51] showed that block or graft copolymers added to concentrated solutions of immiscible polymers could slow down the phase segregation or demixing process by two or more orders of magnitude.

In later work, the morphology of solution cast blends was examined in the bulk state by electron microscopy aided by the osmium tetroxide staining method reported by Kato [56]. Via this route the individual domains may be observed. Inoue has reported extensive observations of particular interest here. He examined the domain morphology of a wide range of block copolymers by this method and what happens when homopolymers are added. Block copolymers may form spherical, rodlike, or lamellar domains whose size, shape, and identity depend on the relative proportions of the two segments. When either or both homopolymers were blended with the block copolymer, it was found that under certain circumstances that the homopolymers were incorporated into the block copolymer domains of its type, as evidenced by the growth in size of these domains in proportion to the amounts added. Stated in the reverse manner, the size of domains in homopolymer blends were reduced by addition of a suitable block copolymer. This provides direct evidence that block polymers may act as an emulsifier, which "solubilizes" the two immiscible homopolymers into the domains of the corresponding block segments. The block copolymer restrains the phase segregation of the homopolymers into their macroscopic domains and therefore facilitates mixing of the two immiscible homopolymers. This does not appear to change the state of immiscibility of these two, and the term "compatibilizing" ability of block or graft copolymers only applies in the technological sense mentioned in Section I.

It is important to note that this "solubilization" does not occur in all cases as shown by every careful investigation of this type. There seems to be universal agreement that solubilization only occurs when the molecular weights of the homopolymers are less than or comparable to the molecular weights of the corresponding segments in the block copolymer [5, 6, 49–55, 57–59]. When the homopolymer molecular weight is larger than that of the corresponding block segment, the homopolymer forms a separate phase and is not solubilized into the domains of the block copolymers (See Chapter 18 for further explanation). This universal finding may be very relevant to design and selection of surfactant block and graft copolymers for blends and would appear to override the requirements on molecular weights given in the previous section. Apparently, this conclusion has been arrived at only by observations on solution cast mixtures, and it would be most important to know if and to what extent it may be altered for melt blended systems, which are of more practical concern. It would also be interesting to know to what extent this apparent rule limits the practical utilization of block and graft copolymers.

Riess *et al.*[5, 49] have made extensive studies on ternary blends of Polymers A and B with Block copolymer AB using other techniques. They found that the ternary triangular diagram can be divided into regions of relative clarity or opacity. These observations can best be understood in terms of the effect the block copolymer has on domain size and the subsequent effect of this on light transmission. Binary blends of immiscible homopolymers are usually opaque due to large domain sizes, whereas block copolymers themselves are usually transparent because of much smaller domains. For the ternary mixtures, there may be a zone of transparent blends near the AB apex, whereas near the A–B binary leg there is a zone of opaque blends. The line of demarcation may be diffuse but its location depends critically on the molecular weights of the homopolymers and the block copolymer segments. Riess' findings may be summarized roughly on the schematic molecular weight map shown in Fig. 3.

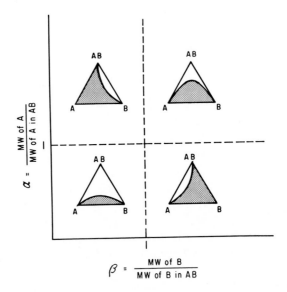

Fig. 3    Effect of relative molecular weights (MW) on clarity, or domain size, of ternary blends of Polymers A and B plus Block Copolymer AB.

The coordinates of this graph α and β are the ratios of the molecular weight of the homopolymer to the molecular weight of the corresponding segment in the block copolymer for Components A and B, respectively. It should be understood that this map is too simplistic to represent all of the important factors, for example, absolute molecular weights, but it does allow a useful visualization scheme that is an adequate summary of the information described previously and shown by Riess. This map is divided into four zones as shown. Each zone contains a schematic triangular diagram whose light and dark regions represent transparent and opaque blends, respectively. For the two diagonal zones, the dividing boundaries are such that transparent blends result no matter what the proportions of A and B are, provided enough AB is added. For $\alpha < 1$ and $\beta < 1$, less AB is required for transparency than for the zone in which $\alpha > 1$ and $\beta > 1$. This agrees with the expectations stated earlier. For the two off-diagonal zones, the ratio of A to B in the blend does matter. The upper left zone, where $\alpha > 1$ and $\beta < 1$, may be visualized by two limiting cases:

$$\overline{M}_A(AB) < \overline{M}_A(A) = \overline{M}_B(B) < \overline{M}_B(AB) \qquad (2a)$$

$$\overline{M}_A(A) > \overline{M}_A(AB) = \overline{M}_B(AB) > \overline{M}_B(B) \qquad (2b)$$

In either of these, copolymer domains of A do not accept Homopolymer A but copolymer domains of B do accept Homopolymer B in agreement with Inoue and others. As a result, clarity only exists near the AB–B binary leg since outside this region rather large domains of homopolymer A are present. An analogous situation occurs in the lower right zone except the roles of A and B are reversed. Riess noted that block copolymers were more efficient in promoting clarity than graft copolymers and that diblocks were more efficient than triblocks [5, 49].

Riess also reports some interesting attempts to determine the extent to which block copolymers are actually located at the blend interface [49]. These techniques are crude but do allow some additional direct confirmation of interfacial activity. It was generally concluded that the optimal situation is near the $\alpha = \beta$ line in Fig. 3. The limited experiments failed to confirm the expectation that interfacial location would be greater in the zone $\alpha = \beta < 1$ than in the zone $\alpha = \beta > 1$. It is significant that the block copolymer was observed to be at the interface in the latter region, since this may not have been expected on the basis of the previous conclusions.

In summary, the best design rule to be extracted from these observations would appear to be that $\alpha$ should be as nearly equal to $\beta$ as possible. It is probably better to have both $\alpha$ and $\beta$ less than one, but this must be weighed against other factors for each situation.

## C.  Relevance of Block Copolymer Theories

In recent years theories relevant to this chapter have begun to emerge. First, a start has been made toward a theory for the polymer–polymer interface [60–63]. This work has shown that interpenetration of segments across the interface will occur which has important ramifications to adhesion. Second, theories for domain formation and the state of the interface in block copolymers have received even greater attention [47,63–73]. The latter results are reviewed here for the relevance they have to the present objective. However, of more direct value would be to combine these two theoretical approaches to treat the ternary system of A, B, plus AB. There is every reason to believe that attempts to do just this will soon be made because of the obvious need for the guidance such a theory could offer.[†]

Theoretical approaches to block copolymer problems have been made using different starting points and simplifications [47,63–73]. Of most importance here is the statistical mechanical approximation of Meier [64]. The present interest is to know under what conditions a block copolymer will or will not phase segregate and what is the nature of the "interface" in intermediate cases. Meier's physical picture of the region between domains of A and B segments of a block copolymer is shown in Fig. 4a. Also seen there is a single AB molecule with a circle denoting the juncture between segment types. The entropy part of the free energy favors locating this juncture at random rather than in a well-defined interfacial plane; however, owing to the usual nature of segmental interactions, the energy part of the free energy would prefer a sharp interface. Consequently, a diffuse interfacial zone

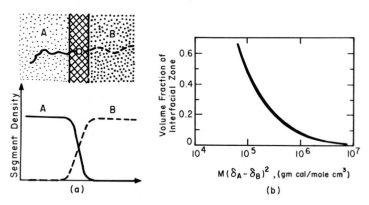

Fig. 4   Physical picture (a) and theoretical estimate (b) of the interfacial region of an AB block copolymer. (After Meier [64].)

[†] *Note added in proof:* The following represent such an effort: D. J. Meier, *Polym. Prepr. Amer. Chem. Soc. Div. Polym. Chem.* **18**(1), 340, 837 (1977).

(shaded area) may result the size of which depends on the compromise created by the relative magnitudes of the entropy and energy contributions. The graph in Fig. 4a shows schematically the density of A and B segments in this zone. The graph in Fig. 4b shows a quantitative estimate of the size of the interfacial zone as its fraction of the system volume. To a first approximation, this is a function only of the product of the copolymer molecular weight $M$, and a measure of the segmental interactions estimated by the solubility parameter difference for the two segment types $(\delta_A - \delta_B)^2$.

The latter graph offers the opportunity to unify, at least conceptually, the diverse observations found in the literature for the phase behavior of block copolymers. It is clear that, when the volume of the interfacial zone is near zero, a sharp interface exists. This has been shown to be the case for many systems [17, 47]. This requires a high molecular weight or a large segmental interaction such that the abscissa in Fig. 4 is very large. If the interfacial volume fraction is near unity, there would be no segregation into domains. This occurs for very low molecular weights or very small segmental interactions, for example, $\delta_A \simeq \delta_B$. Systems of this type are known and examples include polystyrene–poly($\alpha$-methylstyrene) [17], polysulfone–polycarbonate [17], and polycarbonate–polycaprolactone [74]. For systems with intermediate situations, an interfacial zone may exist more or less as a third phase whose properties are intermediate to those of A and B. Experimental evidence for this has been reported [73].

The system styrene–$\alpha$-methylstyrene is interesting because its block copolymers have been reported to have one or two phases depending on structure, composition, and molecular weight [69, 75–78]. The solubility parameter difference $(\delta_A - \delta_B)$ for this system is estimated to be about 0.1 $(\text{cal/cm}^3)^{1/2}$. For $M = 10^5$, the abscissa in Fig. 4 becomes $10^4$, and thus, in this graph it is suggested that the volume of the interfacial zone occupies the entire system. This may be interpreted to mean there is no phase separation within the block copolymer, which is in accord with experimental observations [75, 76]. If the molecular weight were increased by more than an order of magnitude, the data curve in Fig. 4 would suggest that phase separation is likely. More refined analyses have been made for this system [68, 69], but this calculation serves to illustrate the point.

Because of this tendency not to phase segregate, block copolymers of styrene and $\alpha$-methylstyrene would not be expected to have ideal interfacial activity of the type described earlier for blends of the homopolymers. Robeson *et al.* [76] have described experiments, summarized below, that seem to support this. Their diblock copolymers had glass transition temperature $(T_g)$ values intermediate to those for polystyrene and poly($\alpha$-methylstyrene), which is strong evidence that they are homogeneous. A blend of a low molecular weight block copolymer with polystyrene showed a single

$T_g$ intermediate to that of polystyrene and the homogeneous block copolymer. However, a similar blend using a somewhat higher molecular weight block copolymer showed two-phase behavior, and the $T_g$'s observed were those of polystyrene and the homogeneous block copolymer. The latter indicates that the polystyrene part of the block is not incorporated into the polystyrene phase but rather it preferentially forms a homogeneous phase with the poly($\alpha$-methylstyrene) segments. This is clear proof that there is no interfacial activity in this case. All blends of the block copolymers with poly($\alpha$-methylstyrene) were homogeneous, as indicated by one $T_g$. Some ternary blends of polystyrene and poly($\alpha$-methylstyrene) very rich in low molecular weight block copolymer showed a single $T_g$, while lower proportions of the block copolymer resulted in phase separation in which apparently the block copolymer was located in the poly($\alpha$-methylstyrene), that is, no evidence for interfacial activity. In systems of this type a different situation and role of the block copolymer exists than that envisioned in Fig. 1. Evidently addition of the block copolymer causes the blend to become one phase in the same way as addition of solvent can. This type of action is not unique to block copolymers at all. Recent results [79] have shown that a homopolymer miscible individually with each of two immiscible polymers will give some homogeneous ternary blends as suggested.

Two generalizations for this section seem possible. If the two homopolymers are very nearly miscible, addition of their block copolymer to the blend may result in a one-phase system. This would be favored by a low molecular weight for the block copolymer. If the two polymers are far from being miscible, then no block copolymer is likely to cause one-phase mixtures. In this case the role of the block copolymer is to be an interfacial agent and this is only possible if the block copolymer phase segregates readily. The latter is favored by high molecular weight, as shown in Fig. 4. For intermediate-type systems, a segment that is too short may result in a failure of the block or graft copolymer to phase separate or cause the whole block or graft molecule to be miscible in one of the homopolymers phases. Either would detract from its interfacial activity.

## V. SELECTED EXAMPLES OF THE EFFECT OF "COMPATIBILIZERS" ON MECHANICAL BEHAVIOR OF BLENDS

Blends of immiscible polymers may assume phase morphologies ranging from random dispersions to the highly structured laminates of films. Addition of a block of graft copolymer with interfacial activity for the blend components can affect many behavioral characteristics, but since mechanical properties are important to most products, it is pertinent to focus this section of examples in this area. Very often the mechanical behavior of polymer blends is quite poor and limits or prohibits their use. Specific examples in which polymers with interfacial activity have been used to improve blend mechanical properties are discussed in this section. The first set of examples deals with dispersed blends; the second deals with laminates. While adhesion between the two immiscible polymers is a key factor in both sets, the latter allows a more direct observation of this. The important examples of dispersed blends to be found in impact modified plastics are purposely excluded since these are dealt with extensively in Chapters 13 and 14.

In nearly all of the examples described here, the block or graft copolymers employed are not well defined nor have they been designed for this purpose along the lines described in the previous section. Studies of the type one could easily envision, using well-defined block copolymers, have not been reported except in a few cases, and these have not been extensive in scope. The effects shown in these examples, therefore, should be encouraging to those who wish to utilize this concept since one could reasonably expect to find even greater benefits if carefully designed and optimized "compatibilizers" were developed and used.

## A. Dispersed Blends

In extreme cases, dispersed mixtures of two immiscible polymers are weak and brittle and are frequently described as cheesy. Blends of polystyrene and polyethylene are good examples of this [1, 10–12]. The small deformation modulus may obey an appropriate additive mixing rule, but the ultimate properties invariably do not. The ultimate strength and deformation of some blends may be less than that of either pure component in the mixture [1, 10]. It is generally agreed that poor adhesion between the phases plays a significant role in these cases [1, 80]. The inclusion of one phase in a matrix of another can cause very significant stress concentrations [80]. Fracture may initiate at the blend interface, and in any case the fracture path would be expected to follow preferentially the weak interface between the two polymers. These processes are reflected in the ultimate mechanical response of the blend in a very complex manner. Some attempts have been made to quantify this generally understood qualitative model [80].

In all of the following examples a separately formed "compatibilizer" has been added to a blend of two immiscible polymers. Similar situations may be had by (1) graft formation during polymerization of a monomer in the presence of another polymer, (2) block formation by interchange reactions during processing of a blend of condensation polymers, or (3) block or graft formation through chain scission mechanisms in processing operations like mastication [18, 81]. All of these are important possibilities but in most cases the extent of the reactions that produce the "compatibilizer" are undetermined so they do not represent good examples for present purposes.

### 1. Polystyrene–Polyethylene plus Graft Copolymer

Enhancement of the very poor mechanical properties of polystyrene–polyethylene blends by this route has received considerable attention from two groups [10–12]. Both used graft copolymers as the blend additive. In one case, styrene was grafted to low-density polyethylene by a radiation technique [10], while in the other, Friedel–Crafts alkylation of polystyrene (PS) with low-density polyethylene (LDPE) was used [11, 12]. Attempts were made to characterize these grafts, but no definitive measures of the length of the grafts or their frequency on the backbone were found in either case. Cross-linking is a side reaction of grafting [10] that undoubtedly detracts from the efficacy of the interfacial behavior.

All of the results shown here were determined from compression-molded specimens of previously mixed material. Different mixing sequences of the graft copolymer were used in the two studies [10–12]. Figure 5 is a summary of the ultimate tensile strength data reported by Barentsen and Heikens [11]. Comparable but less extensive data were reported by Locke and Paul [10]. The lowest curve is the response for the binary PS–LDPE system, which displays a minimum in the midconcentration range. The uppermost curve

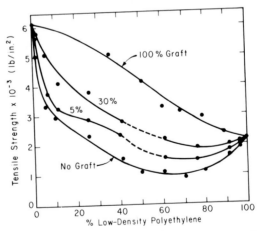

Fig. 5   Tensile strength of PS–LDPE blends and graft copolymer. For the middle two curves the graft was added to the blend in the amounts shown based on the minor component of the blend (LDPE on the left, PS on the right). (After Barentsen and Heikens [11].)

is for a series of grafts with different proportions of LDPE and PS, and it displays a more nearly additive response. The curves in between are for blends of LDPE and PS to which a graft copolymer of nearly equal parts PS and LDPE has been added. The graft was preblended in the amounts shown with the homopolymer present in minor proportion, thus the dispersed phase, in the final blend. The percentages of graft shown thus are based on the minor component, not the final blend. The dotted lines connect the series in which graft was preblended with PS to the series in which graft was preblended with LDPE. Addition of the graft improves the strength of PS–LDPE blends, and the data suggest that the additive response of the graft material will be approached as more graft is added. It would be expected that perfection of the graft structure for this purpose would not alter this trend but might reduce the amount required to accomplish a given degree of improvement.

Locke and Paul simultaneously mixed equal parts PS and LDPE with a graft of this same composition and found that the elongation at break responded as shown in Fig. 6. This blend composition is approximately at the minimum in the curve of elongation at break versus composition [10]. Addition of the graft improves this property in more or less direct proportion

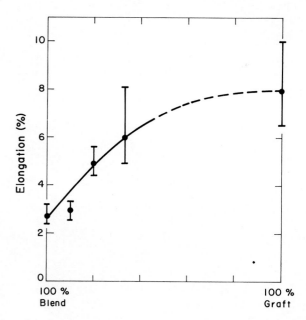

Fig. 6 Effect of graft copolymer on elongation at break of LDPE–PS blends. The graft contains 50/50% LDPE/PS. (From Locke and Paul [10]. Reproduced by permission of *J. Appl. Polym. Sci.*)

to the amount added. While the graft has an elongation approximately four times that of the blend, the absolute value for the graft is not impressive. It would be interesting to learn whether optimization of the graft structure would result in a more ductile material and if this would produce correspondingly greater improvements in the ductility of blends, as Fig. 6 suggests.

Barentsen and Heikens [11, 12, 82] made an interesting study of the fracture surfaces of their mixtures using scanning electron microscopy, and some of their typical photomicrographs are shown in Fig. 7. Figure 7a is for a 75% PS and 25% LDPE blend in which the dispersed phase is LDPE. All domain surfaces are very smooth with no evidence of adhesion between the two. An analogous blend containing graft copolymer is shown in Fig. 7b. In this case, 25% of a premixed blend of 5% graft with LDPE

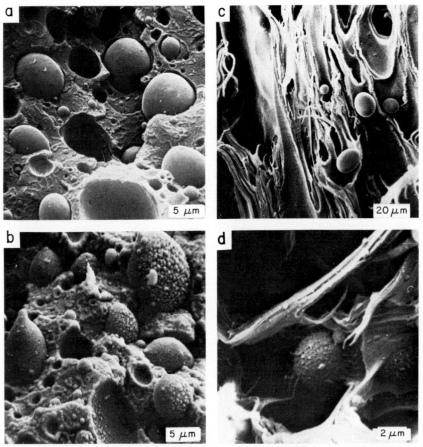

Fig. 7    Scanning electron photomicrographs of fracture surfaces of PS–LDPE blends with and without added graft copolymer: (a) 75/25 PS/LDPE; (b) 75/23.75/1.25 PS/LDPE/graft; (c) 30/70 PS/LDPE; (d) 25/67.5/7.5 PS/LDPE/graft. (Courtesy of D. Heikens.)

was blended with 75% PS (graft content is 1.25% of total blend). Small globules are seen at the surface between the two phases which is presumed to be graft copolymer—the total volume of these irregularities corresponds approximately to the volume of graft added. These results are similar to those for ABS with grafted rubber particles [11, 36]. A 30% PS and 70% LDPE blend is shown in Fig. 7c in which yielding of the LDPE matrix is apparent with attendant void formation. Again no adhesion is evident. The LDPE-rich (75%) blend in Fig. 7d had graft copolymer added (30%) to polystyrene prior to final blending to give 7.5% graft based on the total blend. A quite pronounced coating is seen on the dispersed PS particles with evidence of fibrils connecting to the matrix. The latter undoubtedly result from drawing of the PE during fracture and attest to very strong adhesion between phases as a result of the graft. These photographs definitively prove that the graft does play an interfacial role although this is not as simple as the ideal visualized in Fig. 1. Additional evidence of interfacial activity is the reduction in domain size attending graft addition observed by Locke and Paul [10].

## 2. Polyethylene–Poly(vinyl chloride) plus Chlorinated Polyethylene

Blends of polyethylene (PE) and poly(vinyl chloride) (PVC) are not quite as poor mechanically as the previous system. An interesting "compatibilizer" for this system is some special chlorinated polyethylenes (CPE), first proposed by Schramm [14] and studied more extensively by Paul *et al.* [8, 9]. For these CPEs, chlorination is done via a slurry process in which the high-density polyethylene (HDPE) employed is in the solid state rather than in solution (see Fig. 8). In solution, all hydrogens are equally accessible for

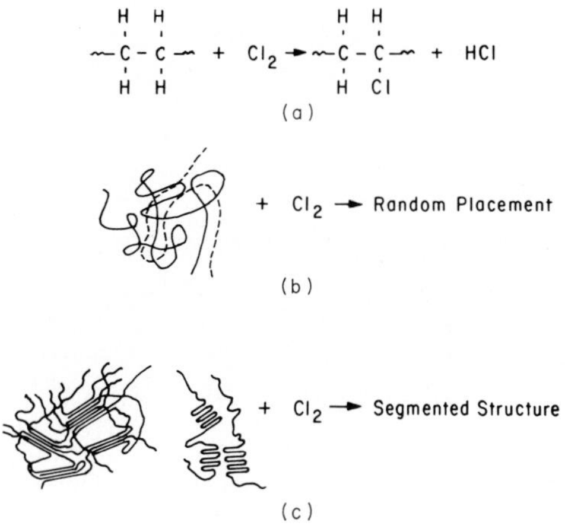

$$\sim\!\!\overset{\text{H}\ \text{H}}{\underset{\text{H}\ \text{H}}{\text{C}-\text{C}}}\!\!\sim \ + \ \text{Cl}_2 \longrightarrow \sim\!\!\overset{\text{H}\ \text{H}}{\underset{\text{H}\ \text{Cl}}{\text{C}-\text{C}}}\!\!\sim \ + \ \text{HCl}$$

(a)

+ $\text{Cl}_2 \longrightarrow$ Random Placement

(b)

+ $\text{Cl}_2 \longrightarrow$ Segmented Structure

(c)

Fig. 8 Schematic illustration of the chlorination of polyethylene: (a) chlorination of polyethylene; (b) solution process; (c) solid state process.

replacement with chlorine so a random chlorine placement occurs. The resulting polymer is similar to a random copolymer of ethylene and vinyl chloride, and crystallinity goes to zero at relatively low chlorine contents. However, in the solid state process only carbon atoms in the amorphous phase get chlorinated since $Cl_2$ cannot diffuse into the crystalline lattice. The result can be a highly segmented structure resembling a block co-polymer, and polyethylenelike crystallinity can be preserved to rather high chlorine contents [9]. The chlorinated parts are similar to PVC, while the unreacted ethylene sequences in the crystalline regions are more like poly-ethylene. A CPE with properly balanced chlorine content and residual crystallinity can be made that will adhere well to both polyethylene and PVC presumably due to this blocklike nature. A particular CPE with 36% chlorine proved most effective of those examined [8, 9]. Stress–strain curves for PVC–PE blends with and without this CPE are shown in Fig. 9. The upper two curves are for HDPE. The unmodified blend is very brittle, but with 20% CPE the blend becomes remarkably more ductile; the specimen exhibited yielding and subsequent neck formation. The modulus and ultimate strength were reduced by the CPE. The effects shown here are quite similar to those seen in impact modification of plastics [35]. The lower two curves are for LDPE and changes similar to the HDPE case are seen, except here the tensile strength is increased by the CPE.

The effect of three different CPEs (the number indicates % Cl) on elongation at break of binary blends of PS–PVC, LDPE–PS, and LDPE–

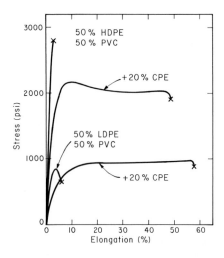

Fig. 9  Effect of CPE (36% chlorine) on stress–strain diagrams for polyethylene–PVC blends. The PVC contained impact modifiers. (From Paul and Locke [8]. Reproduced by permission of *Polym. Eng. Sci.*)

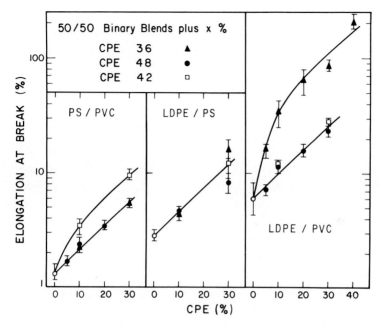

Fig. 10   Effect of various CPEs on elongation to break of binary blends of LDPE, PS, and PVC containing equal parts of each. PVC was a copolymer with vinyl acetate. (From Paul and Locke [8]. Reproduced by permission of *Polym. Eng. Sci.*)

PVC is shown in Fig. 10. One would only expect CPE to play a true interfacial role in the latter system. In each of the nine cases, blend ductility is increased by CPE addition, but the system LDPE–PVC–CPE 36 shows the most dramatic change. Chlorinated polyethylenes 42 and 48 have poorer adhesion to PE because of a lower content of unchlorinated chain segments [9], and as a result they are poorer "compatibilizers" for LDPE–PVC. The elongation for the remaining cases increases in part because a more ductile component has been added to the blend which has some adhesive power for at least one component in the system. Addition of CPE to PVC–PE blends reduces domain size with CPE 36, the most effective of the three. All evidence points to an interfacial role for a properly designed CPE in the PE–PVC system.

These studies were prompted by an interest in upgrading the mechanical properties of blends from mixed scrap plastics [1].

### 3.   Nylon 6–Polypropylene plus Graft

Ide and Hasegawa [13] studied blends of nylon 6 and polypropylene (PP) to which a graft copolymer was added to improve dispersability and mechanical behavior of the two immiscible polymers. They grafted maleic anhydride (MAH) onto polypropylene which on subsequent mixing with nylon 6 could, as they envision it, react as follows:

$$\begin{array}{c}\text{—CH—C}\diagdown \\ \text{O} \\ \text{CH}_2\text{—C} \\ \text{O}\end{array} + \text{H}_2\text{N}\sim\sim \longrightarrow \begin{array}{c}\text{—CH—C—NH}\sim\sim \\ \text{CH}_2 \\ \text{COOH}\end{array} \qquad (3)$$

to give the desired graft. They observed loss of amine groups as proof of the reaction. How the presence of this graft affects the mechanical behavior of a 20% nylon 6 and 80% PP (this includes contribution of graft) blend is shown in Table 1. The amount of PP–*g*–MAH graft added was varied to

Table I

Effect of Polypropylene–Maleic Anhydride Graft Copolymer
on Mechanical Properties of 20% Nylon 6–80% Polypropylene Blend

| PP–*g*–MAH (%)[a] | [N][b] | Yield strength (lb/in.$^2$) | Elongation at break (%) |
|---|---|---|---|
| 0 | 0 | 3300 | 5 |
| 1.8 | 0.5 | 5200 | 13 |
| 3.6 | 1.0 | 5500 | 28 |
| Polymer | | | |
| Polypropylene | | 4600 | >30 |
| Nylon 6 | | 10,500 | >30 |

[a] Based on total blend; percentage of MAH in graft = 1.15% by weight.
[b] [N] = equivalents of MAH/equivalents of —NH$_2$.

yield the ratio of MAH to —NH$_2$ equivalents shown. Ideally when this ratio is one, there would be no free nylon 6 homopolymer in the blend. The blend with no graft has a very low elongation at break and is weaker than the weakest component, PP. Addition of the graft raises the blend yield strength until it approaches the weighted average of the two components, viz. 5780 psi. Addition of graft also makes the blends more ductile and show cold drawing, which the unmodified blend does not. Domain size is greatly reduced by addition of graft. Melt flow properties are also altered. No doubt the chemistry of this system is somewhat more complex than the ideal described above, but nevertheless this seems to be a very good example of the effects possible via "compatibilizers."

### 4. Cellulose Acetate–Polyacrylonitrile plus Graft

Literature from the Soviet Union contains numerous reports on blend fibers made from cellulose acetate (CA) and polyacrylonitrile (PAN) (see Chapter 16). Some of these papers describe the use of a graft of PAN onto CA as an additive for these blends [83, 84]. The graft is reported to contain equal parts CA and PAN. The effect of this graft on the mechanical behavior of blend fibers, which also contain equal proportions of CA and PAN, are shown in Table II. Blends containing graft copolymer can be drawn more and are stronger and more ductile. Addition of the graft also reduced domain size.

Table II

Effect of Cellulose Acetate–Acrylonitrile Graft Copolymer
on Mechanical Properties of Blend Fibers
from 50% Cellulose Acetate–50% Polyacrylonitrile

|  | Graft copolymer (%) | | |
|---|---|---|---|
|  | 0 | 10 | 50 |
| Maximum stretch (%) | 290 | 320 | 500 |
| Strength (gm/den.) | 1.2 | 2.2 | 4.6 |
| Elongation at break (%) | 5.9 | 11.2 | 13.0 |

### 5. Polystyrene–Rubber plus Blocks or Grafts

Blends of PS and various types of rubber with addition of appropriate graft [18, 85] or block [6, 19, 49] copolymers have been studied but these data are not reviewed here. Numerous patents on such systems [86] relate to impact improved PS by postreactor blending rather than through conventional reactor blending.

### B. Laminates

The quantitative influence of interfacial adhesion on the mechanical response of dispersed blends is complex and not well understood. However, laminates of two polymers represent a series arrangement of phases in which direct stressing of the interface may occur and the strength of the composite in certain modes of testing is no greater than the adhesive forces between the two phases. Because different polymer types often do not adhere well to one another, adhesives that adhere to both polymers are needed. In some cases a homogeneous polymer may be a mutual adhesive for these two polymers, but clearly the possible interfacial activity of block and graft copolymers may be advantageous for this purpose. Selected examples that illustrate this are reviewed in this section.

For rather obvious reasons, PVC does not adhere well to natural rubber (NR). An effective and commercially used adhesive for these two is a graft copolymer of methyl methacrylate (MMA) onto NR [87]. The NR backbone of this material will adhere well to NR, while the PMMA graft chains will adhere to PVC because of the partial miscibility of PMMA with PVC [88]. In Fig. 11 are shown results [87] from a 180-degree peel strength test on laminates of PVC and NR bonded together by a layer of the above-mentioned grafts the PMMA content of which was varied. The dashed lines schematically extend these data to show the expected behavior for the extreme cases. When there is no PMMA, the adhesive is simply NR, which will adhere well to the NR sheet but not to the PVC sheet, and the peel strength would be zero. When a pure PMMA adhesive is used, it will stick to the PVC sheet but not to the NR sheet, and peel strength is again zero. It is reasonable that the peel strength would be highest when PMMA and NR are present in nearly equal proportions in the graft. In this example, the graft clearly does not form a monolayer at the interface, as suggested in Fig. 1, but rather exists as a third phase between the two sheets; however, it is believed that the mechanism of adhesion to each sheet by the graft is a simple extension of this concept. The graft can adhere to both NR, and

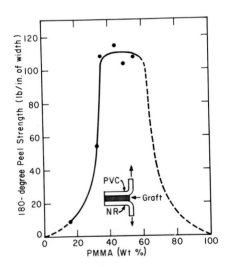

Fig. 11   Use of NR–g–PMMA as an adhesive for natural rubber and poly(vinyl chloride) laminates. The plot shows the effect of the PMMA content of graft on peel strength. Dotted lines drawn to illustrate expected response. (After Pendle [87].)

PVC because its surface can present two different kinds of segments to promote wetting and/or interpenetration of chain segments between the two phases.

Other examples similar to the above have been reported; however, the amount of quantitative data is usually limited. A patent [89] describes bonding an ethylene–propylene–diene terpolymer rubber (EPDM) to styrene–butadiene rubber (SBR) by a graft of styrene and butadiene onto an EPDM backbone. Similarly an SB block copolymer has been used to bond high-impact polystyrene to ABS sheets [90]. Presumably the styrene segments of the SB block penetrate the polystyrene while the butadiene segments of the block penetrate or adhere to the rubber phase of the ABS. Styrene–butadiene blocks in which the butadiene part has been hydrogenated are reported [91] to provide good adhesion between PE and PS. Apparently the hydrogenated polybutadiene is sufficiently similar to PE to be miscible with or adhere to it.

Adhesion of polymers to substrates other than another polymer, for example, glass or metals, may also be improved by this route. For example, it has been shown that poly(diorganosiloxane)–polycarbonate block copolymers are good adhesives for glass–polycarbonate laminates [32]. Jackson and Caldwell [92] have demonstrated a related effect by blending carboxylated polyesters with polymeric substrates such as polycarbonates and polystyrene. This polymeric additive improved the adhesion of the polymer substrate to metals presumably because of the polar interaction of the carboxyl groups with the metal surface. Interestingly, they found that adhesion improved as the carboxylated polyester became less miscible with the polymer substrate. This suggests that the additive goes preferentially to the surface, where it is needed, and performs a better "compatibilizing" role when it is not so soluble in the substrate.

Appropriate blending of polymers also seems to provide the ability to bond simultaneously two dissimilar polymers. For example, PE has been bonded to impact-modified PS with a blend of PE, impact-modified PS, and SB block copolymers [93]. The block copolymer plays a complex role here. There seems to be evidence that dispersed blends of Polymers A and B will adhere to either pure A or pure B. It would be interesting to compare the effectiveness of such a blend with a corresponding block or graft copolymer.

ACKNOWLEDGMENTS

The author extends sincere appreciation of Prof. D. Heikens for numerous scanning electron photomicrographs and to Prof. J. E. McGrath for numerous references.

# REFERENCES

1. D. R. Paul, C. E. Vinson, and C. E. Locke, *Polym. Eng. Sci.* **12**, 157 (1972).
2. A. J. Yu, *in* "Multicomponent Polymer Systems" (N. A. J. Platzer, ed.), Adv. Chem. Ser., Vol. 99, p. 2. Amer. Chem. Soc., Washington, D.C., 1971.
3. N. G. Gaylord, *in* "Copolymers, Polyblends, and Composites" (N. A. J. Platzer, ed.), Adv. in Chem. Ser., Vol. 142, p. 76. Amer. Chem. Soc., Washington, D.C., 1975.
4. O. W. Lundstedt and E. M. Bevilacqua, *J. Polym. Sci.* **24**, 297 (1957).
5. G. Riess, J. Kohler, C. Tournut, and A. Banderet, *Makromol. Chem.* **101**, 58 (1967).
6. J. Kohler, G. Riess, and A. Banderet, *Eur. Polym. J.* **4**, 173, 187 (1968).
7. G. Riess, J. Periard, and Y. Jolivet, *Angew. Chem. Int. Ed.* **11**, 339 (1972).
8. D. R. Paul, C. E. Locke, and C. E. Vinson, *Polym. Eng. Sci.* **13**, 202 (1973).
9. C. E. Locke and D. R. Paul, *Polym. Eng. Sci.* **13**, 308 (1973).
10. C. E. Locke and D. R. Paul, *J. Appl. Polym. Sci.* **17**, 2597, 2791 (1973).
11. W. M. Barentsen and D. Heikens, *Polymer* **14**, 579 (1973).
12. W. M. Barentsen, D. Heikens, and P. Piet, *Polymer* **15**, 119 (1974).
13. F. Ide and A. Hasegawa, *J. Appl. Polym. Sci.* **18**, 963 (1974).
14. J. N. Schramm and R. R. Blanchard, paper presented at SPE RETEC, Cherry Hill, New Jersey (October 1970).
15. M. E. L. McBain and E. Hutchinson, "Solubilization and Related Phenomena." Academic Press, New York, 1955.
16. G. E. Molau, *in* "Block Polymers" (S. L. Aggarwal, ed.), p. 79. Plenum, New York, 1970.
17. M. Matzner, L. M. Robeson, A. Noshay, and J. E. McGrath, *Encyl. Polym. Sci. Technol.* (in press).
18. R. J. Ceresa, *Encycl. Polym. Sci. Technol.* **2**, 485 (1965).
19. M. Morton, *Encycl. Polym. Sci. Technol.* **15**, 508 (1971).
20. S. L. Cooper and A. V. Tobolsky, *Text. Res. J.* **36**, 800 (1966).
21. D. Heinze, *Makromol. Chem.* **101**, 167 (1967).
22. S. F. Edwards, *J. Phys. A* **7**, 332 (1974).
23. L. G. Lundsted and I. R. Schmolka, *in* "Block and Graft Copolymerization" (R. J. Ceresa, ed.), Vol. 2, pp. 1 and 113. Wiley, New York, 1976.
24. W. J. Burlant and A. S. Hoffman, "Block and Graft Polymers." Van Nostrand-Reinhold, Princeton, New Jersey, 1960.
25. S. Marti, J. Newo, J. Periard, and G. Riess, *Colloid Polym. Sci.* **253**, 220 (1975).
26. J. Periard and G. Reiss, *Colloid Polym. Sci.* **253**, 362 (1975).
27. G. Reiss, J. Periard, and A. Banderet, *in* "Colloidal and Morphological Behavior of Block and Graft Copolymers" (G. E. Molau, ed.), p. 173. Plenum, New York, 1971.
28. M. J. Owen and T. C. Kendrick, *Macromolecules* **3**, 458 (1970).
29. A. G. Kanellopoulos and M. J. Owen, *J. Colloid Interface Sci.* **35**, 120 (1971).
30. G. L. Gaines and G. W. Bender, *Macromolecules* **5**, 82 (1972).
31. D. G. LeGrand and G. L. Gaines, *Polym. Prepr. Amer. Chem. Soc. Div. Polym. Chem.* **11**, 442 (1970).
32. General Electric, Netherlands Patent Appl., 07,871 (1974) [*Chem. Abstr.* **83**, 132868 (1975)].
33. S. Wu, *J. Macromol. Sci. Rev. Macromol. Chem.* **C10**, 1 (1974).
34. J. L. Amos, *Polym. Eng. Sci.* **14**, 1 (1974).
35. S. L. Rosen, *Polym. Eng. Sci.* **7**, 115 (1967).
36. R. N. Haward and J. Mann, *Proc. Roy. Soc. London Ser. A* **282**, 120 (1964).

37. D. Hardt, *in* "Block and Graft Copolymerization" (R. J. Ceresa, ed.), Vol. 2, p. 315. Wiley, New York, 1976.
38. F. M. Merrett, *Trans. Faraday Soc.* **50**, 759 (1954).
39. H. Keskkula, S. G. Turley, and R. F. Boyer, *J. Appl. Polym. Sci.* **15**, 351 (1971).
40. H. Keskkula and W. J. Frazer, *Appl. Polym. Symp.* **7**, 1 (1968).
41. G. E. Molau and H. Keskkula, *Appl. Polym. Symp.* **7**, 35 (1968).
42. G. F. Freeguard, *Polymer* **13**, 366 (1972).
43. G. F. Freeguard, *Brit. Polym. J.* **6**, 205 (1974).
44. C. F. Parsons and E. L. Suck, *in* "Multicomponent Polymer Systems" (N. A. J. Platzer, ed.), Adv. in Chem. Ser., Vol. 99, p. 340. Amer. Chem Soc., Washington, D.C., 1971.
45. D. G. Campbell, The Effects of Ethylene–Propylene Block Copolymers on Melt Blended Linear Polyethylene and Isotactic Polypropylene Polyblends. Ph.D. dissertation, Univ. of Maryland, 1974.
46. W. H. Tung, The Influence of Triblock Copolymers upon the Mutual Compatibility of Polystyrene/Polybutadiene Blends. Ph.D. dissertation, Univ. of Maryland, 1974.
47. D. J. Meier, *J. Polym. Sci. Part C* **26**, 81 (1969).
48. J. A. Manson and L. H. Sperling, "Polymer Blends and Composites." Plenum, New York, 1976.
49. G. Riess and Y. Jolivet, *in* "Copolymers, Polyblends, and Composites" (N. A. J. Platzer, ed.), Adv. in Chem. Ser., Vol. 142, p. 243. Amer. Chem. Soc., Washington, D.C., 1975.
50. G. E. Molau, *J. Polym. Sci. Part A* **3**, 1267, 4235 (1965).
51. G. E. Molau and W. M. Wittbrodt, *Macromolecules* **1**, 260 (1968).
52. T. Inoue, T. Soen, T. Hashimoto, and H. Kawai, *Macromolecules* **3**, 87 (1970).
53. M. Moritani, T. Inoue, M. Motegi, and H. Kawai, *Macromolecules* **3**, 433 (1970).
54. T. Inoue, T. Soen, T. Hashimoto, and H. Kawai, *in* "Block Polymers" (S. L. Aggarwal, ed.), p. 53. Plenum, New York, 1970.
55. B. Ptaszynski, J. Terrisse, and A. Skoulios, *Makromol. Chem.* **176**, 3483 (1975).
56. K. Kato, *Polym. Eng. Sci.* **7**, 38 (1967).
57. G. E. Molau, *Kolloid Z. Z. Polym.* **238**, 493 (1970).
58. L. Toy, M. Niinomi, and M. Shen, *J. Macromol. Sci. Phys. B* **11**, 281 (1975).
59. M. S. Akutin, B. V. Andrianov, and V. S. Kulyamin, *Vysokomol. Soedin. Ser. B* **17**, 457 (1975).
60. E. Helfand and Y. Tagami, *J. Polym. Sci. Part B* **9**, 741 (1971).
61. E. Helfand and A. M. Sapse, *J. Chem. Phys.* **62**, 1327 (1975).
62. E. Helfand, *J. Chem. Phys.* **62**, 999 (1975).
63. E. Helfand, *Accounts Chem. Res.* **8**, 295 (1975).
64. D. J. Meier, *Polym. Prepr. Amer. Chem. Soc. Div. Polym. Chem.* **15**, 171 (1974).
65. D. J. Meier, *in* "Block and Graft Copolymers" (J. J. Burke and V. Weiss, eds.), p. 105. Syracuse Univ. Press, Syracuse, New York, 1973.
66. S. Krause, *J. Polym. Sci. Part A-2* **7**, 249 (1969).
67. S. Krause, *Macromolecules* **3**, 84 (1970).
68. S. Krause, *in* "Block and Graft Copolymers" (J. J. Burke and V. Weiss, eds.), p. 143. Syracuse Univ. Press, Syracuse, New York, 1973.
69. D. J. Dunn and S. Krause, *J. Polym. Sci. Part B* **12**, 591 (1974).
70. E. Helfand, *in* "Recent Advances in Polymer Blends, Grafts, and Blocks" (L. H. Sperling, ed.), p. 141. Plenum, New York, 1974.
71. E. Helfand, *Macromolecules* **8**, 552 (1975).
72. D. F. Leary and M. C. Williams, *J. Polym. Sci. Part B* **8**, 335 (1970); *J. Polym. Sci. Polym. Phys. Ed.* **11**, 345 (1973).
73. D. F. Leary and M. C. Williams, *J. Polym. Sci. Polym. Phys. Ed.* **12**, 265 (1974).

74.  J. M. Huet and E. Marechal, *Eur. Polym. J.* **10**, 771 (1974).

75.  M. Baer, *J. Polym. Sci. Part A* **2**, 417 (1964).

76.  L. M. Robeson, M. Matzner, L. J. Fetters, and J. E. McGrath, *in* "Recent Advances in Polymer Blends, Grafts, and Blocks" (L. H. Sperling, ed.), p. 281. Plenum, New York, 1974.

77.  G. S. Fielding-Russell and P. S. Pillai, *Polymer* **15**, 97 (1974).

78.  D. R. Hansen and M. Shen, *Macromolecules* **8**, 903 (1975).

79.  T. K. Kwei, H. L. Frisch, W. Radigan, and S. Vogel, *Macromolecules* **10**, 157 (1977).

80.  W. M. Barentsen and D. Heikens, *J. Mater. Technol.* **1**, 49 (1970).

81.  A. Casale and R. S. Porter, *Adv. Polym. Sci.* **17**, 1 (1975).

82.  D. Heikens, private communication (1976).

83.  N. I. Naimark, B. V. Vasilev, G. S. Zaspinok, N. G. Shcherbakova, V. A. Landysheva, and V. Ye. Lozhkin, *Polym. Sci. USSR* **12**, 1865 (1970).

84.  I. Z. Zakirov, A. A. Geller, Y. B. Monakov, S. I. Slepakova, and B. E. Geller, *Fibre Chem.* **2**, 623 (1971).

85.  G. Natta, M. Pegoraro, F. Severini, and S. Dabhade, *Rubber Chem. Technol.* **39**, 1667 (1966).

86.  R. R. Durst (General Tire and Rubber), German Patent, 2,342,219 (1975) [*Chem. Abstr.* **83**, 29180 (1975)].

87.  T. D. Pendle, *in* "Block and Graft Copolymerization" (R. J. Ceresa, ed.), Vol. 1, p. 83. Wiley, New York, 1973.

88.  J. W. Schurer, A. de Boer, and G. Challa, *Polymer* **16**, 201 (1975).

89.  C. F. Paddock (to Uniroyal), U.S. Patent, 3,758,435 (1973) [*Chem. Abstr.* **80**, 48968 (1974)].

90.  T. Hasegawa, K. Kishida, and K. Tamiya (to Denki Kagaku Kogyo), Japanese Patent, 39,656 (1974) [*Chem. Abstr.* **82**, 18107 (1975)].

91.  K. Bronstert (Badische), German Patent, 2,201,243 (1973) [*Chem. Abstr.* **80**, 16005 (1974)].

92.  W. J. Jackson and J. R. Caldwell, *in* "Multicomponent Polymer Systems" (N. A. J. Platzer, ed.), Adv. in Chem. Ser., Vol. 99, p. 562. Amer. Chem. Soc., Washington, D.C., 1971.

93.  U. Koenig (BASF), German Patent, 2,236,903 (1974) [*Chem. Abstr.* **81**, 50711 (1974)].

# Chapter 13

# Rubber Modification of Plastics

*Seymour Newman*

*Plastics, Paint, and Vinyl Division*
*Ford Motor Company*
*Detroit, Michigan*

## I. INTRODUCTION

Efforts to reduce the abuse resistance of amorphous, glassy polymers by the addition of particulate rubber date to the 1940s and are still in progress. Commercial exploitation has outpaced our understanding of this mechanism of impact enhancement. Early work was directed at modifying polystyrene but has since expanded to include most styrenic polymers; poly(methyl methacrylate); poly(phenylene oxide); cross-linked polymers of many types, for example, epoxies, acrylics, phenolics, and styrenated polyesters; and semicrystalline polymers with amorphous phases possessing low or high glass transition temperatures, for example, polypropylene, polyethylene, nylon, and poly(tetramethylene terephthalate).

In a sense, rubber modification competes with several other chemical methods for achieving improved toughness, elongation to fracture, ductility, or enhanced propensity to yield. These include the closely related block copolymerization, which also almost always yields a two-phase morphology but requires special synthetic procedures; random copolymerization with copolymers with a lower glass transition temperature ($T_g$); and miscibility with compatible liquids (plasticizers) or, in special cases, other polymers (see Chapter 17).

However, the intent of this chapter is a survey, principally, of the more recent efforts at rubber modification by the polyblend approach; it is not my endeavor to review the entire domain of toughness improvement by all possible approaches. Also, a far more exhaustive treatment of the styrenic polymers and poly(phenylene oxide)–high-impact polystyrene blends is discussed in Chapter 14 by Bucknall and therefore will be omitted from this discussion. The use of thin layers of soft polymers, encapsulating fibers, or mineral fillers represents a relatively new field of research and this matter will be touched upon briefly as it represents, in my view, an important and growing activity.

Finally, since the main intent of this chapter is to bring the results of recent investigations of newer elastomer–polymer blends to the reader's attention, I shall only review the theoretical advances sufficiently to permit a sensible interpretation of the experimental information reported.

## II. BACKGROUND

### A. Mechanisms

The various theories proposed thus far as explanations of rubber toughening may be categorized as follows:

1. Energy absorption by the rubber particles.
2. Energy absorption by yielding of the continuous phase; ductility enhanced by strain induced dilatation near the rubber occlusion.
3. Craze formation involving cavitation and polymer deformation within the craze.
4. Shear yielding as a source of energy absorption and crack termination.
5. Stress distribution and relief.
6. Rubber particles acting as craze termination points and obstacles to crack propagation.

No one of these alone should be considered adequate to provide a total theory of rubber toughening. Moreover, the relative role of these elements are expected to vary with the particular system and test conditions (see, for instance, later discussions of the epoxy system in which craze formation and shear yielding are considered; and the PVC blends in which craze formation is rejected as a reinforcing mechanism).

The concept of energy absorption by the rubber was first advanced by Merz *et al.* [1] for the system butadiene–styrene rubber in a polystyrene matrix. These authors took the position that the rubber particles behaved in such a manner as to hold the fracture surfaces of small cracks together. Since the rubber is capable of extensive elongation, it was argued that the rubber phase absorbed more energy during tensile fracture than an equivalent volume of continuous phase.

In 1965, however, Newman and Strella [2] showed that this model was inadequate since the energy absorbed by the rubber represented only one-tenth of the total energy absorbed by the composite. More importantly, these authors asserted that the presence of the dispersed particle served to trigger yielding in the continuous phase, and the enhanced toughness could be attributed to the large energy absorption involved in the local deformation or cold drawing of the matrix. More specifically, the following arguments were advanced to rationalize the energy to rupture in a tensile experiment:

1. Under a tensile strain, an increase in free volume occurs in the matrix near the rubbery particle. Also, because the rubber particle with a Poisson ratio of about 0.5 exhibits no volume change on deformation, then, if the volume expansion of the strained sample is to be maintained at its usual level, an accentuated expansion occurs in the matrix immediately surrounding the rubber particle. This effectively lowers the glass transition temperature $(T_g)$ of the matrix, bringing it closer to the liquid state in which large scale molecular motion is favored.

2. The rubber particles are subjected to combined stresses under which they are sufficiently strong to withstand the dilatational forces involved in the tensile strain.

3. The rubber particles serve to prevent premature catastrophic crack propagation.

Aside from identifying toughness in rubber-reinforced, glassy polymers with matrix yielding, Newman and Strella drew an important analogy to cold drawing in certain unmodified glassy polymers and the relevance of density fluctuations in initiating the onset of this response, along with some "softening" mechanism whereby the effective modulus decreases with increased strain. Thus, a tensile strain $\varepsilon$ does dilate the sample by amount $\Delta = (1 - 2\mu)\varepsilon$, where $\mu$ refers to the Poisson ratio.

Major advances were achieved by Kambour [3] and Bucknall and Smith [4], who identified stress whitening with crazing, which occurs near the rubber–matrix interface. Crazes have been identified by Kambour [3] as regions of porous, plastically deformed matter. Increased energy absorption is associated with the local orientation process that accompanies crazing. According to Bucknall, the rubber particles lower the craze initiation stress by providing sources of local stress concentration and further strengthen the craze by bearing a portion of the triaxial stresses present at the particle–matrix interface. The efforts of these and subsequent workers confirmed that crazes initiate and terminate at rubber particles.

Further attempts to provide a quantitative approach were initiated by Strella [5], who calculated the dilatation or fractional change in volume $\Delta$ outside a rubber inclusion or void in a polyblend under tension. For a model case analogous to an acrylonitrile–butadiene–styrene copolymer blend, the cubical or volume dilation is a maximum at the equator and at the rubber–matrix interface. This dilatation depends on the Poisson ratio of the matrix $\mu_m$, and is given by

$$\Delta_{max} = 2.63(1 - 2\mu_m)\varepsilon$$

Thus, at the equator, $\Delta$ is larger even at a given local stress than the value which would exist in the homogeneous glass (see Table I). Strella asserted that this effect accounts for an increased tendency for cold drawing to occur in the matrix. Furthermore, he took the position that the drawn material must assume a fibrillar structure combined with voids if the width of the

Table I

Effects of Thermal Stresses on Polyblends[a]

| Applied stress (psi) | Dilation ratio[b] $(\Delta_{max}/\Delta_0)$ | Stress ratio[c] $(\sigma_{max}/\sigma_0)$ | $\dfrac{\Delta_{max}/\Delta_0}{\sigma_{max}/\sigma_0}$ |
|---|---|---|---|
| With thermal stresses | | | |
| 500 | 2.271 | 1.096 | 2.072 |
| 1,000 | 2.279 | 1.459 | 1.562 |
| 5,000 | 2.276 | 1.750 | 1.301 |
| 10,000 | 2.269 | 1.786 | 1.267 |
| Without thermal stresses | | | |
| Any value within elastic limit | 2.290 | 2.004 | 1.142 |

[a] Data from Beck *et al.* [8].
[b] Ratio of maximum dilatation at the equator of the rubbery inclusion to the dilatation in the absence of the particle; $\Delta T = 50°C$.
[c] Corresponding ratio of stresses.

undrawn material is to be maintained. In other words, this localized drawing may take the form of crazes.

Most recently shear yielding has been identified as an energy-absorbing process in glassy polymers and even in cross-linked glassy polymers. The occurrence of this mechanism is dependent on temperature, pressure, applied stress conditions, and internal morphology of the polymer. Little stress whitening is associated with this deformation process. Also, according to Bucknall and Street [6, 7], shear bands, aside from acting as an energy-absorbing process, also function as crack stoppers.

Assuming that shear yielding occurs at constant volume, Bucknall and co-workers [6, 7] have assessed the contribution of crazing and shear yielding during creep in various polymers. The relative contributions of these mechanisms may vary over wide ranges, depending on material properties and test conditions.

It is important to note that, while no dilation occurs when a homogeneous isotropic solid is subjected to pure shear, both positive and negative dilations have been calculated [5] in the neighborhood of an occlusion as a function of angle. Therefore, cold drawing may be favored even under conditions of pure shear. Yielding under pure shear does not preclude the free-volume concept. Clearly, this point deserves further investigation.

## B. Dilation and the Thermal Coefficient of Expansion

In general, the coefficient of thermal expansion of rubbery polymers is considerably higher than that of the glassy polymers in which they may be dispersed. Since such polyblends are usually formed or molded at temperatures significantly higher than those at which they are used, the cooling process should result in radial tension. Beck *et al.* [8] and Newman [9] have analyzed the resulting thermal stresses, assuming no loss of interface adhesion.

Their results (Table I) show that the thermal stresses have no appreciable effect on the dilation of the glassy matrix. However, as shown in the third column, the initial compressive tangential stress significantly reduces the maximum stress produced by the external load. Consequently, the dilation required for plastic flow is more likely to be reached well before fracture.

## C. Morphological Features

The detailed structure and performance of each two-phase system ultimately depends on the subtleties of chemical composition and its effect on the spatial arrangement of the constituent polymer species. The latter also depends on processing conditions.

Thus, Baer [10] has shown in rubber-modified polystyrene blends that inadequate cross-linking may lead to particle deformation and disintegration during mechanical working. Excessive cross-linking, however, has been shown by Wagner and Robeson [11] to lead to substantial increases in the particle modulus, which decreases the overall blend impact performance, possibly as a consequence of decreased efficiency in craze initiation and termination.

In high-impact polystyrene, the modulus of the rubber phase may be also influenced by the extent of occlusion of the matrix material within the individual gel particles. For batch processes, this, in turn, is affected by the composition and the conditions of polymerization, for example, agitation during phase inversion. Occlusion simultaneously increases the rubber-phase volume and particle size for a fixed-system rubber content.

Only a glance at one aspect of high-impact polystyrene systems reveals a myriad of competing effects. No single parameter may be altered, therefore, without causing a variety of other changes, which may induce opposing effects. Therefore it is not surprising that in systematic studies, maxima in impact performance and ultimate elongation are often observed, as in Fig. 1, for instance.

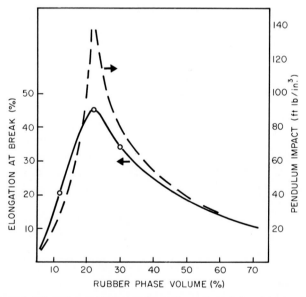

Fig. 1   Effect of rubber phase volume on elongation and pendulum impact on 10-mil film. (Plotted from data of Wagner and Robeson [11].)

Table II

Principal Effect of Rubber Particle Variables

| Particle variable | Physical–chemical factors | Influence on micromechanical behavior |
|---|---|---|
| 1. Size | Interparticle distance smaller with decreasing particle size | With decreasing particle size:<br>Greater density of shorter crazes<br>More efficient craze initiation and termination<br>Reduced strain rate in each craze for a given applied rate<br>May shift principal deformation mode from crazing to shear yielding |
| 2. Modulus<br>a. Solid particles | Principally affected by:<br>Presence of hard inclusions or matrix material<br>Degree of cross-linking<br>Chemical composition of rubber | Low inclusion content increases rubber phase volume and thereby increases craze termination efficiency<br>High inclusion content, however, leads to excessive particle rigidity and brittleness; results in decreased craze initiation and termination efficiency<br>Low cross-linking may allow particle deformation and disruption during processing<br>High cross-linking analogous to high inclusion content |
| b. Liquid droplets and voids | | Increased stress concentration and dilation factor as particle modulus $\rightarrow 0$; but inefficient craze termination<br>May function as site for crack initiation |
| 3. Shape | Nonspherical shapes (elliptical) may be induced by shear history and preparation conditions, e.g., in interpenetrating networks | Little investigated but may be expected to affect craze initiation and termination efficiency; probably deleterious to impact |
| 4. Concentration | Results in increased particle size or number or both | Greater efficiency of craze initiation with increased number of particles<br>For size effects see Merz et al. [1] |
| 5. Poisson ratio | Dependent on chemical structure, temperature, and strain | Little studied; affects stress concentration and dilatation factor; affects craze initiation |
| 6. Interface adhesion | Affected by segment interpenetration, chemical bonding, and wetting | Failure of the bond under combined stresses or chemical degradation results in void formation |

69

It is not my intent here to review the chemical and morphological features of rubber-modified polymers as they have been reported in particular studies on defined systems nor to indicate optimum values for a given system. Nevertheless, a list of the more important morphological features and their relation to the fracture mechanisms previously noted is given in Table II. (For additional information also see Chapter 14.)

### D.  High Rate Impact

Among the principal large strain properties achieved by rubber modification and the energy dissipation mechanisms it provides are improved toughness and elongation to break: These benefits, as discussed elsewhere, depend on a large number of structural and test conditions: rubber concentration, particle size and internal morphology, particle–matrix adhesion, glass transition of the phases, temperature, and rate of loading. With increasing test rate as in a uniaxial stress–strain measurement, the elongation to break will show a rapid decline at some test speed to low values. This is also exemplified by a brittle transition in fracture toughness $K_c$, a critical value of the stress field intensity factor measured experimentally by Marshall et al.[12] for polystyrene under slow crack propagation conditions (Fig. 2). Aside from the effect of test speed on the $T_g(R)$ of the rubber phase which is of particular

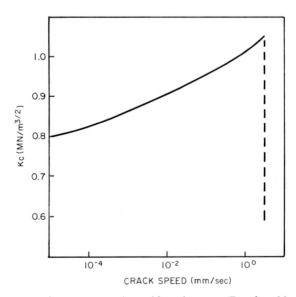

Fig. 2   Fracture toughness versus crack speed for polystyrene. (Data from Marshall et al.[12].)

significance when the test temperature approaches $T_g(R)$, other explanations are needed when $T \gg T_g(R)$. Recent proposals by Doyle [13] are of interest here.

From observations of fracture surface morphology and other considerations, Doyle in referring to Fig. 2 concludes that under high strain rates, that is, at a critical crack velocity, the "stress in the craze layer ahead of the propagation crack exceeds the strength of the craze-bulk boundary" [13, p. 480]. Consequently, "the craze layer splits off the adjacent bulk in a quasi-brittle fracture and the fracture toughness drops" [13, p. 480]. This phenomenon appears to be general for glassy polymers and in some semi-crystalline polymers at low temperatures. Furthermore, the same fracture morphologies observed for cleavage crack propagation are evident in unnotched specimens as well. In the latter, at low strain rates, fracture initiates by nucleation and coalescence of voids within the craze layers. As the strain rate increases, the stress to nucleate cracks within the craze exceeds the craze boundary strength, and fracture initiates by the brittle, low toughness *craze-boundary* fracture mechanism.

For rubber-modified polystyrene, fracture occurs only after crack initiation within the craze and the crack reaches a high velocity.

Therefore, according to Doyle, rubber particles, for example, in high-impact polystyrene, suppress the brittle fracture transition by nucleating a large number of crazes. This has the effect of reducing the strain rate in each individual craze if the craze thickness is not much different from that in the unmodified material. Thus, the strain rate in a craze layer $\dot{\varepsilon}$ is related to the applied strain rates $\dot{s}$ through the number of crazes per unit gauge length, $n$, and their thickness $t$:

$$\dot{\varepsilon} = \dot{s}/nt$$

## III. PROPERTIES OF PARTICULAR SYSTEMS

### A. Cross-Linked Amorphous Glasses

#### 1. General Remarks

Highly cross-linked glassy resins are generally amorphous as a consequence, in part, of the copolymer nature of their chemical structure. More importantly, for this discussion, chain mobility is severely restricted. Therefore, we might expect that molecular deformation associated with cold drawing or crazing should also be severely limited except over relatively short chain segment distances. Polymer networks of extremely high cross-link density, in fact, show only barely detectable glassy transitions and should not respond well to rubber modification. Some rubber toughening can be obtained, however, with appropriate resin systems and carefully selected particulate rubbers, as has been amply demonstrated.

## 2. *Acrylic Systems with "Graded" Rubber Particles*

The patents and publications of Dickie and Newman *et al.* [14–17] deal with a number of cases in which improved elongation, impact strength, and fatigue properties of glassy, thermoset compounds are achieved by incorporating designed rubber particles. To achieve this, rubber particles of controlled shell–core composition were emulsion polymerized so as to be chemically matched to the continuous phase in which they were to be incorporated.

Unique particulate material termed "graded rubber" was prepared by an appropriate sequence of monomers in a two-stage addition process to achieve a rubbery, elastomeric core of cross-linked acrylic polymer and a glassy outer shell containing reactive comonomers in order to provide the particle with the desired surface functionality. Thus, in one example, the core was composed of butyl acrylate cross-linked with a small amount of 1,3-butylene dimethacrylate. The shell was formed after most but not necessarily all of the original monomer charge was consumed, and consisted of a mixture of methyl methacrylate and glycidyl methacrylate. The average particle size thus prepared was 0.1–1.0 $\mu$m.

Particles produced in this manner by emulsion polymerization but without glycidyl methacrylate [17] were found by electron microscopy to be, in general, essentially spherical, with very few new particles formed on addition of the second-stage monomer. Microtomed sections of samples compression molded directly from such particles revealed, on exposures to xylene vapor, a pattern of spherical regions equal approximately in size to the rubber latex particles used as seed in the emulsion polymerization. These observations, combined with the conclusion that the dynamic modulus of molded samples could be well represented by assuming the composite to consist of rubbery particles in a glassy matrix, all lent credence to the assumption of a simple shell–core morphology.

To evaluate the efficiency of "graded rubber" particles, a thermoset prepolymer, which consisted of methyl methacrylate and glycidyl methacrylate, was prepared in emulsion. A blend of this thermoset was prepared with the particulate rubber described above and contained glycidyl methacrylate in the shell for coreaction to the matrix. The tensile properties of this material after molding and curing are shown in Table III.

Table III

Properties of Rubber-Modified Acrylic Thermosets

| Composition | Modulus (psi) | Stress at break (psi) | Elongation to break (%) |
|---|---|---|---|
| Molded epoxy–acrylic thermoset[a] | 460,000 | 8200 | 2.2 |
| Same with 25% rubber[b] | 260,000 | 6100 | 27.0 |

[a] Epoxide cross-linking catalyzed with 2-ethyl-4-methylimidazole.
[b] Based on the core portion of the "graded rubber" particles.

A remarkable increase in elongation to break of the thermoset is observed as a consequence of rubber modification. Extensive investigations also have been reported on rubber-modified epoxy systems, as discussed below. This work is particularly noteworthy for its intense investigation of deformation mechanism on systems that unfortunately did not lend themselves as readily to control of chemical structure as was possible with the particulate graded rubbers.

### 3. Epoxy Systems

On exploring the influence of dispersed elastomer particles on the deformation behavior of cross-linked epoxy resins in the glassy state, McGarry and co-workers [18–20] have reported two flow mechanisms. Both crazing and shear yielding are observed, depending on particle size and the nature of the stress field applied.

Their sample preparation technique consisted of adding low molecular weight (liquid) carboxy-terminated butadiene–acrylonitrile (CTBN) copolymers to a liquid epoxy resin. During polymerization and cross-linking, phase separation occurs together with reaction of the rubber particle to the matrix phase. A variety of chemical factors affect the dispersed particle size, including curing agent reactivity, temperature, molecular weight, and solubility of the CTBN. By an appropriate choice of these parameters, the average particle size could be varied from a few hundred angstroms to several hundred microns.

Particle size effects previously noted with high-impact polystyrene were also observed here, for example, a modest increase in toughness with particles in the range of $10^2$–$10^3$ Å and substantial effects with particles of $10^3$–$10^4$ Å (see, for instance, Table IV and Fig. 3).

By using biaxial stress fields to obtain yield and fracture envelopes, together with scanning electron microscopy of yielded and fractured

Table IV

Fracture Energy of Epoxy Resins Modified with CTBN[a]

| Material | Average particle size (Å) | Fracture surface work ($10^5$ ergs/cm$^2$) |
|---|---|---|
| Epon 828 | None | 1.8 |
| Epon 828 + 10% CTBN R–146 | ~400 | 3.3 |
| Epon 828 + 10% CTBN R–151N | ~12,000 | 15.5 |

[a] Data from Sultan and McGarry [18].

specimens, it was possible for the authors to arrive at the following con-clusions concerning the flow mechanisms that occurred.

When small particles of CTBN were formed in the epoxy and the material tested in a biaxial stress field, the shape of the entire failure envelope was similar to that for an unmodified epoxy. Optical examination showed that shear yielding was predominant in tension and compression. No hole struc-ture or stress whitening was evident. While the fracture energy was greater than that for unmodified epoxy, it was five to ten times less than that obtainable with large particles.

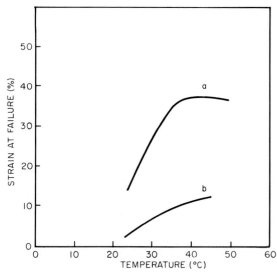

Fig. 3   Tensile strain at failure versus test temperature: (a) Epon 828 modified with small rubber particles; (b) unmodified Epon 828, $\dot{\varepsilon} = 6.3$ in./in./min. (Data from Sultan and McGarry [18].)

With large particle sizes, however, an additional mechanism appears necessary to fully describe the failure envelope. Thus, when large particles are present and the stress field is tensile, crazing appears to be the predominant flow mechanism. This process is highly energy absorbent, as evidenced by a large increase in fracture work. Furthermore, the diameters of holes formed during deformation are larger than the particle size and are attributed to matrix stretching and cavitation around the particle in the triaxial stress field at the crack tip. Coalescence of cavities ultimately leads to fracture without large average strains. If compressive stresses are applied, the shear mechanism appears to dominate, and the material may be macroscopically ductile.

The appearance of both microcavitation and shear banding may be due, in part, to a wide distribution of particle sizes. For a discussion of the adhesive fracture behavior of rubber-modified epoxy resins specifically with carboxy-terminated butadiene–acrylonitrile, the reader is referred to the work of Bascom and Coltington [21].

The dynamic mechanical properties of the epoxy resin of the diglycidyl ether of bisphenol A, containing various concentrations of carboxy-terminated butadiene–acrylonitrile copolymers, have also been investigated by Kalfoglou and Williams [22].

Some evidence for phase inversion at higher levels of CTBN during curing is presented along with evidence for the formation of a rubber–epoxy compound. A correlation of a low-temperature mechanical loss peak height with impact strength under conditions that correspond to the existence of discrete elastomeric particles is reported.

### 4. Styrenated Polyesters

Vinyl-terminated butadiene–acrylonitrile liquid copolymers with pendant vinyl groups (Goodrich's VTBN) can be incorporated in commercial styrenated polyester resin formulations to develop improved toughness characteristics.

Sheet molding compound (SMC) and bulk molding compound (BMC) are rigid, styrenated polyester resin systems reinforced with glass fibers and mineral fillers, generally calcium carbonate, and have widespread commercial use (see Chapter 23). The addition of modest amounts of VTBN, that is, 5–10%, based on the resin content are claimed to noticeably improve impact resistance, crack propagation, and the number and severity of cracks under a given set of test conditions. Some typical results are given in Table V.

While the published results are sketchy, the indications are clear that rubber modification can be achieved even with multicomponent reinforced

Table V

Properties of Modified BMC

|                                  | BMC control | With 10 phr VTBN |
|----------------------------------|-------------|------------------|
| Flexural strength (psi)          | 14,100      | 17,000           |
| Flexural modulus (psi $\times 10^6$) | 1.84        | 1.79             |
| Gardner drop dart impact (in. lb) | 0–2         | 6–8              |

systems. The somewhat improved flex strength also suggests enhanced elongation and delayed fracture.

The detailed micromechanical behavior of such systems, including the interaction of dispersed rubber with fibrous and particulate fillers, deserves further attention.

## B.   Noncross-Linked Glassy Polymers: Crystalline and Amorphous

### *1.   General Remarks*

Historically, the principal efforts of rubber modification to achieve increased toughness and commercial utility have been directed at polystyrene and poly(styrene-*co*-acrylonitrile). It is no surprise then that the literature abounds with reports of investigations of the structure–property relations in high-impact polystyrene (HIPS) and acrylonitrile–butadiene–styrene (ABS). These polymer systems together with their blends with poly(phenylene oxide) are dealt with systematically in other chapters (Volume 1, Chapter 5, and this volume, Chapter 14). In surveying the behavior of particular systems here, I therefore have chosen to omit discussions of these systems to avoid undue overlap with other chapters.

### *2.   Poly(vinyl chloride)*

Among the few studies reported in the open literature on this rigid, moderately crystalline polymer, the work of Petrich [23] is particularly noteworthy. Rigid poly(vinyl chloride) (PVC) appears to be an intrinsically more ductile glassy polymer than polystyrene (see, for instance, the discussion of Newman and Strella [2]). Impact modification is readily achieved, for example, by mechanically mixing, under appropriate conditions, with elastomers such as methacrylate–butadiene–styrene (MBS) copolymers. Carefully engineered grafting procedures such as those used in ABS are not mandatory for the successful rubber modification of PVC.

Incorporation of an MBS-type impact modifier appears to have a

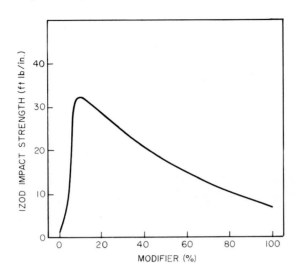

Fig. 4   Plot of Izod impact strength for PVC versus concentration of methacrylate–butadiene–styrene modifier, Acryloid KM-611 of Rohm & Haas Co. (Data from Petrich [23].)

maximum benefit in the 8–25% range (Fig. 4), which is largely attributed by Petrich to a decreased yield stress and an increased elongation to break. A fairly good correlation (Fig. 5) is obtained between impact strength and yield stress, leading to the conclusion that the addition of a rubbery impact modifier is analogous to an increase in temperature and results in a decreased yield stress. This decreased yield stress increases the probability of yielding rather than brittle fracture, except that with dispersed rubber particles the effect is achieved on a local rather than macroscopic scale.

By resorting to electron microscopy and other measurements, Petrich attempted to ascertain whether the mechanical properties observed were more consistent with the cold-drawing mechanism of Newman and Strella [2, 5] or the crazing mechanism of Bucknall [4, 6, 7] previously discussed.

While stress whitening is strongly evident in ABS or HIPS as a result of craze formation, its extent of occurrence is quite variable in unmodified PVC, depending on temperature and the choice of stabilizers and lubricants used. The addition of impact modifiers generally contributes to the degree of stress whitening but may not with appropriate systems, which are, nevertheless, very efficient as impact modifiers, for example, Rohm and Haas's Acryloid KM-607N.

From combined stress, density, and light-transmission measurements, it is found that stress whitening is associated with yielding but that density changes are not the causative factors per se. In fact a densification occurs

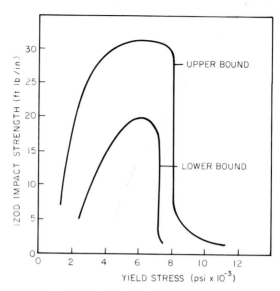

Fig. 5   Correlation of yield strength and Izod impact strength for PVC systems. (Data from Petrich [23].)

around the yield point for both unmodified and MBS-reinforced PVC. This is confirmed by electron microscopy on a variety of tensile and Izod specimens. From these studies [23] to which the reader is referred for greater detail, the author reports that (a) no evidence of voids or crazes is observed in stress whitened samples of MBS-modified PVC; (b) rubber particles are elongated in modified PVC during sample strain; and (c) long fibrils of PVC are observed stretching across the fracture plane. (Presumably, the densification is due to orientation, or possibly crystallization effects.)

Petrich therefore concludes that for these systems stress whitening (light scattering) is associated with birefringence effects and refractive index mismatches between the rubber particle and adjacent matrix. He consequently rejects the view that crazing can be involved as the reinforcing mechanism in all modified-PVC systems.

Overall, then, Petrich maintains that in modified PVC, the rubber particles lower the yield stress, facilitating cold drawing and thereby increasing energy absorption as manifested by enhanced impact. No evidence for voids and crazes is observed in certain systems, so that the stress whitening noted can only be ascribed to refractive index gradients brought about by local deformation of the PVC and rubber particles.

The remarkable fluidity and mobility of the PVC in the fracture plane is consistent, however, with the increased tendency of the PVC matrix to cold draw, as discussed by Newman and Strella, without this drawability taking the morphological form of crazing. This, however, was not proven directly. Optically transparent blends of PVC have been reported by Kenney [24] with terpolymers of α-methylstyrene–methacrylonitrile–ethyl acrylate. A single $T_g$ at 100°C suggests a single-phase system or an extremely fine dispersion. No mechanical properties are reported, however.

### 3. Nylon

Chompff [25] prepared thermoplastic, rubber-reinforced polyamides, nylon 6 and nylon 6,6, containing a cross-linked, elastomeric, dispersed phase grafted to the matrix at their interfaces. Generally, the materials were prepared by reacting a carboxy-bearing polymer ($M = 1500 \rightarrow 10,000$), with a diepoxide to form a hydroxy- and epoxy-functional reaction product that was both self-cross-linkable and capable of reacting with amide groups.

The reaction product of an excess of the diglycidyl ether of bisphenol A with carboxy-terminated butadiene–acrylonitrile represents a typical example. This viscous liquid was dispersable in nylon 6 to about a 1-$\mu$m particle size. With appropriate catalysts added to the system, both self-cross-linking in the particle and reaction with the amide or terminal amine groups of the nylon 6 were achieved. A wide variety of systems was explored, including elastomeric polyester prepolymers with random carboxy functionality.

While the patent contains only a small amount of mechanical property data, substantially improved tensile elongation and impact characteristics are indicated for dry, as-molded specimens.

In a related but older patent [26] grafted polyamide–hydrocarbon polymers are described that have improved toughness and elongation. Thus, an ethylene–ethyl acrylate copolymer ($M = 17,000$) containing 18% ethyl acrylate was added at concentrations of 5, 10, and 25% to ε-caprolactam and reacted at 255°C. The graft polymers are described as having the following repeat unit structure:

$$-(CH_2-CH_2)_n-CH_2-CH_2-$$
$$C{=}O$$
$$NH$$
$$(CH_2)_5$$
$$C{=}O$$
$$NH$$
$$(CH_2)_5$$
$$C{=}O$$
$$OH$$

where $n$ is 0 or a whole number and $x$ is a whole number. No information is given on the morphology of the resultant polymers, but it is presumed that the polycaprolactam crystallizes and the amorphous phase itself is multiphase. The graft polymers were purified, extruded into pellets, and molded.

High-impact and elongation values are reported in Table VI and compared with nylon 6 or straight blends of nylon 6 of ethylene–ethyl acrylate copolymer.

Table VI

Physical Properties of Grafted Caprolactam

| Sample | Notched Izod impact (meter gm/cm) | Elongation to break at 23°C (%) |
|---|---|---|
| 25% Graft | 1430 | 280 |
| Nylon 6 | 65.1 | 190 |
| 25% Blend (control) | 65.1 | 60 |

### 4. Poly(methyl methacrylate)

Rubber-modified poly(methyl methacrylate) resins are currently available as commercial products. The Rohm and Haas Company, for instance, has been granted a substantial number of patents in this area. Unfortunately, as a result of proprietary concerns, the amount of published information correlating the mode of preparation, composition, morphological relationships, and final properties is extremely limited. Moreover, there is no readily discernible connection between the patent literature and existing products. The patent by Dickie and Newman [27] illustrates, however, the properties achievable by carefully controlled preparative methods.

Emulsion polymerized shell–core polymers have already been described in Section III.A. on cross-linked amorphous glasses. In the above patent an elastomeric core is emulsion polymerized generally by using butyl acrylate with minor amounts of 1,2-butylene dimethacrylate. A variety of compositions and conditions are described, including the procedure of initiating the shell polymerization prior to complete conversion of the core structure. This presumably results in the formation of an intermediate layer and consequently a particle of "graded" composition. Uniform particle sizes are realized with a particle size falling within the range of .04–1 $\mu$m. Modifying comonomers may be used to influence the refractive indexes of core and shell to achieve a high degree of transparency.

The properties of these "multistage" polymerization products vary, of course, with composition and timing of the sequenced operations. The limited tensile data reported lie within the ranges shown in Table VII; the elongation to break is seen to be profoundly altered by rubber modification with designed elastomer particles.

Table VII

Properties of Rubber-Modified Methyl Methacrylate
(Graded Shell–Core Structures)[a]

| Butyl acrylate/ methyl methacrylate (ratio) | Yield stress (psi) | Ultimate elongation (%) | Modulus (psi) |
|---|---|---|---|
| 2.7 | 690 | 150 | 28,000 |
| 0.92 | 3000 | 88 | 128,000 |
| 0.30 | 5700 | 38 | 248,000 |
| 0.19 | 7200 | 14 | 310,000 |
| 0.10 | 8260 | 12 | 360,000 |

[a] Data from Dickie and Newmann [27].

## 5. Polysulfone

While shell–core, graded emulsion, polymerized particles and rubber grafted by addition polymerization are commonly used as types of rubber modifiers, the technique of additive blending of block copolymers (see Chapter 12) recently has become of greater significance. This method is uniquely suited to condensation polymers, for example, polysulfone, a generally amorphous thermoplastic with a high glass transition temperature.

A careful investigation of this system has been conducted by Noshay et al. [28], who achieved remarkable toughness improvement with the addition of polysulfone–poly(dimethylsiloxane) block copolymers. Fortunately, these block copolymers form two-phase systems of high melt viscosity and elasticity so that particle integrity appears to be largely retained during processing, even in the absence of cross-linking. The polysulfone segments are believed to be miscible with the polysulfone matrix, resulting in a dispersion of poly(dimethylsiloxane).

Table VIII

Effect of Block Copolymer Composition on Impact Strength[a,b]
5% Block Copolymer Added

| Block $\overline{M}_n$ | | Poly(dimethylsiloxane) (wt %) | Notched Izod impact (ft lb/in.) |
|---|---|---|---|
| Polysulfone | Poly(dimethylsiloxane) | | |
| 6,600 | 1,700 | 20 | 1.6 |
| 5,000 | 5,000 | 50 | 20.4 |
| 5,000 | 10,000 | 67 | 20.5 |
| 10,000 | 5,000 | 33 | 15.9 |
| 10,000 | 10,000 | 50 | 18.6 |

[a] Compression-molded polysulfone–block copolymer blend.
[b] Data from Noshay *et al.* [28].

Table IX

Effect of Block Copolymer Particle Size
on Impact Strength[a,b]
5% Block Copolymer Added

| Maximum particle size of 5000/5000 copolymer | Izod impact (ft lb/in.) |
|---|---|
| <0.5 | 1.5 |
| <3.0 | 10.5 |
| <8.0 | 16.5 |

[a] Injection-molded polysulfone–copolymer blend.
[b] Data from Noshay *et al.* [28].

Block copolymer composition, concentration, and particle size show a profound effect on impact strength with optimum properties indicated with blocks of (5,000–10,000) molecular weight and particle sizes in the range of 0.5–3.0 μm (see Tables VIII and IX). No fracture studies are reported.

## C. Semicrystalline Polymers of Low Glass Temperature

### 1. General Remarks

Semicrystalline polymers at temperatures above the $T_g$ of their amorphous phase also exhibit impact improvement through modification with dispersed rubber. Moreover, the magnitude of the effect can be similar to that achieved with glassy, amorphous polymers. It should be noted at this point that crazing or crazelike bodies have been reported in a variety of crystalline thermoplastics including polyethylene and polypropylene. Van den Boogart [29], for instance, reported the existence of crazes in polypropylene with their planes oriented normal to the applied stress. Other related evidence for a variety of crystalline thermoplastics is summarized by Kambour [3]. In essence then, polymers like polypropylene ($T_g \simeq 0°C$) may experience localized deformation under tensile stresses so as to result in regions of low polymer density containing fibrillated matter. Phenomenologically, one may therefore expect to observe effects with rubber modification analogous to glassy polymers. Detailed morphological features and the mechanism of local deformation may be expected to be different, however.

## 2. Polypropylene

Speri and Patrick [30] have explored to a limited degree the behavior of polypropylene mechanically blended with poly(ethylene–propylene) elastomers. Impact properties of such blends are strongly dependent on rubber particle size and size distribution. These particle parameters reflect, in part, shear developed during melt compounding and therefore are dependent on the melt viscosity of the polypropylene. (For a general discussion of processing parameters and viscoelastic properties as they influence the mode of dispersion, the reader is referred to Volume 1, Chapter 7 by Van Oene.) An optimum enhancement is achieved as the rubber particle size approaches 0.5 $\mu$m with a narrow particle size distribution.

The increase of impact strength reported by Speri and Patrick with increasing elastomer content in polypropylene with a melt flow index of 5 is given in Table X. For reasons that are not explained by these writers, only the notched Izod values exhibit a monotonic increase with elastomer content, whereas the unnotched do not. However, an increase of yield strain is suggested by the data. Of particular interest is the dramatic improvement noted with the amorphous ethylene–propylene–diene terpolymer (EPDM)–HDPE blend. This presumably is associated with the smaller particle size of about 0.5 $\mu$m achieved by compounding with this elastomer as compared with 1.5 $\mu$m for the crystalline EPDM.

Speri and Patrick acknowledge that dispersed rubbers are known to cause a decrease in average crystalline size in polymers like polypropylene, and thereby result in an improvement in impact strength. Nevertheless, the more likely mechanisms operating here are those of craze initiation and prevention of catastrophic craze and/or crack growth. The severe dependence of impact strength on the size of the dispersed rubber-phase particles as well as the observed stress whitening on deformation provide further, but not conclusive, evidence of the local yielding or craze mechanism.

Table X

Impact Properties of Injection-Molded Polypropylene Homopolymer[a]
Modified with Ethylene–Propylene Rubbers

| EPDM rubber[b] (%) | EPDM–HDPE rubber[c] (%) | Izod impact (ft lb/in.) | |
| --- | --- | --- | --- |
| | | Notched | Unnotched |
| 0 | — | 0.6 | 23.1 |
| 5 | — | 0.8 | 32.7 |
| 10 | — | 1.6 | 31.3 |
| 15 | — | 2.2 | 31.9 |
| 20 | — | 2.5 | 30.0 |
| — | 20 | 11.0 | 29.6 |
| 25 | — | 3.8 | 31.1 |

[a] Resin with melt flow index of 5; see Speri and Patrick [30].
[b] Crystalline EPDM hydrocarbon rubber.
[c] Amorphous EPDM preblended with high-density polyethylene.

## IV. SOFT INTERLAYERS IN FIBER AND PARTICULATE FILLED COMPOSITES

### A. Coated Fibers

Those workers familiar with ABS technology are naturally prone to speculate about whether an elastomeric skin over an inexpensive core material would suffice to provide toughening. As mentioned previously, however, it has become recognized that particles, for instance, with an excess of hard inclusions in the core, are inefficient as regards craze initiation and termination. Nevertheless, the presence of a soft interlayer between a hard core and the matrix has been demonstrated to provide useful property modification over that achievable with the particles themselves. This is documented by the evidence cited below.

The strength of composites containing well-aligned, continuous fibers is affected by stress concentrations. These may arise because of (a) the presence of the fibrous inclusion, which causes a nonuniformity in the stress field, or (b) shrinkage stresses caused by either resin curing or differential thermal contraction. Under transverse loading (direct or resolved) such stress concentrations may result in premature debonding or resin cracking.

Calculations of the radial, transverse, and shear stresses (polar co-ordinates) in a matrix containing a fibrous (cylindrical) inclusion have been made and reviewed by Arridge [31], who has demonstrated further that the presence of an interlayer on the fiber can affect the stress concentrations in the matrix significantly.

The analysis of real composites is exceedingly complex and depends on a number of factors, including fiber concentration, modulus ratio $E_f/E_m$, fiber packing, and interlayer thickness. Nevertheless, on theoretical grounds, Arridge clearly shows that for interlayer thicknesses, which are 1–2% of the fiber radius, a considerable reduction in the stress concentrating effects of the fibers is possible. In particular, for a steel wire embedded in an epoxy resin, the transverse (hoop) stress at 90 degrees to the direction of loading may vary from negligible values to about three times the applied stress, as the shear modulus of the interlayer is altered from equivalence to the matrix to low values characteristic of a cavity. The radial stress exhibits a different dependence on the shear modulus of the interlayer. Consequently, by an appropriate choice of interlayer shear modulus, either the radial or transverse stress or both may be minimized.

In general, for hard fibers in brittle matrixes, an optimized system is obtained if the shear modulus of the interlayer is smaller than that of the matrix and the Poisson ratio of the interlayer is considerably less than 0.5. A practical composite should be one in which the interlayer–fiber combination, that is, the interlayer with its fiber reinforcement, is matched to the composite. This requires an interlayer with a shear modulus about one-tenth that of the matrix.

Using a finite element analysis, Broutman and Agarwal [32] have also achieved a theoretical analysis of the effect of interfacial layers on the properties of spherical particle and oriented short fiber composites. For glass fiber-reinforced resins the following typical properties assumed were

Matrix:

$$\langle E_m = 4 \times 10^5 \quad \text{psi}, \quad \mu_m = 0.35 \rangle$$

Fiber:

$$\langle E_f = 11.8 \times 10^6 \quad \text{psi}, \quad \mu_f = 0.197, \quad L/D = 10.375 \rangle$$

For a low interface modulus, the strength of the composite along the fiber axis is relatively low due to high stress concentrations at the fiber end. With an increasing interface modulus, the composite strength is predicted to increase until the interface modulus reaches a value of $\sim 10^4$ psi; thereafter, the composite strength remains essentially constant (Fig. 6). Also, the strain energy absorbed, a measure of impact strength, peaks at an interface modulus of $10^4$ psi; consequently, a good combination of tensile strength and toughness may be expected with an interlayer modulus about one-tenth that of the matrix. On the other hand, the modulus of the composite reaches its maximum

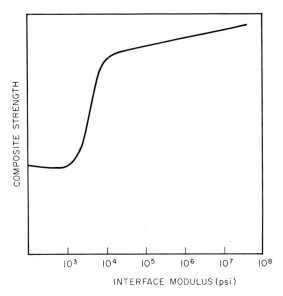

INTERFACE MODULUS (psi)

Fig. 6   Composite strength as a function of interface modulus, as qualitatively represented by Broutman and Agarwal [32].

between $10^5$ and $10^6$ psi as a consequence of improved efficiency of load transfer from matrix to fiber.

Experimental confirmation that coating of continuous fibers with thin layers of soft materials reduces stresses in the matrix and improves the off-axis performance of composites has been demonstrated by Lavengood and Michno [33]. In this study, oriented "E" glass fiber composites in an epoxy matrix were prepared with glass fiber coatings applied in thicknesses of 1, 2, and 4% of the fiber radius. The coatings consisted of Versamid-modified epoxies, which exhibited a glass transition temperature of 70°C, compared with 165°C for the unmodified matrix resin. Unfortunately, a detailed characterization of the interlayer properties was not possible since the interlayer material in situ differed from the same composition in bulk; however, the interlayer material was certainly more flexible than the matrix material.

Using a three-point flexure test, it is shown that the transverse strengths of dry samples is substantially increased from 5 ksi for uncoated fibers to 10 ksi for samples with coated fibers. A 2% coating thickness is optimum. Exposure to boiling water for 2 hr causes all samples to lose strength, but here too the coated fiber specimens exhibit a much improved retention of properties. This effect, not discussed by the authors, may also be due to simple protection of the glass by the interlayer, however.

Most importantly, the fracture surface of uncoated fiber samples show failure at the glass resin interface, whereas, for interlayered samples fracture occurs away from the fiber surface with evidence of flow, drawing and tearing in the residual polymer attached to the fiber.

It is of interest to note (Fig. 7) that with increasing temperature, the transverse strengths of the coated fiber samples fall off rapidly, especially in the neighborhood of $70°C$, the measured $T_g$ of the interlayer. With increasing temperature, the shear modulus of the interlayer decreases while its Poisson ratio may be expected to increase. Evidently, the combination of these parameters becomes increasingly less optimum in the temperature range investigated, that is, 25–90°C, thereby leading to higher stress concentration factors. Certainly, in the limit of a liquid interlayer, conditions of a cavity may be approached.

As previously mentioned, a practical composite should be chosen so that the interlayer–fiber combination is matched to the matrix. This calls for an interlayer of a particular modulus about one-tenth that of the matrix. One may speculate that in the temperature range shown the modulus of the interlayer has fallen well below this level.

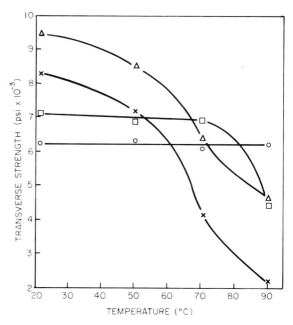

Fig. 7  Transverse strength of composites at temperatures above and below the glass temperature of the inner layer. Inner-layer thicknesses are ($\square$) 1, ($\triangle$) 2, and ($\times$) 4% of the fiber radius; ($\bigcirc$) control. (Replotted from data of Lavengood and Michno [33].)

However, some improvement in the toughness of fiber-reinforced composites has been reported by coating fibers with viscous fluids [34]. This is attributed to energy dissipation through fiber pullout and the increased viscous contribution of the liquid layer.

### B.   Coated Particulate Mineral Fillers

Unfortunately, few definitive published experimental studies are available to substantiate in detail the quantitative predictions of the theoretical studies of short fiber-reinforced composites. This is similarly true of particulate mineral-reinforced composites.

Particulate mineral fillers are widely used to effect property modification and reduced cost in commercial plastic systems. Enhancement of modulus, hardness, creep resistance, heat distortion temperature, etc. are generally realized, however, with an attendant loss in elongation and toughness. A substantial recovery of toughness may be expected and, in fact, has been achieved with the use of appropriately selected interlayers. Unfortunately, there is a paucity of such information in the open literature, and documentation of such studies are, therefore, difficult to provide. (See, however, Fig. 8.)

With semicrystalline polymer matrixes, soft inclusions may be achieved simply by controlled reduction in the degree of polymer crystallinity in the neighborhood of a hard inclusion.

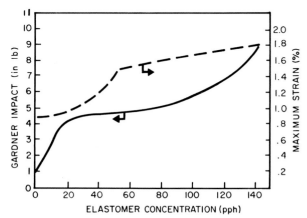

Fig. 8   Gardner impact and maximum strain for an elastomer coated mineral filler in a thermoset matrix. (Data courtesy of B. F. Goodrich Co.)

# REFERENCES

1. E. H. Merz, G. C. Claver, and M. J. Baer, *J. Polym. Sci.* **22**, 325 (1956).
2. S. Newman and S. Strella, *J. Appl. Polym. Sci.* **9**, 2297 (1965).
3. R. P. Kambour, *Macromol. Rev.* **7**, 134 (1971).
4. C. B. Bucknall and R. R. Smith, *Polymer* **6**, 437 (1965).
5. S. Strella, *J. Polym. Sci.* **4**, 527 (1966); paper presented at the Amer. Chem. Soc. Meeting, Chicago (Sept. 1967); *Appl. Polym. Symp.* No. 7, p. 165 (1968).
6. C. B. Bucknall and D. G. Street, *SCI Monograph No. 26* p. 272 (1967).
7. C. B. Bucknall, *Brit. Plast.* **40** (11), 118; (12) 84, 1967.
8. R. H. Beck, S. Gratch, S. Newman, and K. C. Rusch, *Polym. Lett.* **6**, 707 (1968).
9. S. Newman, *Polym. Plast. Technol. Eng.* **2**(1), 67 (1973).
10. M. Baer, *J. Appl. Polym. Sci.* **16**, 1125 (1972).
11. E. R. Wagner and L. M. Robeson, *Rubber Chem. Technol.* **43**, 1129 (1970).
12. G. P. Marshall, L. E. Culver, and J. G. Williams, *Int. J. Fract. Mech.* **9**, 295 (1973).
13. M. J. Doyle, *SPE Annu. Tech. Conf., 34th Atlantic City, New Jersey* p. 480 (1976).
14. R. Dickie and S. Newman, Rubber-Modified Thermosets and Process, U.S. Patent 3,833,682 (September 3, 1974).
15. R. Dickie and S. Newman, Rubber-Modified Thermosets and Process I, U.S. Patent 3,833,683 (September 3, 1974).
16. R. Dickie and S. Newman, Graded Rubber Particles Having Hydroxy Functionality and a Polymeric Crosslinking Agent, U.S. Patent 3,856,883 (December 24, 1974).
17. R. Dickie, M. F. Cheung, and S. Newman, *J. Appl. Polym. Sci.* **17**, 65 (1973).
18. J. N. Sultan and F. J. McGarry, *Polym. Eng. Sci.* **13**, 29 (1973).
19. F. J. McGarry, *Proc. Roy. Soc. London A* **319**, 59 (1970).
20. J. N. Sultan, R. C. Laible, and F. J. McGarry, *Appl. Polym. Symp.* No. 16, 127 (1971).
21. W. D. Bascom and R. L. Coltington, *J. Adhes.* **7**, 333 (1976).
22. N. K. Kalfoglou and H. L. Williams, *J. Appl. Polym. Sci.* **17**, 1377 (1973).
23. R. P. Petrich, *Polym. Eng. Sci.* **13**, 248 (1973).
24. J. F. Kenney, *J. Polym. Sci.* **14**, 123 (1976).
25. A. J. Chompff, Rubber-Modified Polyamides, U.S. Patent 3,880,948 (April 29, 1975).
26. R. J. Kray and R. J. Bellet, Grafted Polyamide Hydrocarbon Polymers and Processes to Prepare Them, French Patent 1,470,255 (February 17, 1967).
27. R. A. Dickie and S. Newman, Acrylate Polymer Particles Comprising a Core, an Outer Shell, and an Intermediate Layer, U.S. Patent 3,787,522 (January 22, 1974).
28. A. Noshay, M. Matzner, B. P. Barth, and R. K. Walton, *Div. Organ. Coatings Plast. Chem. Prepr. Amer. Chem. Soc.* 217 (September 1974).
29. A. Van den Boogart, *in* "Physical Basis of Yield and Fracture" (*Conf. Proc. Oxford*) p. 167. Inst. Phys. and Phys. Soc., London, 1966.
30. W. M. Speri and G. R. Patrick, *Polym. Eng. Sci.* **15**, 668 (1975).
31. R. G. C. Arridge, *Polym. Eng. Sci.* **15**, 757 (1975).
32. L. J. Broutman and B. D. Agarwal, *Polym. Eng. Sci.* **14**, 581 (1974).
33. R. E. Lavengood and M. J. Michno, *Soc. Plastic Eng. Div. Tech. Conf. Additives, Hudson, Ohio* p. 126 (1975).
34. T. Jones, N. R. Suh, and N. Sung, A method of improving the fracture toughness of fiber reinforced composites, *Soc. Plastic Eng., 34th Annu. Tech. Conf., Atlantic City, New Jersey* p. 458 (1974).

# Chapter 14

# Fracture Phenomena in Polymer Blends

## C. B. Bucknall

*Cranfield Institute of Technology*
*Cranfield, Bedford*
*England*

## I. INTRODUCTION

The principal application of polymer blends is in the manufacture of fracture-resistant plastics, a fact that testifies to the practical importance of fracture properties to the user of plastics products. Over 50 years ago it was

91

discovered that the fracture resistance of polystyrene could be increased substantially by adding a small proportion of rubber, and the principle has been employed commercially since 1948 in making high-impact polystyrene (HIPS). Manufacture of acrylonitrile–butadiene–styrene polymer (ABS) began a few years later, and the technology has since been extended to a wide range of other polymers: there are now rubber-toughened grades of poly(vinyl chloride) (PVC), poly(methyl methacrylate) (PMMA), polypropylene (PP), polycarbonate (PC), polysulfone (PSF), poly(2,6-dimethyl-1,4-phenylene oxide) (PPO), epoxy resin, and a number of other polymers and copolymers.

General features of a variety of rubber-toughened plastics are reviewed in Chapter 13. In this chapter another aspect of rubber toughening is concentrated upon, the relationship between structure and fracture resistance. A more extensive review of the whole subject is contained in a recently published book [1a].

A qualitative study of rubber-toughened plastics shows that phase separation is an essential feature: the rubber must be dispersed randomly as small discrete particles in the continuous-matrix phase if satisfactory resistance is to be achieved. Other necessary features of the rubber phase are: (a) low shear modulus in relation to the matrix polymer; (b) good adhesion to the matrix; (c) adequate crosslinking; (d) average particle diameter near the optimum value for the material; and (e) low glass transition temperature. The reasons for these requirements are discussed below. The structure of the matrix polymer also affects fracture resistance: molecular weight is a fundamental consideration, and toughness is also dependent upon the presence of antiplasticizers and other additives, the degree of cross-linking (in thermosetting resins), and the crystalline morphology (in semicrystalline polymers).

## II. QUALITATIVE THEORIES OF TOUGHENING

The eventual aim of any study of rubber toughening must be to develop a quantitative understanding of structure–property relationships. The osmium-staining technique[†] provides a wealth of quantitative information about structure, and the methods discussed in the latter part of the present chapter help to satisfy the requirement for quantitative data on the deformation and fracture behavior of rubber-modified plastics. However, in order to interpret these data, it is necessary to have a qualitative picture of the phenomena

---

[†] This technique was developed by K. Kato and has been described in a number of his publications (e.g., [1b–d]). [—EDITORS]

contributing to rubber toughening. A brief outline of current theories is presented below.

## A. The Multiple-Crazing Theory

Early theories of toughening tended to concentrate upon the rubber phase, and its role in preventing brittle fracture. The rubber particles were considered to act as obstacles to crack propagation. This idea was developed by Merz *et al.* [2], who suggested that the rubber particles held together the opposite faces of a growing crack, being stretched out as the matrix material parted, and thus inhibiting brittle crack propagation; they suggested that the stabilizing effect of the rubber permitted the formation of a large number of microcracks instead of a single catastrophic crack. This theory explained the characteristic stress whitening observed in rubber-toughened plastics under tensile stress, and accounted for the ability of these materials to deform to high tensile strains before fracture. It did not, however, give a satisfactory account of a number of features of rubber toughening. The theory did not explain precisely how energy was absorbed in an impact test, nor did it account, for example, for the observed differences in fracture behavior between HIPS and toughened PVC.

It is now generally recognized that energy absorption occurs almost entirely in the matrix polymer. The function of the rubber particles is to promote and control deformation in the matrix, by providing a large number of stress concentrations, where localized deformation can be initiated. Shear yielding has some part to play in the process, but the dominant mechanism of toughening is crazing. Numerous experiments have shown that the familiar phenomenon of stress whitening is due to multiple-craze formation rather than microcracking [3–7]. Because of their low refractive indexes, crazes scatter light in the same way as cracks.

Multiple crazing explains many of the other observed features of rubber toughening. Fracture mechanics measurements show that a large amount of energy is absorbed in forming a single craze: fracture surface energies are of the order of $1 \text{ kJ/m}^2$, which is very high in relation to the amount of material contained in a craze [8, 9]. It follows that by increasing the number of crazes formed during fracture, crazes increase the energy absorption very considerably. The open, voided structure of crazes explains the decrease in density produced by tensile deformation of HIPS [2], and the resulting permeability of the material [10]. The low-modulus and high-mechanical hysteresis of stress-whitened HIPS and ABS [11] are directly attributable to the distinctive viscoelastic properties of crazes [12].

Crazing is the first stage of fracture in glassy polymers [13]. Under a sufficiently high stress, the fibrillar structure of the craze breaks down, and a true crack forms. When the crack reaches a critical size, the material breaks. In a rubber-toughened polymer, the function of the rubber particles is not only to initiate crazes, but also to prevent, or at least delay, the formation of a crack of critical length. The latter function is extremely important. Other types of inclusion, including glass beads, are able to initiate multiple crazing [14]. Rubber particles are distinguished by their greater ability to control the growth of individual crazes.

Crazing is not the only deformation mechanism operating in rubber-toughened plastics under tensile loading. In addition to elastic deformation, there is usually some shear yielding, in the form of shear bands or more diffuse shear yielding. The stress concentrations produced by the rubber particles are initiation sites for shear bands as well as crazes. Shear yielding is not simply an additional deformation mechanism, but appears to be an integral part of the toughening mechanism. The molecular orientation within the shear zones is approximately parallel to the applied tensile stress, and therefore normal to the planes in which crazes are formed. Since both initiation and propagation of crazes are inhibited by orientation in this direction, shear bands have the effect of limiting craze growth. As the number of shear bands increases, the length of newly-formed crazes decreases.

The multiple crazing theory outlined above, with its emphasis upon multiple craze initiation by rubber particles, and control of craze growth by the rubber particles and by any shear bands that may be formed, appears to be widely accepted. Nevertheless, attempts have been made, notably by Bragaw [15] and by Hagerman [16, 17], to modify the theory substantially. Both authors suggest that the rubber particles act not by initiating crazes, but in some other way. Neither presents a convincing case.

Bragaw [15] notes that craze branching is sometimes observed in rubber-toughened plastics, and argues that this branching is the essential feature of the multiple-crazing mechanism. Referring to Goodier's analysis of the stresses surrounding an isolated spherical inclusion, he concludes that craze branching cannot occur in the matrix polymer, and must therefore originate within the rubber particles. In order to explain branching within the particles, Bragaw quotes Yoffe's classic paper [18] on branching of cracks propagating in a homogeneous medium at speeds approaching the terminal velocity, and points out that terminal velocities are lower in rubbers that in glassy polymers. This set of hypotheses ignores several obvious features of rubber toughening. First, craze branching does not easily account for the general pattern of craze formation, for example, along the length of a HIPS tensile bar. Second, rubber particles in a typical HIPS or ABS are very far from being isolated spherical inclusions; there can be little doubt that the observed branching occurs because rubber particles are close together in random positions throughout the matrix polymer, so that the stress field in the matrix is quite different from that predicted by Goodier. Third, the stress field within a small rubber particle is unlikely to bear much resemblance to the stress field in a homogeneous medium, especially if the rubber particle is surrounded by other similar particles; the case for applying Yoffe's analysis to a rubber particle in ABS is simply not argued. Fourth, rubber particles in HIPS are themselves inhomogeneous, consisting largely of polystyrene subinclusions, with a minor proportion of rubber; many ABS polymers have a similar structure. In short, Bragaw's hypothesis is ingenious, but is based upon false premises, and is not supported by the available evidence. Craze branching is an incidental feature of rubber toughening rather than a fundamental requirement.

Hagerman suggested that individual rubber particles were not important in generating crazes, since scanning electron micrographs of fracture surfaces showed no obvious evidence of rubber particles [16]. This suggestion is contradicted by numerous transmission electron micrographs of crazed polymers [4–7]. Hagerman's proposal that rubber particles produce a state of plane stress in the styrene–acrylonitrile copolymer (SAN) matrix of ABS polymers is based upon an erroneous analogy with stress conditions in thin plates.

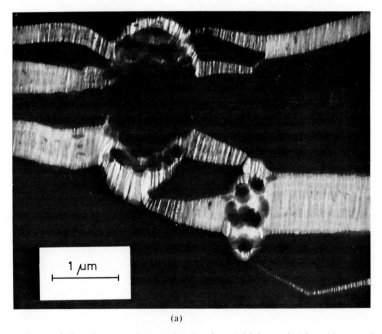

(a)

Fig. 1 Transmission electron micrography showing multiple crazing in rubber-toughened plastics: solvent-case films of (a) HIPS; (b) ABS, stretched in the vertical direction. (Micrographs by courtesy of M. Bevis, P. Beahan, and A. Thomas, Liverpool University.)

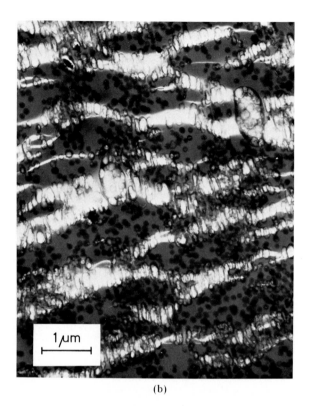

(b)

Fig. 1   Transmission electron micrographs showing multiple crazing in rubber-toughened plastics: solvent-cast films of (a) HIPS; (b) ABS, stretched in the vertical direction. (Micrographs by courtesy of M. Bevis, P. Beahan, and A. Thomas, Liverpool University.)

## B.   Experimental Evidence

There is ample experimental evidence to support the multiple craze theory of toughening. The first studies were made by stretching thin sections of HIPS on the stage of an optical microscope, and examining them between crossed polars [3]. Subsequent electron microscope studies have confirmed the formation of crazes in HIPS [5, 6], ABS [7], toughened PVC [4], and toughened epoxy resins [1, 19]. In Fig. 1 are shown two transmission electron micrographs obtained by solvent casting thin films of HIPS and ABS, stretching the films, and then immersing them in osmium tetroxide, which is a preferential stain for the rubber phase [20]. The micrographs show that large numbers of crazes are formed in the glassy matrix, and that the rubber particles encountered by the crazes also undergo substantial deformation. The rubber phase extends by fibrillar deformation, as does the matrix, so that the crazes can be considered as propagating into, and in many cases through, the rubber particles. The subinclusions of PS or SAN present in the larger rubber particles do not appear to deform, nor is there any substantial deformation of rubber particles that are not embedded in crazes.

Evidence for the simultaneous formation of crazes and shear bands in a rubber-toughened plastic is presented in Fig. 2. The specimen is a blend of HIPS with PPO, which was deformed in tension, polished by ultra-microtoming parallel to the tensile direction, and etched with a mixture of chromic and phosphoric acids [21]. The acid mixture preferentially attacks the rubber particles and crazes, and also etches the shear bands to a lesser extent. After coating with gold–palladium, the specimen was examined in the scanning electron microscope, which reveals the presence of large numbers of short crazes, many of them terminating at shear bands. Similar interactions probably occur in other rubber-toughened plastics based on relatively ductile matrix polymers.

Figures 1 and 2 emphasize the points made in Section II.A. The function of the rubber particles is to ensure that large deformations can occur in the matrix without producing a crack that is large enough to cause fracture. In HIPS at room temperature, multiple crazing is the only mechanism by which

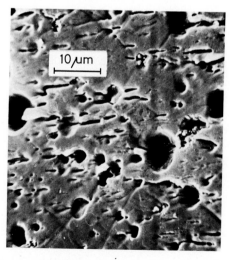

Fig. 2  Scanning electron micrograph showing crazes and shear bands formed simultaneously in a HIPS–PPO blend under tensile stress (vertical direction). Specimen etched in $H_2SO_4$–$H_3PO_4$–$CrO_3$, and coated with Au–Pd.

large tensile strains can be achieved. The rubber particles have low shear moduli in relation to the matrix polymer, and therefore provide stress concentrations from which crazes can grow. The experimental evidence shows that the rubber particles also limit the growth of individual crazes. There are several ways in which this control of craze size may be exercised: large particles may simply act as barriers to advancing crazes, smaller particles embedded in crazes probably have a stabilizing effect, and mutual termination of crazes through the overlap of stress fields is another possibility. In more ductile polymers such as toughened PVC and PPO–HIPS blends, rubber particles nucleate shear bands, which are extremely effective in controlling craze growth. Providing that the loading conditions allow time for shear bands to develop, a very tough product is obtained.

## C.  Role of Adhesion

The need for good adhesion between rubber particle and matrix is adequately demonstrated in Fig. 1a. If the bond between the rubber and the polystyrene were not a strong one, a void would form at the interface, and a crack would be initiated. Once such a crack developed, it would run from one rubber particle to the next with little hindrance from the poorly anchored particles in its path. The problem can be demonstrated by making a latex blend of polybutadiene with PS or SAN polymer [22]; in commercial HIPS and ABS polymers it is avoided by grafting (see Chapter 12).

Because of poor adhesion, mechanical blends of polystyrene with polybutadiene rubber have very low impact strengths. However, the addition of a graft copolymer of the two materials produces a very dramatic increase in impact strength [23]. A similar effect can be demonstrated in blends of polystyrene with low-density polyethylene [24, 25]: addition of a graft copolymer of styrene and ethylene greatly increases the fracture resistance of the blend. Since polyethylene is a highly deformable polymer of low shear modulus, it fulfils the function of a rubber in this blend system, and the graft copolymer ensures good adhesion despite the incompatibility of the two constituents (this system is discussed more fully in Chapter 12).

## D.  Role of Cross-Linking

Why cross-linking is desirable in the rubber particles also is indicated in Fig. 1a. The rubber phase is subjected to very large tensile strains, giving a crazelike structure. In PS and SAN, craze fibrils are stabilized by molecular entanglements, and can sustain high stresses for long periods. In uncross-linked polybutadiene, on the other hand, molecular entanglements are unable to prevent rapid flow and fracture in response to an applied stress: at room temperature, the rubber is over 100°C above its $T_g$, and relaxation times are extremely short. A moderate degree of cross-linking allows the rubber to reach high strains by fibrillation, and at the same time confers mechanical strength upon the fibrils.

There are other reasons for preferring a cross-linked rubber. Uncross-linked rubber particles tend to break down during melt processing, and can become so finely dispersed that the particle size is below the critical diameter for toughening. Even if the critical particle size is not reached, problems can occur in extrusion or injection molding: uncross-linked rubber particles become smeared out into flat sheets, which give rise to laminar tearing [26, 27].

## E. Effect of Particle Size

It is generally recognized that the size of the rubber particles has a strong influence upon the toughness of rubber-modified polymers. For each type of material, there appears to be an optimum particle size for toughening. Impact strength falls drastically if the average particle diameter is reduced below a critical value, and there is evidence for a rather slower fall in fracture resistance as the particle size is increased above the optimum diameter. The literature contains numerous examples of the critical particle size effect. For example, Durst *et al.* [23] obtained an Izod impact strength of 7.5 ft-lb/in. notch for a blend of polystyrene with 25% styrene–(1,4-butadiene)–styrene (SBS) block copolymer, which was dispersed as particles having an average diameter of 1.0 μm; but a blend of the same composition with an average particle size of 0.2 μm had an impact strength of 1.0 ft lb/in. notch. Parsons and Suck varied particle size in ABS emulsion polymers between 0.1 and 0.3 μm, and observed an approximately linear increase in impact strength from 1.9 to 4.5 ft lb/in. notch in polymers containing 20% polybutadiene [28]; parallel experiments on ABS, using different grafting procedures, showed a similar trend but with a different slope to the curve [29]. A further example of the particle size effect is given in Fig. 3, which is taken from Purcell's work on rigid PVC blends containing 9% MBS (methyl methacrylate–butadiene–styrene) graft copolymer [30]: reducing the particle size from 0.2 to 0.05 μm causes a marked drop in the impact strength of standard ¼-in. specimens, and a much smaller drop in the impact strength of ⅛-in. specimens.

In studying the literature on particle size, it is necessary to examine the evidence carefully. Some of the effects attributed to particle size are due, at least in part, to other factors. The problem is that it is difficult to alter rubber particle size without at the same time altering the structure and morphology of the polymer in other ways. For example, in the bulk polymerization of HIPS,

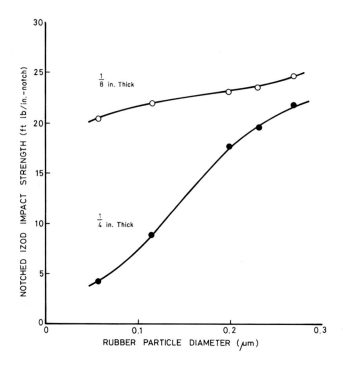

Fig. 3   Relationship between particle size and impact strength in PVC–MBS blends, tests at two different specimen thicknesses (see Section IV.B.3). (After Purcell [30].)

rubber particle size is reduced by increasing stirrer speeds during the period immediately following phase inversion [31, 32]. However, the reduction in size is accompanied by a reduction in the volume fraction of the particles; stirring decreases the amount of polystyrene occluded by the particles. Since the volume fraction of the disperse phase is known to be an important factor in rubber toughening, it is difficult to establish from this type of experiment whether there is also a particle-size effect. Other factors that are known or suspected to affect toughening include molecular weight of matrix, degree of grafting, volume ratio of subinclusions to rubber, size of subinclusions, glass transition temperature of rubber, and degree of cross-linking in the rubber. Experiments on particle size effects should preferably be designed to ensure that each of these factors is held constant. Measurements of the modulus provide a useful check on the volume fraction of rubber particles, which can be determined accurately by electron microscopy.

Despite the difficulties listed above, one point emerges clearly from the literature: critical particle sizes are not the same for all materials. The best available estimates are 0.8 $\mu$m for HIPS, 0.4 $\mu$m for ABS, and 0.2 $\mu$m for toughened PVC. Polycarbonate is toughened by the addition of a silicone block copolymer which forms domains only 5nm in diameter [33]. These results indicate that the critical particle size for toughening decreases with increasing ductility of the matrix polymer. Evidence to support this view was obtained by Bucknall et al. in experiments on HIPS–PS–PPO blends [21]. The particle size of the HIPS was small, so that HIPS–PS blends were not tough. However, on substituting PPO for PS in the blend, a much tougher product was produced, although the particle size and concentration were unchanged.

There are three possible explanations for the particle size effect: (a) small particles are unable to control craze growth effectively, (b) small particles are unable to initiate crazes efficiently, and (c) both the initiation and control of craze growth are ineffective in materials containing small particles. All three mechanisms produce the same overall result: instead of a large number of small crazes, the material forms a small number of large crazes, and the fracture resistance is thereby reduced. A simple tensile test is sufficient to determine whether craze initiation is affected by particle size, since a reduction in the rate of initiation leads to an increase in yield stress. The same test also provides information about craze growth. A reduction in elongation at break indicates that the rubber particles are less effective in controlling craze propagation. Experiments on ABS by Grancio show that reducing particle diameter from 0.56 to 0.10 $\mu$m both increases yield stress and reduces elongation at break [34], indicating that explanation (c) applies to this material.

In relatively ductile polymers, craze growth is inhibited by shear yielding, as explained in Section II.A. The rubber particles may be too small to control crazing directly, but large enough to control it indirectly, by initiating shear bands. This line of reasoning would explain why the critical particle size decreases with increasing ductility of the matrix. Shear yielding compensates for a lower rate of craze formation and at the same time prevents any crazes that are formed from growing too large. It is perhaps significant that toughened PVC materials can sometimes exhibit ductile failure in impact tests without stress whitening, under conditions that cause brittle failure in unmodified PVC. Multiple crazing is suppressed almost completely, while shear yielding is promoted by the rubber particles, so that a very tough product is obtained. Another significant result comes from Grancio's work on cold rolling of ABS [34]: the material containing 0.10 $\mu$m particles became very much tougher on rolling to 85% of its previous thickness. The elongation at break increased from 10 to 130%, and whereas the compression-molded specimens showed only local stress whitening, the rolled specimens whitened along the entire gauge length. These changes can be related directly to shear yielding during rolling. Although the rubber particles were very small, they were effective in toughening the ABS after rolling because the shear zones controlled the growth of individual crazes.

Most of the published studies have been concerned with reducing particle size. Large particles impair surface appearance, and are therefore of little interest within the industry. The limited amount of information that is available indicates that toughness decreases if the particle size becomes too large. Grancio's experiments showed that small particles were more effective than large particles in toughening cold-rolled ABS, although the reverse was true of compression-molded ABS [34]. In other words, there appears to be an optimum particle size for toughening, which decreases on cold rolling. At a constant volume fraction of rubber, the effect of increasing particle size is to increase the interparticle spacing. If the particles are above the average diameter, the predicted result of increasing particle size is an increase in average craze size, which is directly related to the distance between neighboring rubber particles. The basis of the argument is that craze sizes are at a minimum when the particles have an optimum diameter for toughening: larger particles are too far apart, whereas smaller particles are unable to control craze growth.

Particle sizes are not fully characterized by the average diameter, and in discussing the relationship between structure and mechanical properties, it is a logical step to consider the effects of particle size distribution. Experiments on HIPS bulk polymers show that the number of particles of diameter $d$ follows a gaussian dependence on $\log d$ [35]. Different distributions may be obtained quite readily. For example, bimodal distributions are produced by blending bulk polymers with emulsion polymers, or by mixing two emulsion polymers of differing particle size. Some commercial ABS polymers contain bimodal distributions of rubber particle sizes, and there is evidence that this type of distribution gives a tougher product than one having a normal distribution. Grancio prepared ABS polymers containing 14% rubber by mixing materials of differing particle sizes, and found that a 3/1 mixture of small (0.1-$\mu$m) and large (0.56-$\mu$m) particles was considerably tougher than either of the parent materials [34]. Substituting small particles for large ones resulted in a rapid increase in yield stress, which was accompanied by a much slower decline in elongation at break over the range of compositions up to the 3/1 mixture. The low elongation at break of the material containing only small particles has already been discussed. This experiment again points to an optimum condition, in this case an optimum particle size distribution, which minimizes craze sizes. The larger particles appear to cooperate with the smaller particles, providing a stabilizing influence on craze growth, and thus increasing the energy at break of the ABS.

## III. VOLUMETRIC STRAIN MEASUREMENTS

The qualitative studies discussed in the preceding section show that both crazing and shear yielding play a significant part in rubber toughening, and it is clear that a quantitative study of these two mechanisms would greatly advance our understanding of the phenomenon. The aim of such a study should be to measure rates of crazing and of shear yielding, and to relate these to the structure of the polymer and the conditions of testing. Volumetric strain measurements provide the basis for a quantitative approach of this kind. The principle of the method is simple: crazes produce a large increase in volume, which can be measured, whereas other deformation mechanisms occur essentially at constant volume.

### A. Analysis of Strain

It is convenient to divide the tensile deformation of a toughened polymer into three components: (a) elastic deformation; (b) crazing; and (c) shear yielding. Elastic deformation may be defined as the deformation occurring instantaneously on loading the specimen, including both extensional and volumetric components of strain. A typical figure for the Poisson ratio of a polymer would be 0.4, which means that an elastic extension of 1.0% is accompanied by a lateral contraction of 0.4%, and therefore by a volume increase of 0.2%. The initial elastic volume strain is inversely proportional to the bulk modulus of the polymer.

Any subsequent increase in volume strain can be identified with craze formation. Thus, by measuring volume strains it should be possible to follow rates of crazing in a stressed specimen. Void cracks can also contribute to the measured volume strain, but the problem is not a serious one, since the formation of a crack large enough to be detected in this way will usually lead immediately to fracture. Ideally, the specimen should be examined by microscopy, to confirm that crazes rather than cracks are responsible for the measured volume strain.

In a tensile deformation of an isotropic material, crazing causes the material to extend in the tensile direction, but has no detectable effect upon lateral dimensions. Any decrease in the cross-sectional area of the specimen must therefore be due to shear yielding, so that measurements of lateral strain provide direct information about rates of shear deformation.

If we consider a cube of unit dimensions, subjected to tensile stress in the three direction, as illustrated in Fig. 4, the volume change $\Delta V$ is given by

$$\Delta V = \lambda_1 \lambda_2 \lambda_3 - 1$$
$$= \lambda_1^2 \lambda_3 - 1 \qquad (1)$$

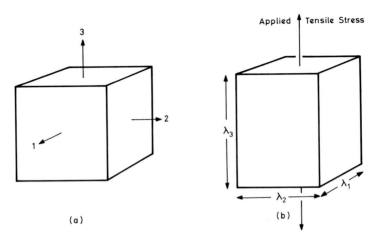

Fig. 4  Tensile deformation of a cube of unit dimensions: (a) unstressed; (b) under uniaxial tension.

where $\lambda$ is the natural draw ratio, defined as current length divided by original length. Since the material is isotropic, $\lambda_1 = \lambda_2$. Substituting $e_1$ for $(\lambda_1 - 1)$ and $e_3$ for $(\lambda_3 - 1)$, Eq. (1) gives

$$\Delta V = 2e_1 + e_3 + e_1^2 + 2e_1e_3 + e_1^2 e_3 \qquad (2)$$

Provided that $e_1$ is small, Eq. (2) reduces to

$$e_3 = \Delta V - 2e_1 \qquad \text{for small} \quad e_1 \qquad (3)$$

In other words, the extension is the sum of volume strain and lateral contraction terms. On subtracting the instantaneous elastic deformation from each term, we are left with the statement that the time-dependent part of the extension is the sum of the volume strain due to crazing and the lateral strain due to shear, yielding

$$e_3(t) - e_3(0) = \Delta V(t) - \Delta V(0) - 2[e_1(t) - e_1(0)] \qquad (4)$$

The more complete Eq. (2) must be used if $e_1$ is not small. The terms involving $e_1$ in Eqs. (3) and (4) are negative because the lateral strain is a contraction. At small strains, the term $-2e_1$ is equal to the fractional decrease in cross-sectional area of the specimen $-\Delta A$.

This analysis of strains as a means of differentiating between the deformation mechanisms operating in rubber-toughened plastics was first developed by Bucknall and Clayton, and applied to creep of HIPS and other polymers [21, 36–38]. Experiments consisted of accurate measurements of $e_1$ and $e_3$ using specially designed extensometers. The technique has since been extended to tensile impact tests by Fenelon and Wilson, who employed high-speed cinematography to determine $e_1$ and $e_3$ [39], and ordinary tensile tests by Truss and Chadwick, who determined $\Delta V$ directly by displacement dilatometry [40, 41].

## B.  Factors Affecting Deformation Mechanism

The quantitative approach described in the preceding section has revealed a number of interesting and in some cases unexpected features in the deformation behavior of toughened plastics, and especially of ABS. It is now clear that both the rate and mechanisms of deformation vary with stress, strain, temperature, and polymer structure. The effects of stress upon the mechanism of tensile creep are illustrated in Fig. 5, which shows creep and recovery data for a typical ABS emulsion polymer [38]. At the moderate stress of $26.5 \, \text{MN/m}^2$, there is very little time-dependent volume change, and no detectable residual volume strain after unloading, showing that crazing makes a negligible contribution to creep at this stress. Creep is dominated by the

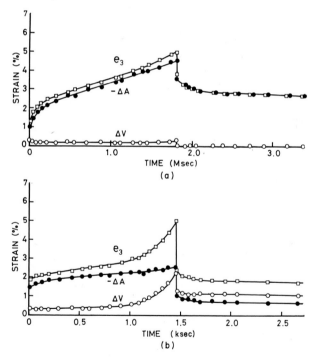

Fig. 5  Creep and recovery of ABS polymer at two stress levels: (a) $26.5 \, \text{MN/m}^2$; (b) $34.5 \, \text{MN/m}^2$. ($\square$) Elongation $e_3$; ($\bullet$) area strain $-\Delta A$; ($\bigcirc$) volume strain $\Delta V$. (From Bucknall and Drinkwater [38].)

shear mechanism, which proceeds at a decreasing rate, and leaves a residual strain of approximately 3% after unloading. The creep behavior of the ABS changes considerably on raising the stress to 34.5 $MN/m^2$; after a slow initial period, the volume strain rises rapidly, indicating craze formation at an accelerating rate. The total creep strain $e_3$ is dominated by the contribution of crazing, and the high residual strain after unloading is due largely to the fact that crazes close up only partially on removal of the stress. Because of the difference in time scale between Fig. 5a and b, it might appear that the shear mechanism is partially suppressed on increasing the stress. This is not the case, however. Raising the applied stress increases the rate of shear deformation by a factor of approximately 500, but the effect is overshadowed by the concurrent increase in the rate of crazing. Because of differences in kinetics between the two mechanisms, crazing becomes more dominant as the stress is increased.

A more explicit representation of the relationship between deformation mechanism and stress or strain is given by plotting volume strain $\Delta V(t)$ at time $t$ against the corresponding elongation $e_3(t)$, as shown in Fig. 6. At 26.5 $MN/m^2$, the curve has zero slope, showing that there is no crazing, and that creep is due entirely to shear yielding. At higher stresses, the mechanism begins to change: shear mechanisms still predominate at elongations up to 2.5%, but at higher strains crazing becomes the principal mechanism of creep. The slope increases with increasing applied stress.

Truss and Chadwick obtained similar results in tensile tests on transparent ABS emulsion polymer [41]. The relationship between $\Delta V$ and $e_3$ is shown in Fig. 7 as a function of strain rate in tensile tests performed at room temperature; the slope increases with increasing strain rate, showing that the mechanism changes from shear yielding, with a 10% contribution from crazing, to 100% crazing over the range of strain rates studied. The results also show a change in mechanism with increasing strain. At elongations above

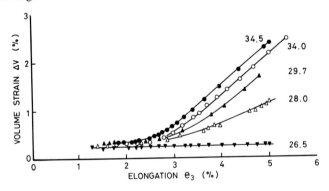

Fig. 6    Relationship between $\Delta V$ and $e_3$, showing mechanism of creep at 20°C as a function of elongation, at five different stresses ($MN/m^2$). (From Bucknall and Drinkwater [38].)

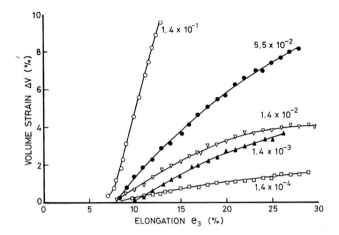

Fig. 7 Relationship between $\Delta V$ and $e_3$, showing mechanism of tensile deformation as a function of strain and strain rate. (After Truss and Chadwick [41].)

15%, the slope begins to decrease. Fenelon and Wilson observed a similar decrease in slope at high strains in tensile impact tests on ABS polymer [39].

The simplest approach to these changes in mechanism with stress and strain is to treat crazing and shear yielding as separate and independent mechanisms. The kinetics of the two processes are completely different, so that changes in test conditions alter the balance between the two. Table I is a summary of the principal differences.

Table I

Differences in the Kinetics of Crazing and Shear Yielding

|  | Rate of crazing | Rate of shear yielding |
|---|---|---|
| Primary stage | Slow (induction period) | Decelerating |
| Secondary stage | Approximately constant | Approximately constant |
| Tertiary stage | Slowly decelerating | Accelerating |

Crazing begins slowly. Immediately after loading, there is an induction period during which the volume of the specimen remains almost constant. In contrast, shear yielding is relatively rapid at first, but becomes slower after a time. During the secondary stage of creep, both processes proceed at approximately constant rate. The final, tertiary stage of creep is again marked by differences between the two mechanisms: rates of crazing tend to decrease, perhaps because the number of sites available for craze initiation becomes depleted; rates of shear yielding, on the other hand, tend to increase, partly as a result of strain softening, and partly because the cross-sectional area of the specimen is decreasing. Thus, shear yielding tends to predominate at the beginning of a creep test, before the end of the induction period for crazing. During the secondary stage of the test, crazing usually makes a larger contribution to creep than it does during the primary stage, and the ratio of craze to shear contributions remains almost constant, so that $\Delta V$ increases linearly with $e_3$. Finally, during the tertiary stage, the acceleration of shear yielding and deceleration of crazing results in an increased contribution from shear yielding.

Increasing the applied stress increases the rates of both processes, but the effect upon crazing is greater. The induction period is shortened, and crazing therefore appears at an earlier point in the creep curve. This effect is apparent in Fig. 5. On raising the stress, the induction period is drastically reduced, and the rate of crazing is increased both absolutely, and in relation to the rate of shear yielding. At sufficiently high stresses, crazing is so rapid that it accounts for almost all of the time-dependent deformation of the material; the contribution of shear yielding is negligible. This point is particularly relevant to the discussion of impact behavior.

The deformation of isotropic ABS at room temperature is of particular interest because crazing and shear yielding occur at comparable rates over the normal time span of tensile and tensile-impact tests, so that changes in mechanism with stress and strain are easily observed. Some toughened plastics show no significant change in mechanism under the same conditions of testing. For example, isotropic HIPS shows no sign of shear deformation at 20°C when subjected to creep or tensile tests at comparable strain rates [36]. Volumetric strain measurements show that creep of HIPS is due almost exclusively to crazing, as illustrated in Fig. 8. Results for polypropylene, which are included in the same illustration, are in complete contrast: over the same range of strain rates, deformation is dominated by shear yielding, with only a small contribution from crazing.

Another contrasting pair of polymers is compared in Fig. 9. The toughened PVC, which was made by blending rigid PVC with 5% of an ABS master-batch concentrate, deforms largely by shear yielding. On the other hand, the ASA (acrylonitrile–styrene–acrylate) polymer, which consists of a poly-(styrene-*co*-acrylonitrile) matrix containing acrylic rubber particles, deforms largely by crazing, with only a small contribution from shear yielding [42]. Unlike ABS, neither polymer shows any significant change in mechanism with strain.

The effects of matrix composition on deformation mechanisms are most clearly demonstrated in blends of HIPS with PPO. A blend of HIPS with PS is

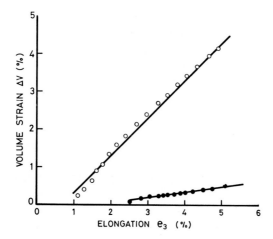

Fig. 8    Mechanisms of creep at 20°C in (○) HIPS at 23.4 MN/m² and in (●) toughened polypropylene at 17.0 MN/m².

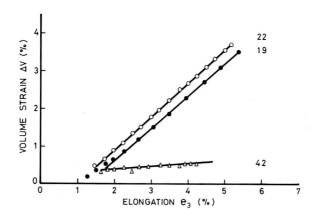

Fig. 9    Mechanisms of creep in (○) ASA polymer and in (●) toughened PVC at 20°C. Stress is given in MN/m².

simply a diluted HIPS, which deforms entirely by multiple crazing. A blend of the same HIPS with PPO, however, has a much more ductile matrix, and therefore deforms by a combination of crazing and shear yielding [21]. Because the two mechanisms interact strongly, as demonstrated in Fig. 2, the total creep response is not simply the sum of the contributions that each mechanism would make if acting alone. With increasing PPO content, the kinetics of crazing become dominated by the kinetics of shear yielding.

These quantitative studies of tensile deformation mechanisms help to clarify the problem of rubber toughening. Isotropic polystyrene is brittle at room temperature at all strain rates of practical interest, and the function of the rubber particles is to stimulate crazing in such a way as to enable the polymer to yield rather than fracture. To a slightly lesser extent, the same comment applies to poly(styrene-*co*-acrylonitrile): ABS is in a sense simply a superior type of HIPS. Rigid PVC and polypropylene are not in the same category. At low rates of strain, these polymers are ductile, and have no need of toughening. At high rates of strain, and at low temperatures, on the other hand, both PVC and PP are brittle. The tendency to brittle fracture is especially evident in notched impact tests, and the function of the rubber particles is to reduce this tendency. Conditions that produce a ductile-to-brittle transition in polypropylene produce a ductile-to-multiple-crazing transition in toughened polypropylene. The experiments on ABS illustrate the way in which the mechanism changes with increasing strain rate.

Any factor that affects creep rate is likely to affect crazing and shear yielding differently, and thus to alter the balance between the two mechanisms, either increasing or decreasing the contribution of crazing to the total strain. The following factors are known to affect the mechanisms of deformation of rubber-toughened plastics: (1) stress and loading history; (2) strain; (3) temperature; (4) rubber-phase volume; (5) matrix composition; (6) molecular

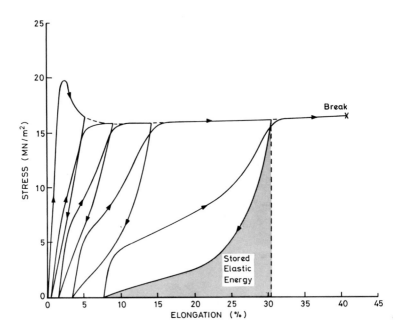

Fig. 10   Cyclic stress–strain curve for HIPS at 21°C, showing increase in hysteresis and stored energy as a result of multiple crazing.

orientation; (7) degree of craze–shear band interaction. The techniques of volumetric strain determination provide a satisfactory quantitative basis for the study of these factors, and therefore for a more complete understanding of rubber toughening.

## C.  Properties of Crazed Polymers

The mechanisms of deformation of toughened plastics have a practical as well as a theoretical importance. Crazes are porous, allowing rapid passage of gases, and the ingress of liquids, whereas shear yielding has little effect upon permeability. For this reason, shear yielding is preferred to multiple crazing in materials to be used for food packaging, and especially in carbonated drink containers. In some applications, changes in color or transparency resulting from stress whitening are a problem in themselves. Changes in mechanical properties are of more general significance.

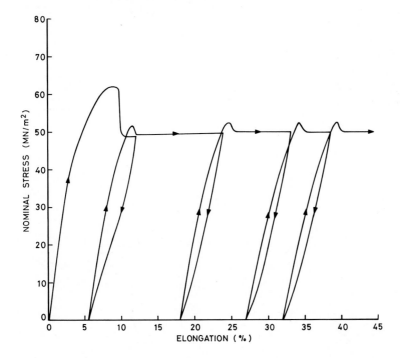

Fig. 11   Cyclic stress–strain curve for polycarbonate at 21°C. The specimen necked at an elongation of 8%, beyond which the true stress in the neck was approximately 70% higher than the nominal stress.

Crazes have much lower moduli than the parent bulk polymer, and exhibit a considerably greater degree of mechanical hysteresis. These properties are reflected in the cyclic stress–strain curve for HIPS presented in Fig. 10 [11]. As the specimen extends, the secant modulus falls and the size of the hysteresis loop increases. There is a strong similarity between the properties of a highly crazed HIPS specimen and the properties of a single craze [12]. The cyclic stress–strain behavior of polycarbonate, illustrated in Fig. 11, contrasts quite strongly with that of HIPS: polycarbonate exhibits only a small change in properties with increasing strain. Yielding occurs by the formation of a neck, which propagates along the gauge portion as the test proceeds.

The cyclic stress–strain curve shows that the stored elastic energy in a HIPS specimen increases with strain beyond the yield point. The relationship between stored elastic energy and strain for both HIPS and polycarbonate is given in Fig. 12. The difference between the two is particularly important in the discussion of fracture in highly strained tensile specimens, since it is elastic strain energy that provides the driving force for crack propagation. This problem is discussed further in Section V.A.

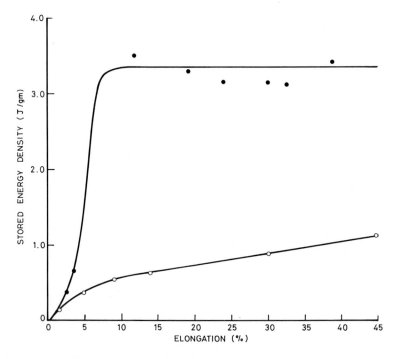

Fig. 12   Calculated changes in stored elastic energy with elongation in (●) polycarbonate (neck region) and (○) HIPS.

## IV. FRACTURE MECHANICS

The preceding sections have dealt with the deformation of rubber-toughened plastics, and the mechanisms that enable these materials to reach high strains before fracture. An understanding of deformation mechanisms provides the necessary background to the discussion of criteria for fracture.

The most frequently used fracture tests are tensile tests at constant strain rate and impact tests of various types. While these tests are rapidly and easily performed, they are not always as informative as they might be. Conventional impact tests are particularly difficult both to interpret and to relate to service performance. In recent years, there has been an increasing interest in fracture mechanics as a technique for resolving some of the problems inherent in fracture tests. The principles of fracture mechanics are reviewed briefly below. For a more extended treatment, the reader should consult more specialized texts [1, 43].

### A. Principles of Fracture Mechanics

Fracture mechanics begins with the assumption that all materials contain intrinsic cracks, flaws, and other defects, even when apparently free from surface damage and other flaws. Under an applied stress, these intrinsic flaws provide the nuclei for crack growth, which leads eventually to fracture of the material. The existence of such flaws may now be taken as established. The problem is to define the conditions under which a crack or flaw of a given size will propagate.

A solution to this problem was first proposed by Griffith, for the case of an ideally brittle, linearly elastic material [44], that is a material in which fracture occurs by bond rupture in the crack surface, and all other response to stress is restricted to elastic deformation according to Hooke's law. Griffith's criterion for an incremental increase in crack length was that the stored elastic energy released by the specimen should be greater than the energy absorbed in forming new crack surface. In an ideally brittle material, the energy absorbed in crack propagation is proportional to the number of bonds broken, and therefore to the area of new fracture surface. For a wide and relatively thick plate containing a central crack of length $a$, the Griffith energy-balance criterion predicts catastrophic fracture when the stress $\sigma$ reaches a critical value $\sigma_c$ given

$$\sigma_c{}^2 = 2E\gamma/\pi a(1 - \mu^2) \quad \text{in plain strain} \quad (5)$$

where $E$ is Young's modulus, $\mu$ Poisson's ratio, and $\gamma$ energy absorbed per unit area of fracture surface.

Subsequent research has shown that ideally brittle fracture is rarely if ever observed. There is always some plastic or viscoelastic deformation round the crack tip. The Griffith criterion can still be used provided that the total energy absorbed in crack propagation is directly proportional to the fracture surface area, and is independent of specimen geometry. The only modification necessary is to include the plastic work term in the definition of $\gamma$. The standard practice is to express fracture resistance in terms of the critical strain energy release rate, measured in joules per square meter of the crack, and given the symbol $\mathcal{G}_{IC}$. As there are two fracture surfaces, $\mathcal{G}_{IC}$ is equal to $2\gamma$. The Griffith criterion in its modified form then becomes

$$\sigma_c^2 = E\mathcal{G}_{IC}/\pi a(1 - \mu^2) \quad \text{in plane strain} \tag{6}$$

This equation is the basis of linear elastic fracture mechanics (LEFM). It is based on the assumption that $\mathcal{G}_{IC}$ is a materials property, and is independent of specimen geometry and loading history. There is experimental evidence to show that $\mathcal{G}_{IC}$ is not strictly a materials parameter in this sense [45], but in the absence of a satisfactory alternative the concept has sufficient validity to be useful. The LEFM approach is not applicable when there is a large plastic zone around the crack tip.

The treatments developed for linearly elastic materials can be applied to nonlinear elastic materials with only minor modifications. In the case of linear materials, it is possible to calculate the strain energy release rate using Hooke's law. For non-Hookean materials, it is generally necessary to measure the change in stored elastic energy $U$ as a function of crack length $a$ experimentally. Instead of $\mathcal{G}_{IC}$, the critical value of stored energy release rate is given the symbol $\mathcal{J}_{IC}$. Rubber physicists use the symbol $\mathcal{T}$ (tearing energy) for the same quantity. The condition for crack propagation in non-linear elastic materials is a more general form of the energy-balance criterion, as follows:

$$-\partial U/\partial a \geqslant \mathcal{J}_{IC} \tag{7}$$

which reduces to Eq. (6) for a linear material. Nonlinear elastic fracture mechanics has been used with considerable success to characterize the tearing resistance of rubbers.

The extension of fracture mechanics to materials in which there is a substantial degree of plastic deformation presents considerable practical and theoretical difficulties. In the general case, it is not possible to characterize the conditions for crack propagation by means of a single parameter. Nevertheless, attempts have been made to do just that. Two parameters have been proposed: the crack opening displacement (COD), represented by the symbol $\delta$, and the generalized strain energy release rate $\mathcal{J}$. The alternative criteria for crack propagation in a ductile material are then (a) that $\delta$ reaches a critical value $\delta_{IC}$, or (b) that $\mathcal{J}$ reaches a critical value $\mathcal{J}_{IC}$. Before attempting to apply these criteria, it is necessary to test their validity experimentally. Under certain conditions neither criterion is applicable. For example, yielding fracture mechanics is not applicable to a specimen in a state of general yield, that is, one in which the yield zone occupies the complete cross section of the specimen.

Measurements of $\mathscr{G}_{IC}$ are made using a variety of specimen types, each containing a sharp crack of known length. The simplest type of specimen in principle is a flat plate containing a central planar crack; the tensile stress on the plate is increased until the specimen fractures, and $\mathscr{G}_{IC}$ is calculated using Eq. (6). Alternatively, the specimen may have a single edge crack, or two opposing edge cracks of equal length. Other specimen geometries in frequent use include the double-cantilever beam, wedge opening, three-point bending and compact tension specimens. In principle, $\partial U/\partial a$ can be calculated for each of these specimens using elasticity theory. In practice, however, it is usually preferable to determine $\partial U/\partial a$ experimentally. Similar techniques are used to determine $\mathscr{J}_{IC}$.

In the plate specimens described above, the strain energy release rate $-\partial U/\partial a$ increases with crack length $a$, and the crack therefore tends to accelerate as it extends. In other types of specimen, for example, the double-cantilever beam, $-\partial U/\partial a$ may increase at first, and then begin to decrease because the total strain energy in the specimen has fallen. If the strain energy release rate falls below the critical value $\mathscr{G}_{IC}$ (or $\mathscr{J}_{IC}$), the crack will stop. In order to maintain crack propagation, additional energy must be supplied to the specimen.

## B. Application to Rubber-Toughened Plastics

Under some conditions, especially at low temperatures, rubber-toughened plastics are sufficiently brittle to permit a valid measurement of $\mathscr{G}_{IC}$: yielding is confined either to a single craze or to a very small area of stress whitening around the crack tip. At room temperature, the yielded zone is usually too large to allow LEFM techniques to be applied, and the investigator is left with the choice of employing yielding fracture mechanics or abandoning all attempt to apply fracture mechanics. Despite the difficulties, fracture mechanics measurements have been made on a number of rubber-modified plastics, providing more quantitative information than was previously available on the factors affecting fracture resistance. In most cases, the results are stated in terms of $\mathscr{G}_{IC}$, although it would be more correct to refer to $\mathscr{J}_{IC}$ in experiments that produce substantial amounts of stress whitening.

### 1. Rubber-Phase Volume

The volume fraction of the rubber phase $\phi_2$ is a factor of obvious importance in determining the fracture resistance of a rubber-toughened polymer. A clear distinction should be made between the rubber content, which is simply the concentration of polybutadiene (or other rubber) in the material, and the rubber-phase volume $\phi_2$, which includes both the rubber itself and any occluded hard phase trapped within the rubber particles. The point that $\phi_2$ may be several times greater than the rubber content is illustrated in Fig. 1.

The relationship between $\phi_2$ and $\mathcal{G}_{IC}$ is illustrated in Fig. 13, which presents results for a series of epoxy resins [19]. Each of the materials tested contained 8.7% carboxyl-terminated butadiene acrylonitrile (CTBN) rubber, but the formulation and curing conditions of the resin were varied in such a way as to vary the degree of phase separation between wide limits. In the more rapidly curing formulations, phase separation was prevented completely, while in some of the slowly curing systems the rubber not only separated, but also trapped subinclusions of epoxy resin within the rubber particles. The results emphasize the point that a small amount of rubber, properly incorporated into the structure, can increase toughness by an order of magnitude. Bascom *et al.* obtained similar results for a series of epoxy resins containing varying similar results for a series of epoxy resins containing varying amounts of CTBN

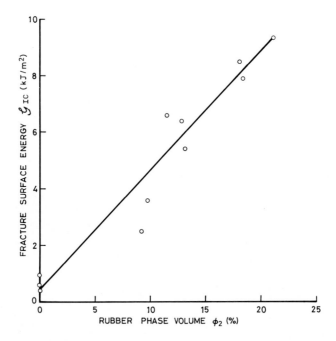

Fig. 13   Relationship between fracture resistance and rubber phase volume in epoxy resins containing 8.7% CTBN rubber. (After Yoshii [19].)

rubber [46], as did Kobayashi and Broutman in experiments on acrylonitrile–methacrylate–butadiene–styrene (AMBS) polymers containing 0–16% polybutadiene [47].

### 2. Crack Speed

Experiments on glassy polymers, notably PS and PMMA, show that $\mathscr{G}_{IC}$ is not a single-valued quantity, but varies with crack speed $\dot{a}$. In a fracture mechanics study of PMMA, Broutman and Kobayashi found that $\mathscr{G}_{IC}$ increased with $\dot{a}$ to a maximum at a speed of 7 cm/sec, and then fell sharply. There was a similar maximum at 20 m/sec. The authors identified both peaks with viscoelastic processes in the glassy polymer [48]. Similar variations in $\mathscr{G}_{IC}$ (or more correctly $\mathscr{J}_{IC}$) occur in rubber-toughened plastics. Data on AMBS, also from the work of Broutman and Kobayashi, that illustrate this point [47] are presented in Fig. 14. The peaks are probably viscoelastic in origin, as in PMMA, but have not been positively identified. The curves all pass through minima at about 1 m/sec, but even at that speed the materials with added rubber are considerably tougher than those without. A decreasing fracture toughness with increasing crack speed means, of course, that the crack tends to accelerate rapidly, meeting less resistance from the material as it does so.

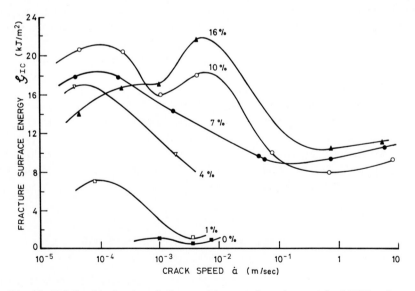

Fig. 14   Relationship between fracture resistance and crack speed in AMBS polymers containing 0–16% rubber. (After Kobayashi and Broutman [47].)

### 3. Constraints

It is well known that the fracture toughness of a metal is higher if the measurement is made on a thin plate than if it is made on a thick plate. Similar observations have been made on ductile polymers (see Fig. 3). The effect is described as the plane stress to plane strain transition: in a thin plate under tensile loading, the component of stress in the through-thickness direction is zero, whereas in a thick plate there is an appreciable tensile stress in the through-thickness direction within a small region near the root of a notch or crack. The material near the notch in a thick plate is constrained from contracting in the thickness direction by the surrounding material, so that a state of plane strain is established in the central part of the plate.

Because yielding depends upon differences between principal stresses, the effect of raising the through-thickness component of stress is to inhibit shear yielding. Yielding can still occur, but at a higher level of applied tensile stress. Fracture stresses are not raised in the same way, so that the net effect of the constraint is to promote brittle fracture. In a rubber-modified plastic, a constraint would be expected to raise the stress required to cause shear yielding without a corresponding increase in the critical stress for crazing, so that a transition from shear yielding to multiple crazing would be predicted for PVC–ABS blends, or PPO–HIPS blends, which show an appreciable amount of shear yielding in tension. On the other hand, increasing the plate thickness or decreasing the notch radius would not be expected to affect the fracture behavior of HIPS, which deforms almost entirely by multiple crazing, since crazing is itself a plane-strain process.

In view of this analysis, the results obtained by Parvin and Williams in experiments on HIPS sheet are somewhat unexpected [49]. These authors found that the toughening mechanism was inoperative in fully constrained regions of the HIPS, which contained 4% rubber. A sharp crack was cut into the surface of a thick sheet of the HIPS, and the severity of the constraint in this type of specimen was sufficient to reduce $\mathscr{G}_{IC}$ to the level of polystyrene. Single-edge notch tests on the same material yielded very different results. At temperatures above $-60°C$, there was sufficient craze yielding around the notch to render LEFM techniques invalid.

## V.  TENSILE AND IMPACT BEHAVIOR

The understanding of deformation and fracture mechanisms provided by electron microscopy, creep studies, and fracture mechanics is of great value in interpreting the results of standard fracture tests, including tensile and impact tests. The value of the approach is further enhanced by the fact that the same analysis can be extended to the behavior of rubber-toughened plastics under service conditions.

## A. Tensile Properties

In standard tensile tests at room temperature, rubber-toughened plastics yield at strains of about 5%, and extend to high elongations before breaking. Typical elongations at break are between 10 and 70%. Although the stress–strain curves may be very similar for many rubber-toughened plastics, there are significant differences in tensile behavior, which arise from differences in deformation mechanism.

A comparison of HIPS with ABS illustrates this point. Isotropic HIPS shows no significant change in the cross-sectional area of the tensile specimen between the onset of yielding and the final fracture of the material. Necking is not observed because there is little or no shear yielding: crazing is the dominant mechanism of deformation. In ABS, on the other hand, shear yielding does contribute to the deformation, with the result that a neck is formed soon after the specimen has yielded. The increased stress in the neck accelerates shear yielding in this region, and also causes intense stress whitening. Meanwhile, the remainder of the specimens is relatively unaffected. Eventual failure occurs by fracture in the neck region, at a much lower total elongation than a comparable HIPS. For this reason, elongations at break are an unreliable guide to the toughness of ABS polymers. Paradoxically, increasing the strain rate may increase the elongation at break, by promoting multiple crazing at the expense of shear yielding (see Section III.B), so that necking is delayed or prevented.

Raising the temperature or introducing molecular orientation increases the contributions of shear yielding to the tensile deformation of HIPS. At temperatures above about 60°C, HIPS behaves in much the same way as ABS at room temperature, forming a neck soon after yielding. Injection-molded HIPS tensile bars will neck at room temperature if the molding conditions are so chosen as to give a high level of molecular orientation in the surface layers of the bars.

In well-prepared specimens, free from surface defects, eventual fracture occurs in quite a reproducible manner at high strains. Crack propagation in a stress-whitened material involves extending and then relaxing large numbers of crazes in the crack-tip region, a process requiring a large expenditure of energy because of the high mechanical hysteresis of crazes. In addition, energy may be absorbed in the formation of new crazes. As explained in Section IV.A, the driving force for crack propagation is the elastic energy stored in the material surrounding the crack tip. Usually the stored energy density increases because the stress increases: this is the basis of Griffith's treatment, and the principle applies equally to nonlinear elastic materials. However, in Figs. 10 and 12 the point is made that HIPS extends at almost constant stress, and that the stored elastic energy increases because dense crazing increases the recoverable strain of the material.

## B. Notched-Impact Tests

The Charpy and Izod impact tests, illustrated in Fig. 15, are the most widely used methods for measuring the fracture resistance of rubber-toughened plastics. The two tests are similar in principle, and differ only in detail: both measure the energy abstracted from a pendulum in breaking a standard bar in flexure. The standard forms of specimen have rounded notches, which make them unsuitable for fracture mechanics studies, but LEFM analyses have been applied with some success to PS and PMMA using razor-notched specimens [50–52]. Experimental programs have tended to concentrate upon the Charpy specimen, as the stress concentrations near the notch are complicated in the Izod specimen by clamping forces.

The Charpy test is a three-point bending test conducted at a relatively rapid rate. In brittle fracture, the pendulum deforms the specimen elastically until a crack begins to propagate from the notch. Once this point has been reached, the stored elastic energy is sufficient to support crack propagation, and no

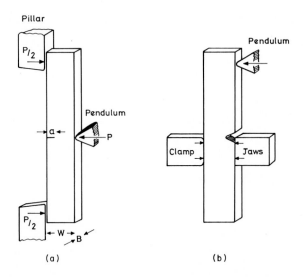

Fig. 15   Schematic diagrams illustrating (a) the Charpy impact test on a razor-notched specimen, and (b) the Izod impact test on a standard specimen with a rounded notch.

further energy is abstracted from the pendulum. The impact strength $I$, which measures the total energy abstracted from the pendulum, is therefore equal to the peak value $U_c$ of stored energy. More precisely, $I$ is the sum of $U_c$ and $U_{ke}$, the kinetic energy of the specimen at the point of crack initiation [51]. The condition for crack propagation is therefore

$$I - U_{ke} = U_c = BWZ\mathscr{G}_{IC} \tag{8}$$

where $B$ is specimen thickness, $W$ specimen width, and $Z$ a geometrical factor that relates specimen compliance $C$ to crack length $a$. If a force $P$ exerted by the pendulum produces a deflection $\delta$, then

$$C = \delta/P \quad \text{and} \quad Z = C\frac{d(a/W)}{dC}$$

Thus in brittle fracture, the impact strength is governed by the geometry of the specimen and by $\mathscr{G}_{IC}$, which represents the resistance of the material to crack initiation. Under these conditions, $I$ increases linearly with $BWZ$, and is not related to the fracture surface area $B(W - a)$.

The foregoing analysis is applicable to rubber-toughened plastics at temperatures below the glass transition of the rubber. Fracture is essentially brittle because the rubber does not fulfil its intended role, and values of $\mathscr{G}_{IC}$ are comparable for PS and HIPS. Viscoelastic energy dissipation by the rubber phase near $T_g$ makes some additional contribution to the fracture toughness of HIPS, but the effect is comparatively small [49].

At temperatures immediately above the glass transition of the rubber, the resistance to crack initiation, as measured by $\mathscr{G}_{IC}$, begins to increase. The rubber particles are able to act as stress concentrators, generating multiple crazing at the root of the notch, and leaving a stress-whitened area on the fracture surface near the notch. As noted earlier, a very severe triaxial constraint can suppress this stress whitening, but this condition is unusual. However, once the crack has started to propagate, there is no further stress whitening. The fracture surface remote from the notch has the rough, broken appearance characteristic of brittle fracture by a rapidly propagating crack. This type of behavior is to be expected when $\mathscr{G}_{IC}$ decreases sharply with increasing crack speed.

At a temperature well above the glass transition of the rubber, there is a further change in fracture behavior. The energy absorbed in impact is no longer represented by Eq. (8). Instead, the impact strength is proportional to the ligament area, as follows:

$$I = B(W - a)\mathscr{J}_{IC} \tag{9}$$

The entire fracture surface is stress whitened, showing that substantial amounts of energy are absorbed by multiple crazing throughout the fracture process, and not simply at the initiation stage. The implication of Eq. (9) is that $\mathcal{J}_{IC}$ does not fall sharply with increasing crack velocity, as $\mathcal{G}_{IC}$ does at lower temperatures. Owing to the increase in fracture resistance with temperature, the elastic energy stored in the specimen at the point of crack initiation is no longer sufficient to maintain the propagating crack: unless additional energy is supplied by the pendulum, the crack slows down and eventually stops. This ability of the material to arrest the progress of a propagating crack is clearly of practical importance.

The variation of fracture resistance with temperature in HIPS is illustrated in Fig. 16. Impact tests show that other rubber-toughened plastics behave in a very similar way [1, 53]. There are three distinct types of fracture behavior, as follows:

I. *Low temperature.* No stress whitening. Impact strength given by Eq. (8). $\mathcal{G}_{IC}$ low. Completely brittle fracture.

II. *Intermediate temperatures.* Stress whitening around root of notch, and increasing in area with increasing temperature. Impact strength given by Eq. (8). $\mathcal{G}_{IC}$ for crack initiation increasing with temperature as shown in Fig. 16, but falling sharply with increasing crack speed, so that resistance to rapid crack propagation is very low.

III. *High temperatures.* Stress whitening over entire fracture surface. Impact strength given by Eq. (9). $\mathcal{J}_{IC}$ increasing with temperature. Crack arrest is possible.

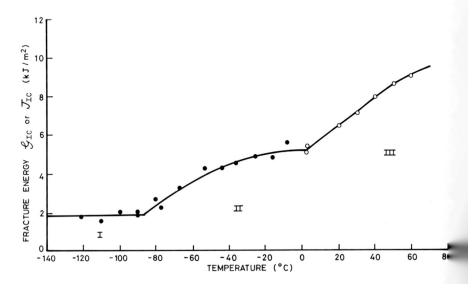

Fig. 16    Fracture resistance of HIPS as a function of temperature: ($\bullet$) $\mathcal{G}_{IC}$ calculated from Eq. (8); ($\bigcirc$) $\mathcal{J}_{IC}$ calculated from Eq. (9).

The change from Temperature Region I to Temperature Region II is clearly due to the glass transition in the rubber, which allows the multiple crazing mechanism to operate. Raising the temperature further causes an increase in the rate of crazing, with the result that the stress-whitened region expands from the root of the notch. The increase in $\mathscr{J}_{IC}$ with temperature in Temperature Region III can also be attributed to an increase in the rate of crazing, with perhaps a contribution from shear yielding at high temperatures.

The significant feature of Temperature Region II is the decrease in $\mathscr{G}_{IC}$ with increasing crack speed. This decrease appears to be another consequence of the glass transition in the rubber phase. The observed modulus of a rubber is a function of the time scale of the experiment. At very short times, or equivalently at high crack speeds, the rubber has a modulus comparable with a glassy polymer, and is therefore ineffective in initiating crazes. In accordance with the time–temperature equivalence principle, the transition from glassy to rubbery properties occurs at shorter times, or higher crack speeds, as the temperature is raised. At sufficiently high temperatures, the rubber is rubbery in its response at all attainable crack speeds, and the sharp fall in $\mathscr{G}_{IC}$ or $\mathscr{J}_{IC}$ with crack speed is no longer observed. This condition appears to be fulfilled in Temperature Region III, as described above.

This analysis explains why a low glass transition temperature is such an important feature of the rubber phase in a rubber-toughened polymer (see Section 1). A low $T_g$ is beneficial in increasing $\mathscr{G}_{IC}$ at low temperatures, but this aspect of the problem is relatively unimportant; very few users are interested in fracture resistance at $-80°C$. The main reason for preferring polybutadiene, which has $T_g$ at approximately $-90°C$, to a 75/25 butadiene/styrene co-polymer with $T_g$ at $-50°C$ is that the polybutadiene gives greater toughness at room temperature when crack speeds are high.

It was emphasized in Section IV.B.1 that $\mathscr{G}_{IC}$ and $\mathscr{J}_{IC}$ are determined by the rubber-phase volume, as well as by the temperature. The effect of reducing $\phi_2$, and therefore reducing $\mathscr{J}_{IC}$, is to extend Temperature Region II to higher temperatures. The condition for crack arrest is that the strain energy release rate should be less than $\mathscr{J}_{IC}$, and reducing the rubber content obviously makes this condition more difficult to satisfy.

## C. Aging

Rubber-toughened plastics tend to become severely embrittled on exposure to sunlight, an effect known as aging. The basic cause of embrittlement is photooxidation of the rubber phase, initiated by the ultraviolet component of sunlight. Attempts have been made to substitute oxidation-resistant rubbers such as the acrylates for the diene polymers usually employed as toughening agents, but with limited success. Polybutadiene is particularly suitable as a toughening rubber because of its low $T_g$ and the ease with which it participates in grafting and cross-linking reactions with styrene, methyl methacrylate, and other monomers.

Oxidation causes a marked change in the properties of the rubber: the secondary loss peak marking the glass transition broadens and shifts to higher temperatures. Ghaffar *et al.* irradiated HIPS for 14 hr under a sun lamp, and found that the loss peak due to the rubber phase shifted from $-85$ to $+20°C$ [54]. The oxidized material becomes extremely brittle, and this embrittlement is especially noticeable in impact tests. However, the embrittlement is limited to a relatively thin surface layer [55], and the impact strength of an aged sample may be restored completely by machining away the exposed surface to a depth of about $100 \, \mu$m.

The effects of aging upon fracture properties can best be understood in terms of fracture mechanics. Oxidation reduces $\mathcal{G}_{IC}$ in the surface layer to the level of the homopolymer, or perhaps lower. In Temperature Region II, where the impact strength is proportional to $\mathcal{G}_{IC}$ in accordance with Eq. (8), the embrittlement of the surface affects the fracture resistance of the whole

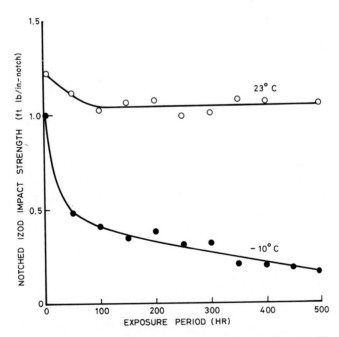

Fig. 17    Artificial aging of HIPS under a xenon lamp: impact tests performed at 23 and $-10°C$ after exposure to ultraviolet light. (After Bucknall and Street [56].)

specimen. In Temperature Region III, on the other hand, the impact strength is proportional to the average value of $\mathscr{J}_{IC}$ over the entire fracture surface, and is therefore relatively little affected by the properties of the oxidized surface layer of the specimen.

The differences between the two types of fracture behavior are amply illustrated in Fig. 17, which shows the effects of aging on the notched Izod impact strength of HIPS [56]. At 23°C, the material is in Temperature Region III, and the impact strength decreases only slightly on aging; at −10°C, on the other hand, the material is in Temperature Region II, and the impact strength falls rapidly as a result of the identical exposure. This comparison makes the point that aging trials can easily give misleading results unless the factors involved are clearly understood. Tests at room temperature appear to show that the HIPS was resistant to aging, whereas it is obvious from the lower curve of Fig. 17 that aging was quite rapid. If a component is to be used for outdoor applications, where it will be exposed to both sunlight and low temperatures, the difference between real and apparent resistance to aging is an important one.

## VI. CONCLUDING REMARKS

Over the past decade, four important techniques have been developed for studying rubber toughening. Each of these techniques is capable of yielding quantitative information, and therefore placing the subject on a proper scientific foundation. They are as follows:

1. *osmium staining*, which provides direct quantitative information about structure and morphology, including particle size, particle size distribution, subinclusion size, and volume fraction of rubber particles;
2. *scanning electron microscopy*, which reveals the positions and lengths of crazes and shear bands;
3. *volumetric strain measurement*, which provides quantitative information about rates of crazing and rates of shear yielding, and characterizes the contribution of each mechanism to a given deformation;
4. *fracture mechanics*, which defines fracture resistance in terms of $\mathscr{G}_{IC}$ and $\mathscr{J}_{IC}$, placing fracture on a more quantitative basis than is possible using conventional impact tests.

Each technique has contributed individually to our present knowledge of structure–property relationships in rubber-toughened plastics. There is much to be gained from coordinated studies, involving all four techniques, which would finally answer some of the outstanding questions, and open the way to the development of a new range of materials with novel properties.

# REFERENCES

1a. C. B. Bucknall, "Toughened Plastics." Applied Science Publ., London, 1977.
1b. K. Kato, *J. Electron Microsc.* (*Jpn.*) **14**, 220, (1965).
1c. K. Kato, *Polym. Lett.* **4**, 35 (1966).
1d. K. Kato, *Polym. Eng. Sci.* **7**, 38 (1967).
2. E. H. Merz, G. C. Claver, and M. Baer, *J. Polym. Sci.* **22**, 325 (1956).
3. C. B. Bucknall and R. R. Smith, *Polymer* **6**, 437 (1965).
4. M. Matsuo, *Polym. Eng. Sci.* **9**, 206 (1969).
5. R. J. Seward, *J. Appl. Polym. Sci.* **14**, 852 (1970).
6. R. P. Kambour and R. R. Russell, *Polymer* **12**, 237 (1971).
7. P. Beahan, A. Thomas, and M. Bevis, *J. Mater. Sci.* **11**, 1207 (1976).
8. J. P. Berry, *J. Polym. Sci.* **50**, 107 (1961).
9. L. J. Broutman and F. J. McGarry, *J. Appl. Polym. Sci.* **9**, 585 (1965).
10. E. Drioli, L. Nicolais, and A. Ciferri, *J. Polym. Sci. Polym. Chem. Ed.* **11**, 3327 (1973).
11. C. B. Bucknall, *J. Mater.* (Amer. Soc. Test Mater.) **4**, 214 (1969).
12. R. P. Kambour and R. W. Kopp, *J. Polym. Sci. Part A-2* **7**, 183 (1969).
13. R. P. Kambour, *J. Polym. Sci. Part A* **2**, 4165 (1964).
14. R. E. Lavengood, L. Nicolais, and M. Narkis, *J. Appl. Polym. Sci.* **17**, 1173 (1973).
15. C. G. Bragaw, *Amer. Chem. Soc. Adv. Chem. Ser.* **99**, 86 (1971).
16. E. M. Hagerman, *Amer. Chem. Soc. Polym. Prepr.* **15**(1), 217 (1974).
17. E. M. Hagerman, *J. Appl. Polym. Sci.* **17**, 2203 (1973).
18. E. H. Yoffe, *Phil. Mag.* **42**, 739 (1951).
19. T. Yoshii, PhD Thesis, Cranfield, England (1975).
20. P. Beahan, A. Thomas, and M. Bevis, *J. Mater. Sci.* **11**, 1207 (1976).
21. C. B. Bucknall, D. Clayton, and W. E. Keast, *J. Mater. Sci.* **7**, 1443 (1972).
22. R. N. Haward and J. Mann, *Proc. Roy. Soc. London Ser. A* **282**, 120 (1964).
23. R. R. Durst, R. M. Griffith, A. J. Urbanic, and W. J. van Essen, *Amer. Chem. Soc. Adv. Chem. Ser.* **154**, 239 (1976).
24. W. M. Barentsen and D. Heikens, *Polymer* **14**, 579 (1973).
25. W. M. Barentsen, D. Heikens, and P. Piet, *Polymer* **15**, 119 (1974).
26. C. B. Bucknall, *Trans. IRI* **39**, 221 (1963).
27. S. Koiwa, *J. Appl. Polym. Sci.* **19**, 1625 (1975).
28. C. F. Parsons and E. L. Suck, *Amer. Chem. Soc. Adv. Chem. Ser.* **99**, 340 (1971).
29. C. F. Parsons, Private communication (1976).
30. T. O. Purcell, *Amer. Chem. Soc. Polym. Prepr.* **13**(1), 699 (1972).
31. E. R. Wagner and L. M. Robeson, *Rubber Chem. Technol.* **43**, 1129 (1970).
32. G. F. Freeguard, *Brit. Polym. J.* **6**, 205 (1974).
33. B. M. Beach, R. P. Kambour, and A. R. Shultz, *Polym. Lett.* **12**, 247 (1974).
34. M. R. Grancio, *Polym. Eng. Sci.* **12**, 213 (1972).
35. H. Willersin, *Makromol. Chem.* **101**, 296 (1967).
36. C. B. Bucknall and D. Clayton, *J. Mater. Sci.* **7**, 202 (1972).
37. C. B. Bucknall, D. Clayton, and W. E. Keast, *J. Mater. Sci.* **8**, 514 (1973).
38. C. B. Bucknall and I. C. Drinkwater, *J. Mater. Sci.* **8**, 1800 (1973).
39. P. J. Fenelon and J. R. Wilson, *Amer. Chem. Soc. Div. Org. Coat. Plast. Prepr.* **34**(2), 326 (1974).
40. R. W. Truss and G. A. Chadwick, *J. Mater. Sci.* **11**, 111 (1976).
41. R. W. Truss and G. A. Chadwick, *J. Mater. Sci.* **11**, 1385 (1976).
42. C. B. Bucknall, C. J. Page, and V. O. Young, *Amer. Chem. Soc. Adv. Chem. Ser.* **154**, 179 (1976).

43. J. F. Knott, "Fundamentals of Fracture Mechanics." Butterworth, London, 1973.
44. A. A. Griffith, *Phil. Trans. Roy. Soc.* **A221**, 163 (1921).
45. D. P. Clausing, *Int. J. Fracture Mech.* **5**, 211 (1969).
46. W. D. Bascom, R. L. Cottington, R. L. Jones, and P. Peyser, *Amer. Chem. Soc. Div. Org. Coat. Plast. Prepr.* **34**(2), 300 (1974).
47. T. Kobayashi and L. J. Broutman, *J. Appl. Polym. Sci.* **17**, 2053 (1973).
48. L. J. Broutman and T. Kobayashi, *in* "Dynamic Crack Propagation" (G. C. Sih, ed.), p. 215. Noordhoff, Leyden, 1973.
49. M. Parvin and J. G. Williams, *J. Mater. Sci.* **11**, 2045 (1976).
50. H. R. Brown, *J. Mater. Sci.* **8**, 941 (1973).
51. G. P. Marshall, J. G. Williams, and C. E. Turner, *J. Mater. Sci.* **8**, 949 (1973).
52. E. Plati and J. G. Williams, *Polym. Eng. Sci.* **15**, 470 (1975).
53. C. B. Bucknall and D. G. Street, *SCI Monogr.* **26**, 272 (1967).
54. A. Ghaffar, A. Scott, and G. Scott, *Eur. Polym. J.* **11**, 271 (1975).
55. E. Priebe and J. Stabenow, *Kunststoffe* **64**, 497 (1974).
56. C. B. Bucknall and D. G. Street, *J. Appl. Polym. Sci.* **12**, 1311 (1968).

# Chapter 15

# Coextruded Multilayer Polymer Films and Sheets

## W. J. Schrenk and T. Alfrey, Jr.

*The Dow Chemical Company*
*Midland, Michigan*

## I. INTRODUCTION

Physical properties of polymers are modified and extended by several techniques, including copolymerization; formulation with plasticizers,

129

stabilizers, and fillers; rubber modification; and polymer blending. Sometimes all of these approaches are employed to tailor polymer properties to fit an end use application. These powerful tools for modifying polymers have broadened applications for plastics tremendously.

Frequently, however, no single polymer can satisfy all of the end use requirements economically, and the engineer then considers composites, such as laminations and coatings of different plastics to themselves or to other materials. The use of layered plastic composites for packaging and industrial uses developed rapidly in the 1950s. Each ply in the composite was chosen to provide a specific end-use characteristic to the product, for example, heat sealability, barrier, and chemical resistance.

In the early 1960s coextrusion emerged as a new fabrication method to make multilayer plastic film and sheet by direct simultaneous extrusion of two or more polymers through a single die. Coextrusion avoids many of the manufacturing steps required by conventional lamination and coating processes, such as making and handling of individual films, application of coatings and primers, and solvent drying. Because of these economic and technical advantages, the use of coextrusion is growing rapidly.

Both polymer blending and multilayer coextrusion are used to combine the properties of two or more polymers by melt fabrication to obtain desired end-use characteristics in a product. While polymer blending randomly mixes the components together, coextrusion has an advantage of yielding a layered morphology that can be an optimum "dispersion" geometry for many applications. Coextrusion shares many of the same technical considerations with polymer blending: adhesion (see Volume 1, Chapter 6), compatibility (see Volume 1, Chapters 1–5), mechanical interaction between phases (layers), (see Chapters 12–14), and rheology (see Volume 1, Chapter 7). Also, some early publications described extensive mixing of viscous fluids by deforming layers until they become "vanishingly thin" [1–4]. Coextruded films are being made that have individual layer thickness less than the wavelength of light and can approach molecular dimensions.

In this chapter we review coextrusion technology, some properties of multilayer films, and several applications that illustrate how coextrusion is being used to obtain composite properties economically.


## II. METHODS OF COEXTRUSION

Coextruded multilayer plastic films are being produced by two processes: (1) the tubular-blown film process, and (2) the flat-die chill-roll casting process. A variety of multilayer die designs can be used for both processes, and we only review the more common methods, most of which are in commercial use.

## A. Tubular Coextrusion Dies for Blown Film

The blown-film process is the most widely used method for manufacturing plastic film, and many of the early two- and three-layer coextrusions were produced by this process. A tubular die geometry is used, and the extrudate is inflated with air to stretch and cool the plastic bubble. Tubular die designs for coextrusion require the formation of uniform concentric layers in the die annulus.

Several dies for tubular coextrusion are illustrated in Figs. 1, 2, and 3. In Fig. 1 Polymers A and B are joined in an adaptor ahead of the single-manifold tubing die, and the concentric melt stream is pierced by the die mandrel to produce a two-layer annular extrudate [5]. The ratio of layer thickness is determined by the volumetric pumping rate of each extruder for each polymer.

A second tubular die design (Fig. 2) uses individual manifolds, which distribute the layers concentrically prior to final joining ahead of the die land [6]. The manifold for each layer must be designed carefully to obtain satisfactory layer uniformity.

A third design (Fig. 3) stacks toroidal distribution manifolds for the

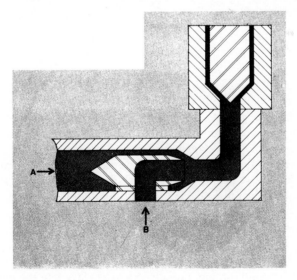

Fig. 1   Adaptor connected to single-manifold blown-film die that arranges Polymers A and B into a concentric melt stream. (Reprinted with permission from J. E. Johnson [6], *Plastics Technology* **22**, 45, February 1976.)

Fig. 2    Multimanifold blown-film die with individual annular channels for each layer that are joined before the final die land. (Reprinted with permission from J. E. Johnson [6], *Plastics Technology* **22**, 45, February 1976.)

desired number of layers [7]. As the melt stream flows toward the die exit, layers are extruded on one another sequentially.

Such tubular die designs become quite complicated for coextrusions exceeding about three layers, but many two- and three-layer packaging films are produced by these blown-film dies.

A unique blown-film die has been developed that can generate a large number of layers of two or more polymers [8, 9]. This tubular die system can make multilayer blown films having several hundred layers with individual layer thicknesses less than 1000 Å. A similar approach has been used to make multilayer bicomponent fibers [10]. The principle involves the generation of layers in an annular die having rotating elements. As shown in Fig. 4, the individual polymers are extruded through a feed port, which arranges them into alternating, radially extending layers in the die annulus. The layers then can be deformed into long, thin spirals by rotation of the annular die boundaries. Basically, four layer patterns, as shown in Fig. 5, can be generated, depending on the mode of die element rotation. Either the

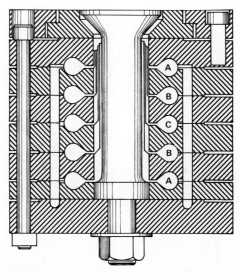

Fig. 3    Toroidal-distribution manifolds stacked to provide the desired number of layers by the sequential addition of polymers.

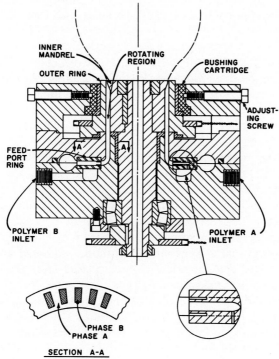

Fig. 4    Rotating annular die elements capable of generating large numbers of layers of Polymers A and B.

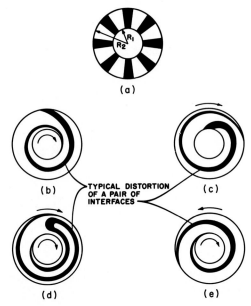

Fig. 5   Four modes of rotation of annular die boundaries can generate different layer patterns: (a) feed distribution entering annular die; (b) Case I; (c) Case II; (d) Case III; (e) Case IV. (From Schrenk and Alfrey [8].)

inner or outer annular die boundaries may be rotated individually, or simultaneously in the same direction or counterrotating. Each mode of die rotation will generate a different layer thickness distribution. The actual layer thickness will depend on the extrusion rate of each polymer, the rotational speed of the annular die elements relative to the feed ports, the length of the annular channel, and the radial location within the annulus.

The relationship between these operating parameters has been analyzed mathematically [3] and was found to be in excellent agreement with experimental observations. The experimentally measured layer distribution as a function of position within a 1-mil film to the calculated distribution for a Case 2 layer pattern are compared in Fig. 6 [9].

The agreement between experiment and theory is even more remarkable when one considers that a Newtonian-flow model was used and that the calculations involved determination of the displaced layer position in the final film after leaving the die annulus [8].

Microlayer coextruded films made by this die system can approach layer dimensions of a few hundred angstroms. The films are, in a sense, polymer blends having a layered morphology of molecular dimensions. The process can reduce the layer thickness to a point where the layers become discontinuous, as illustrated in the electron photomicrograph in Fig. 7. In this

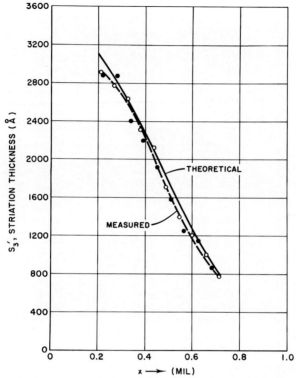

Fig. 6  Comparison of measured striation thickness with calculated values for a Case 2 layer pattern. Striation thickness is the thickness of a pair of adjacent A and B layers. (From Schrenk and Alfrey [9].)

coextruded film of polystyrene (PS) and polymethylmethacrylate (PMMA), the PMMA layers were thinned to where they break into spherical particles. Wherever "layer scission" has occurred, the ends are larger in diameter, presumably due to relaxation of the molecules. As these dubbell-shaped layer segments are sheared further, ruptures continue until spherical particles are formed.

### B.  Flat Die Coextrusion

Coextruded film and sheet is also being produced using flat die geometry, that is, where the multilayer melt is extruded through a wide narrow slit. The film can be drawn to varying thickness while still hot and then cooled below its solidification temperature on a large diameter chill roll.

The flat-die coextrusion geometry is used in:

1.  chill-roll cast film process, which can run at high speeds since cooling is more rapid than air cooling by the blown film process;

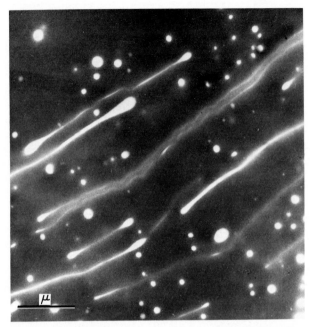

Fig. 7   Electron micrograph showing "layer scission" where PMMA layers become discontinuous. (From Schrenk and Alfrey [9].)

2.   in-line coextrusion coating process in which a multilayer film is coextruded directly onto a nonextrudable substrate, such as paper or cellophane; and

3.   coextrusion of thick film or sheet (>10 mil) in which the tubular blown-film process is impractical.

Two flat-die coextrusion methods are used commercially. In early technology a multimanifold die is used, as illustrated in Fig. 8, with individual manifolds extending across the width of the die for each layer. The manifolds combine the melt streams just ahead of the die lips.

The newer and most widely used flat-die coextrusion method uses a conventional single-manifold die in conjunction with a feed block at the die inlet that introduces a prearranged multilayer melt stream into the die, as shown in Fig. 9. This coextrusion method involves deformation of the joined multilayer melt stream within the die by squeezing and spreading while maintaining layer integrity.

Each system has its advantages and disadvantages [11]. The major advantage of the multimanifold die is its ability to coextrude polymers of more widely different rheological properties. Each layer is independently extruded

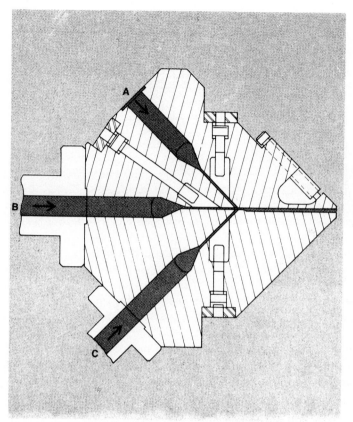

Fig. 8   Multimanifold flat die with individual manifolds for each layer that extend across the width of the die. The layers are joined ahead of the final die land. (Reprinted with permission from J. E. Johnson [6], *Plastics Technology* **22**, 45, February 1976.)

to the width of the die prior to combining, in contrast to the feed-block method in which all layers must flow together, thus requiring more careful matching of viscosities. The major disadvantage of the multimanifold die system is its complexity, cost, and operating difficulties when used for more than three layers.

The feed-block method, while requiring a closer match of polymer viscosities to obtain uniform layers, has greater versatility since a single-manifold die can be used to make a wide variety of products having from two to as many as several hundred layers, including thermally degradable polymers.

Figure 10 is a schematic illustration of the feed-block method [12]. A feed port meters uniform layers of two or more polymers into a die channel. The feed port can subdivide and arrange the melt streams from each extruder

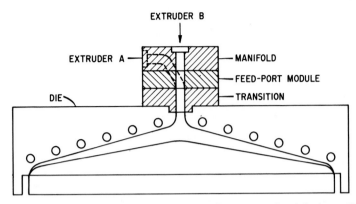

Fig. 9    Feed-block method of flat-die coextrusion using a conventional single-manifold die in conjunction with a feed block that arranges a multilayer melt stream ahead of the die inlet. (From Schrenk [13].)

into almost any layer distribution. It is frequently possible to change the layer arrangement by simply changing a feed-port module. Several layer arrangements are illustrated in Fig. 11. At the exit of the feedports the polymer streams combine into a multilayer melt stream, which flows through a die laminarly and decreases the layer thickness while spreading them to the width of the die.

Commercially, the feed-block method has been used to make five- and six-layer film and sheet from as many as five different polymers (using five

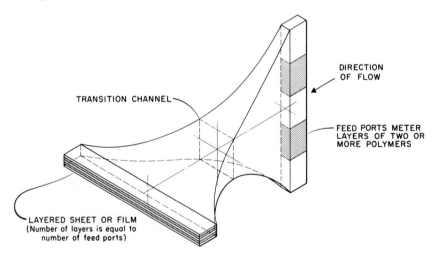

Fig. 10    Schematic illustration of the feed-block method. The feed ports arrange two or more polymers into the desired layer configuration, and the layers are subsequently thinned and spread during extrusion through a single-manifold die. (From Schrenk [13].)

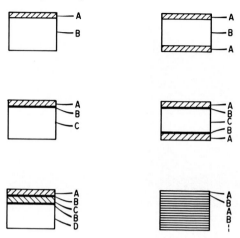

Fig. 11 Wide variety of layer arrangements obtainable by the feed-block method. Frequently the layer distribution can be changed quickly by inserting a different feed-port module, as shown in Fig. 9. (From Schrenk [13].)

extruders connected to the feed block) [13]. Also, unique films having over 100 layers of two polymers (using two extruders) are being made by this flat-die method. Figure 12 is a composite cross-sectional electron micrograph of a 249-layer film having alternating layers of PS and PMMA made by this method [14].

The feed block may be adapted to layer-multiplying devices to increase the number of layers further [15]. The principle of one such device, illustrated in Fig. 13, subdivides, stacks, and recombines the layered melt stream from the feed ports one or more times to increase the number of layers (and inter-facial surface area) exponentially. Thus, it is possible to manufacture unique microlayer films containing many hundreds of layers by the flat-die co-extrusion method. As in the case using the rotating, annular coextrusion die discussed previously, these microlayer films can approach a "layered polymer blend" of molecular dimensions.

## C. Polymer Selection for Coextrusion

Since coextrusion relies on laminar flow of a multilayer melt stream through a die, the proper selection of polymers is crucial. A variety of factors must be considered:

1. Obviously the polymers must be selected to provide the desired end-use properties for the intended application, for example, mechanical properties, permeability, heat sealability, and chemical or weathering

Fig. 12   Composite electron micrograph of 249-layer PS–PMMA film made by the feedblock method. Films with uniform layers as thin as 300 Å have been made.

resistance. Also, postextrusion operations must be considered, for example, if the multilayer is to be thermoformed, it must have sufficient melt strength. Other properties, such as satisfactory surface characteristics, and coefficient of friction, must be specified for machinability on high-speed packaging machines.

2. Frequently, adequate adhesion must be obtained for the proposed end use. A "glue layer" polymer, which adheres to both of two other nonadhering polymers, may be coextruded as an intermediate layer. The degree of adhesion and the selection of glue layer polymers is discussed later.

3. Related to adhesion is the ability to recycle multilayer scrap, which is an important economic consideration. This also is discussed later.

4. A final important criterion for selection of polymers for coextrusion

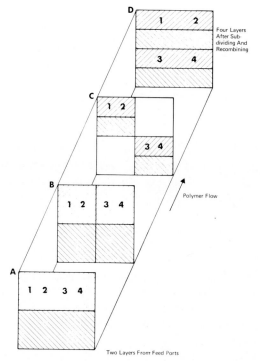

Fig. 13   Illustration of the principle of layer multiplication by using interfacial surface-generating devices.

is their rheological properties. It is particularly true when using the feed-block method of flat die coextrusion since the layers must spread uniformly in the die by laminar flow. This does not mean that all polymers must have identical viscosity; in fact, polymers with viscosity mismatches of three- or fourfold have been coextruded successfully [13]. Frequently, the melt temperature of the polymers entering the feed block may be adjusted to compensate partially for viscosity mismatch; however, this approach is limited since heat transfer takes place between layers. Also, the extrusion temperature of one polymer cannot exceed the short-term thermal degradation temperature of an adjacent layer.

Further, certain coextrusions can exhibit interfacial flow instability, which can cause layer nonuniformity, such as ripples and intermixing of layers. This type of flow instability is not unique to feed-block coextrusion; it can be observed in tubular die coextrusion and multimanifold, flat-die coextrusion as well.

The flow of non-Newtonian multilayer melts through a die is a complex problem. The shear rate (and apparent viscosity) is a function of layer position within the die channel. Therefore, a mathematical model is desirable to assist in selecting the correct polymers for coextrusion.

## III.  FLOW ANALYSIS OF NON-NEWTONIAN MULTILAYER COEXTRUSION

A mathematical analysis of multilayer flow has been done for simple geometric channels, such as concentric flow in pipes and annuli and parallel flow in wide rectangular slits. In these simple channels the analysis involves a one-dimensional flow problem. The analysis is useful in predicting pressure drop and shearing stresses for die design, and also for correlation of laboratory data in selecting polymers for large-scale production lines.

Figure 14 is an illustration of the velocity distribution in a multilayer fluid flowing through a slit die of gap $x = h$, which consists of roughly parabolic segments joined at interfaces $a_{12}$, $a_{23}$, etc. (This is also a good approximation for narrow annular channels, such as tubular dies for blown film.) The boundary conditions are that the velocity is zero at the die walls, and adjacent fluid layers have equal velocity at their interface (no slippage). The shear stress is continuous across the interfaces.

The shearing stress $\tau_{xz}$ (for brevity subscripts will be omitted hereafter) and the pressure gradient are related by Eq. (1):

$$-\partial p/\partial z + \partial \tau/\partial x = 0 \qquad (1)$$

$$\tau = p'(x - c) \qquad (2)$$

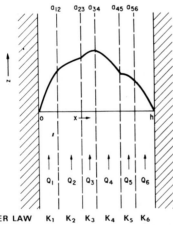

POWER LAW       $K_1$  $K_2$  $K_3$  $K_4$  $K_5$  $K_6$
MAT'L CONST.    $n_1$  $n_2$  $n_3$  $n_4$  $n_5$  $n_6$

Fig. 14  Schematic velocity profile of a non-Newtonian multilayer flow through a wide, narrow slit, where $dv/dx = K\tau^n$.

where $p'$ is the pressure gradient $\partial p/\partial z$, and $c$ denotes the location where $\tau = 0$. (In the case of symmetrical multilayer flow, $c = h/2$.) If each fluid can be approximated by the power law model, then

$$\dot{\gamma}_i = K_i \tau^{n_i} \tag{3}$$

When negative shearing stresses are encountered, Eq. (3) should be written

$$|\dot{\gamma}_i| = K_i |\tau|^{n_i} \tag{4}$$

or more specifically

$$\dot{\gamma}_i = \text{sign}(\tau) K_i |\tau|^{n_i}, \quad \text{where} \quad \text{sign}(\tau) = +1 \quad \text{for} \quad \tau > 0,$$
$$\text{sign}(\tau) = -1 \quad \text{for} \quad \tau < 0 \tag{5}$$

If $v_i(x)$ is the velocity profile in the $i$th layer,

$$dv_i/dx = \dot{\gamma}_i = \text{sign}(p') \quad \text{sign}(x-c) K_i |p'|^{n_i} |(x-c)|^{n_i} \tag{6}$$

Equation (6) can be integrated to yield $v_M(x)$, the velocity profile in the $M$th layer. (Total number of layers $= N$.)

$$v_M(x) = -\text{sign}(p') \left\{ \sum_{j=1}^{M-1} \frac{K_j |p'|^{n_j}}{n_j+1} [|a_{j-1,j} - c|^{n_j+1} - |a_{j,j+1} - c|^{n_j+1}] \right.$$
$$\left. + \frac{K_M |p'|^{n_M}}{n_M+1} [|a_{M-1,M} - c|^{n_M+1} - |x-c|^{n_M+1}] \right\} \tag{7}$$

The total volumetric flow rate of the $M$th layer can be obtained by integrating Eq. (7) between the two interfaces bounding that layer:

$$|Q_M| = \sum_{j=1}^{M-1} \frac{K_j |p'|^{n_j}}{n_j+1} \{|c - a_{j-1,j}|^{n_j+1} - |c - a_{j,j+1}|^{n_j+1}\} (a_{M,M+1} - a_{M-1,M})$$

$$+ K_M \frac{|p'|^{n_M}}{n_M+1} \{|c - a_{M-1,M}|^{n_M+1}\} (a_{M,M+1} - a_{M-1,M})$$

$$+ \frac{K_M |p'|^{n_M}}{(n_M+1)(n_M+2)} \{\text{sign}(c - a_{M,M+1}) |c - a_{M,M+1}|^{n_M+2}$$

$$- \text{sign}(c - a_{M-1,M}) |c - a_{M-1,M}|^{n_M+2}\} \tag{8}$$

If $p'$, $c$, $a_{12}$, $a_{23}$, ..., $a_{N-1,N}$ are known, these equations can be directly applied to yield the velocity profile and the volumetric rates of each layer. However, normally the problem is encountered in inverse form; the volumetric rates $Q_M$ are known (controlled), and we wish to calculate the pressure gradient $p'$, the location of the zero shear plane $c$, and the locations of the fluid interfaces ($a_{12}$, etc.). If a trial value is selected for $p'$ and for $c$, Eq. (8) can be solved sequentially for $a_{12}$, $a_{23}$, ..., $a_{N, N+1}$, and a velocity profile is obtained. The calculated value of $a_{N, N+1}$ does not match the true location $h$, and the calculated velocity at this interface does not match the true value (zero). Iteration, with new trial values for $p'$ and $c$, can remove these mismatches, and converge on the correct solution.

The problem is somewhat simpler with symmetrical multilayer systems, since the location of the zero-shear plane is known at the outset to be at the center of the channel ($c = h/2$). A trial value of $p'$ is selected, Eq. (8) is solved, and the correct value of $p'$ determined by an iteration procedure that removes any mismatch with the boundary at $x = h$. The solution provides the complete velocity profile, locations of the fluid interfaces, and the shear stress at each interface.

## IV. FLOW INSTABILITIES IN MULTILAYER COEXTRUSION

Several types of flow instabilities can occur in a multilayer melt stream while coextruding a plastic film or sheet. Nonuniform layer thickness may be caused by surging in the extruder output rates or by poor melt temperature uniformity within individual resin streams. Southern and Ballman [16] and others [17–19] have reported on another type of instability that occurs in layered melts of substantially different viscosities. The lower-viscosity resin migrates to the region of highest shear stress and will tend to encapsulate the higher-viscosity polymer if the viscosity difference and residence time are sufficient. This effect, which has been observed in capillaries and rectangular orifices, can lead to layer thickness variation and is discussed in greater detail by D. R. Paul in Chapter 16.

### A. Interfacial Flow Instability in Multilayer Coextrusion

Another type of flow instability can occur at the interfaces of a multilayer stream, which in the extreme case can cause intermixing of the layers. Under stable flow conditions, the fluid interfaces within the die and the solid interfaces in the final multilayer film are smooth, flat, and parallel. As output rate is increased, a point is reached at which the outermost interface in the solid multilayer film exhibits a wavelike form. Beyond this point we do not have a quantitative understanding of the fluid velocity pattern within the die.

A low-amplitude waviness of the interface may not be noticeable to the eye, and may not interfere with the functionality of the multilayer film. At still higher output rates, the layer distortion becomes more severe. If a large-amplitude, transverse waveform develops in the flowing multilayer stream within the die, the velocity gradient can carry the crest forward and convert it into a fold. Multiple folding can result in an extremely jumbled interface, and a film with marbleized appearance (which may be desirable for decorative applications). Figure 15 is a schematic illustration of a two-layer film with varying degrees of interfacial flow instability. In Fig. 16 two photomicrographs through the cross section of a three-layer coextruded sheet of acrylonitrile–butadiene–styrene copolymer (ABS) 213–Styron†470–ABS 213 produced under conditions of stable coextrusion and severe instability are shown. The change from stable to unstable multilayer flow was obtained by changing melt and die temperatures. Higher temperatures (lower viscosities) were required to obtain the stable flow pattern.

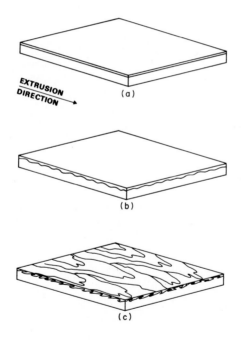

Fig. 15    Illustration of film or sheet appearance under (a) stable flow conditions, (b) incipient interfacial flow instability, and (c) severe instability.

† Trademark of The Dow Chemical Company.

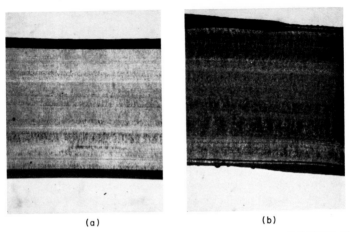

(a)                                                          (b)

Fig. 16  Cross-sectional photomicrographs through a three-layer ABS–HIPS–ABS sheet made under (a) stable and (b) unstable coextrusion conditions. Thicknesses were 0.0013-in. ABS–0.0165-in. HIPS–0.0013-in. ABS.

## B.   Location of Interfacial Flow Instability

To determine the source of instability, a small, single-manifold, flat-film die with a three-layer feed block was used to coextrude an ABS 213–Styron 470–ABS 213 sheet. Extruders deliver individual melts into the feed block, where they can be subdivided and combined into the desired layered structure. The combined melt stream then flows to the die, where the layers are spread and thinned to the final die land dimensions.

A set of coextrusion conditions were chosen in which incipient inter-layer instability was observed in the three layer sheet. At this point the extruders were stopped and the die orifice blocked to prevent drool. The die was allowed to cool, and was then disassembled. The solidified plastic was sectioned to study the interfaces throughout their flow path. It was found that the layer interfaces remained uniform and clearly defined at the exit of the feed block and up to the die land. Within the die land a wavelike distortion of the interfaces started to develop.

These findings, plus other observations, lead us to the hypothesis that there is a "critical interfacial shear stress" at which interfacial flow instability may occur in a given pair of polymers.

It is well known that melt fracture in a single polymer is related to a "critical" shear stress at the wall of a die. In polyethylene, melt fracture can occur at about $10^6$ dyn/cm$^2$. While many factors can affect melt fracture, the roughness of the extrudate is associated with a "stick-slip" instability of the polymer velocity near the rigid die wall. In multilayer coextrusion, it appears that an interfacial flow instability can occur at an even lower "critical shear stress" than that required for melt fracture at the wall.

## C. Experiment

A fractional factorial experiment was designed to test the validity of our hypothesis that interfacial flow instability develops at a certain critical interfacial shear stress for a given pair of polymers. Again, we chose the three-layer coextrusion of ABS–HIPS–ABS. The first experiments were designed to ascertain which variables were most important. The independent variables considered were: (1) skin-layer temperature; (2) core-layer temperature; (3) die temperature; (4) ratio of skin-layer to core-layer thickness; (5) total extrusion rate; and (6) die gap. It was found that the most important variables were the skin-layer temperature and viscosity, the ratio of skin-to-core layer thickness, total extrusion rate, and die gap. The die temperature and core-layer temperature were only significant as they affected the skin-layer temperature (viscosity) through heat transfer.

## D. Determination of "Critical Interfacial Shear Stress"

The four most important variables (skin-layer viscosity, layer thickness ratio, total rate, and die gap) were varied independently over a range of shear stress that gave stable to unstable flow in the three-layer ABS 213–Styron 470–ABS 213 structure.

It was necessary that the point of incipient instability be determined by a rather arbitrary visual judgment. Six persons evaluated all the samples, where each variable spanned the range from stable to unstable flow while maintaining other conditions constant. The consensus of the judges on the occurrence of incipient instability was good.

The interfacial shear stress was calculated from the known coextrusion conditions at incipient instability for each variable and the two material constants, $K$ and $n$, for each polymer as determined by Instron rheometer data. It was determined that interlayer melt fracture developed at an interfacial shear stress of approximately 500,000 dyn/cm$^2$, regardless of which independent variable was varied.

These results support our hypothesis that interfacial flow instability is related to a "critical interfacial shear stress."

The value of 500,000 dyn/cm$^2$ for the "critical interfacial shear stress" refers to the specific system studied (ABS 213–Styron 470–ABS 213). The transition to unstable flow may occur at a different stress level in other systems.

A sheet produced at the point of incipient instability would probably still be commercially acceptable, since very critical visual judgment was employed. Severe instability occurred at higher shear stresses. (The steady-state analysis used in this work of course cannot provide a meaningful value for shearing stress in the regime of severe instability.)

Interfacial flow instability can be reduced or eliminated by increasing the skin-layer thickness, increasing the die gap, reducing the total extrusion rate, or decreasing the viscosity of the skin-layer polymer. (The last remedy for the interfacial instability problem, in some cases, may introduce a systematic nonuniform layer distribution because of viscosity mismatch, as described earlier.)

## V.  ADHESION IN COEXTRUSION

### A.  Interfacial Adhesion

When a coextruded sheet or film is cooled, and the fluid layers solidify (by crystallization or by vitrification), the solidified layers may hold together tenaciously, or they may peel apart easily. Some pairs of polymers exhibit strong interfacial adhesion; other pairs exhibit weak interfacial adhesion. Accumulated experience provides a reservoir of knowledge regarding the degree of adhesion between various thermoplastic materials, but the reasons behind these facts are only partially understood, and predictability of new systems is low (see Volume 1, Chapter 6).

When two molten polymer streams join, a fluid interface is created and is transported for some distance under pressure before leaving the die. These conditions favor the establishment of complete and intimate interfacial contact of the two fluids. This is accompanied by some molecular inter-penetration, as polymer segments of one phase diffuse across the interface into the other phase. However, since dissimilar polymers are usually thermo-dynamically immiscible, this interpenetration is limited in extent. The localized variation in composition in the vicinity of the interface is indicated in Fig. 17. The thickness of the "interphase" increases with time of contact, approaching a limiting equilibrium thickness. The limiting thickness is presumably greater with polymers that are nearly compatible, and smaller with highly incompatible polymers.

The extent of molecular interpenetration of the molten polymer streams must be one of the factors that determine the interlayer adhesion of the final product, but the solidification processes are probably also involved. Two adjacent layers may solidify at different temperatures, at different times,

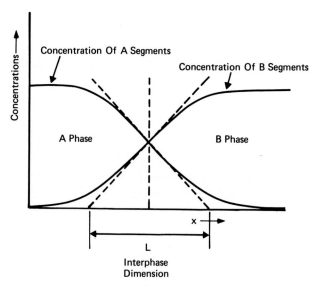

Fig. 17 Concentration of Polymer Segments A and B in the "interphase" region at the interface of A and B.

and by different mechanisms, thereby affecting the structure and properties of the interface. For example, if layer A crystallizes while layer B is still a viscoelastic fluid, penetrating segments of A chains might be drawn back from the "interphase" in the A layer.

Yamakawa [20] has studied the closely related problem of hot-melt adhesive bonding of polyethylene with ethylene copolymers. The observation of cross sections of bonded joints with an interference microscope showed the presence of an interfacial boundary layer and a mixed layer of polyethylene and the copolymer. The relation between peel strength and the character of the two layers was discussed.

Another closely related problem is the interpenetration of polymer chains across a weld line, to develop a bond between two converging streams of the same molten polymer [21]. If adjacent layers of general-purpose polystyrene and rubber-reinforced polystyrene are coextruded, the interfacial bonding is similar to that of a polystyrene weld line. The continuous phase in the HIPS layer is polystyrene. Molecular interpenetration across the fluid interface is thermodynamically allowed. The two layers solidify together, by vitrification, at nearly the same temperature, so that the solidification process should not destroy the bond. In practice, this simple system behaves as expected. In the general case of the coextrusion interface between two arbitrarily chosen polymers, however, we do not know of any quantitative theory that can predict the degree of interfacial adhesion in the final solidified film.

## B.   Use of "Glue Layer" Polymers in Coextrusion

It sometimes develops that a desired multilayer arrangement in sheet or film exhibits poor adhesion between a pair of adjacent layers, with consequent delamination. This weakness can be remedied by introducing a thin layer of a third polymer that shows good adhesion to both of the materials in question ("glue layer").

An obvious candidate for a glue layer between nonadhering Polymer A and Polymer B would be an AB block copolymer, which could penetrate strongly into both phases (see Chapter 12). However, in most cases this is only a hypothetical solution, since the indicated block copolymers are not available. In practice, the choice of a glue layer polymer depends upon the accumulated experience regarding the interlayer adhesion referred to above. Table I is a qualitative guide to the degree of adhesion that can be obtained in coextrusion of several different polymers [6].

## C.   Polymer Blends for Adhesion

Occasionally polymer blends are used as coextruded glue layers between dissimilar, nonadhering layers. Blends are used for:

1.   optimizing adhesion by combining the molecular functionality of two polymers,
2.   improving coextrudability of some glue-layer polymers by incorporating a polymeric "extrusion aid" to adjust viscosity, and
3.   obtaining "mechanical" adhesion by a glue-layer blend of incompatible polymers.

Frequently adhesion between two layers can be optimized by a glue-layer blend of two generally compatible polymers. For example, styrene–butadiene block copolymer adheres well to polystyrene and its copolymers, but its adhesion is borderline with polyethylene. A blend of styrene–butadiene with ethylene–vinyl acetate can be an effective glue layer to adhere a polystyrene layer to a polyethylene layer [22]. Also, a blend of ethylene–acrylic acid copolymer with ethylene–vinyl acetate copolymer can adhere nylon to poly(vinyl chloride) and vinylidene chloride–vinyl chloride copolymers [23].

Similar blends are used to improve coextrudability, especially where viscosity must be adjusted to obtain satisfactory layer uniformity. Some glue-layer polymers are thermally degradable, and blends are used as polymeric lubricants or "extrusion aids" to improve extrudability.

Table I

Qualitative Degree of Adhesion between Resins[a]

| Material: | Olefins | | | Styrenes | | | | | | Vinyl | | | Misc. | | | | | Adhesives | | | | |
|---|---|---|---|---|---|---|---|---|---|---|---|---|---|---|---|---|---|---|---|---|---|---|
| Material: | LDPE | HDPE | PP | Low-impact PS | Medium-impact PS | High-impact PS | Crystal PS | ABS[b] | ABS | Rigid PVC | Flexible PVC | Vinylidene chloride–vinyl chloride copolymer | Polycarbonate | Polyurethane | Acrylic | Nitriles | Nylon 6 | EAA | EVA | Ionomers | SBS[c] | CPE |
| LDPE | G | G | P | P | P | P | P | P | P | P | P | P | P | F | P | P | P | G | G | G | F | G |
| HDPE | G | G | P | P | P | P | P | P | P | P | P | P | P | F | P | P | P | G | G | G | F | G |
| PP | P | P | G | P | P | P | P | P | P | P | P | P | P | F | P | P | P | F | G | P | G | G |
| Low-impact PS | P | P | P | G | G | G | G | G | P | P | P | P | P | F | P | P | P | P | G | P | G | F |
| Medium-impact PS | P | P | P | G | G | G | G | G | P | P | P | P | ? | F | P | P | P | P | G | P | G | F |
| High-impact PS | P | P | P | G | G | G | G | G | P | P | P | P | ? | F | P | P | P | P | G | P | G | F |
| Crystal PS | P | P | P | G | G | G | G | G | P | P | P | P | ? | F | P | P | P | P | G | P | G | F |
| ABS[b] | P | P | P | G | G | G | G | G | G | G | G | G | ? | F | G | P | P | P | G | P | G | G |
| ABS | P | P | P | P | P | P | P | G | G | G | G | G | ? | F | G | P | P | P | G | P | G | G |
| Rigid PVC | P | P | P | P | P | P | P | G | G | G | G | F | U | G | G | P | U | P | G | P | G | G |
| Flexible PVC | P | P | P | P | P | P | P | G | G | G | G | F | U | G | G | P | U | P | G | P | G | G |
| Vinylidene Chloride–Vinyl Chloride copolymer | P | P | P | P | P | P | P | G | G | F | F | G | U | G | G | ? | P | P | G | P | G | G |
| Polycarbonate | P | P | P | P | ? | ? | ? | ? | ? | U | U | U | G | ? | G | P | ? | ? | ? | ? | ? | ? |
| Polyurethane | F | F | F | F | F | F | F | F | F | G | G | G | ? | G | ? | P | ? | F | G | ? | F | G |
| Acrylic | P | P | P | P | P | P | P | G | G | G | G | G | G | ? | G | P | ? | ? | ? | ? | ? | F |
| Nitriles | P | P | P | P | P | P | P | P | P | P | P | ? | P | P | P | G | P | P | F | ? | F | ? |
| Nylon 6 | P | P | P | P | P | P | P | P | P | U | U | P | ? | ? | ? | P | G | G | P | G | P | ? |
| EAA | G | G | F | P | P | P | P | P | P | P | P | P | ? | F | ? | P | G | G | G | G | F | G |
| EVA (ethylene–vinyl acetate) | G | G | G | G | G | G | G | G | G | G | G | G | ? | G | ? | F | P | G | G | ? | G | G |
| Ionomers | G | G | P | P | P | P | P | P | P | P | P | P | ? | ? | ? | ? | G | G | ? | G | ? | G |
| SBS[c] | F | F | G | G | G | G | G | G | G | G | G | G | ? | F | ? | F | P | F | G | ? | G | G |
| CPE (chlorinated polyethylene) | G | G | G | F | F | F | F | G | G | G | G | G | ? | G | F | ? | ? | G | G | G | G | G |

G = good, F = fair, P = poor, ? = unknown, U = undesirable for coextrusion. (Reprinted with permis-
from J. E. Johnson [6], *Plastics Technology* **22**, 45, February 1976.)
ess than 20% acrylonitrile content.
tyrene–butadiene–styrene.

Finally, blends of two incompatible polymers have been used as an intermediate glue layer between layers of the same incompatible polymers [24]. While this approach provides a convenient place for recycle of scrap, the adhesion is usually poorer than that possible with other glue-layer polymers. Delamination at the interface usually exhibits a fibrillar appearance where portions of the mechanically interlocked blended phases adhere to their homogeneous-layered counterparts and are elongated during delamination. In some cases this degree of "mechanical" adhesion is satisfactory.

It is certain that the use of polymer blends will play an increasing role in obtaining adhesion in coextruded products.

## VI.   PROPERTIES OF MULTILAYER FILMS

### A.   Permeability of Multilayer Films

The permeation of gases and water vapor through a homogeneous polymer film depends upon the permeability of the polymer and the thickness of the film (see Volume 1, Chapter 10):

$$\text{Flux (steady state)} \ = P(p_1 - p_2)/l \tag{9}$$

or

$$\text{film permeance} = P/l \tag{10}$$

where $P$ is permeability, $l$ thickness, and $p_1$ and $p_2$ the partial pressures of the diffusing gas on the two sides of the film. The permeability, in turn, is the product of a solubility coefficient $S$ and a diffusion coefficient $D$:

$$P = DS \tag{11}$$

In a multilayer film, each polymer exhibits its own values of $S$, $D$, and $P$. The permeance of the composite film is calculated by treating it as a series of resistances, with each ply contributing to the barrier:

$$1/\text{Permeance} = l_1/P_1 + l_2/P_2 + l_3/P_3 + \cdots + l_N/P_N \tag{12}$$

In the steady state, the diffusion flux is uniform through the film, but the concentration of diffusing gas exhibits a discontinuity at each interface.

Actually, the permeability of a given polymer can depend upon the degree of crystallization, and the crystalline morphology. Very thin layers, unable to accommodate the normal morphology, might be expected to exhibit different values of permeability from thick specimens of the same polymer. However, multilayer films and polystyrene–polyethylene and polystyrene–polypropylene containing layers a few hundred angstroms thick have been reported to follow Eq. (12) [25].

Films containing multiple-barrier layers can retain their barrier properties after severe crumpling. Randomizing of flaws in multiple, sandwiched, barrier layers forces the permeating gas through a tortuous path. For example, a film containing multiple layers of Saran[†] can retain barrier properties after severe crumpling at low temperature [26a]. Because of low-temperature brittleness, a single Saran layer of equivalent thickness may show a catastrophic loss of barrier properties.

Also, the water vapor transmission rates through 125-layer PE–PS and PP–PS films were measured before and after crumpling. It was found that crumpling did not appreciably change the water vapor permeation rate [25].

Many applications for coextruded plastic film and sheet are designed to give gas- and moisture-barrier properties to the layered product. The permeability coefficients for oxygen transmission through several different polymers is given in Table II.

High-barrier Saran, hydrolyzed ethylene–vinyl acetate copolymers, high nitrile copolymers, rigid PVC, nylon, and polyesters all have relatively good oxygen barrier properties compared to polyethylene (PE), polypropylene (PP), polystyrene (PS), etc. One mil of Saran provides a barrier equivalent to about 4–5 mils of nitrile barrier resins, 10 mils of nylon, 40 mils of rigid PVC, and over 1000 mils (1 in.) of PS, PE, or PP. Thin layers of the high-barrier

Table II

Approximate Oxygen Transmission Rates
through Several Plastics

| Resin | Permeability $(cm^3\text{-mil}/100\ in.^2\text{-day-atm})$ (at 23°C) |
|---|---|
| Hydrolyzed ethylene–vinyl acetate | <0.02 dry gas |
|  | 0.08 moist gas |
| High-barrier Saran | <0.2 |
| Nitrile-barrier resin | 0.6 |
| Nylon | 2.6 |
| Polyester | 7 |
| Rigid PVC | 14 |
| Polypropylene | 150 |
| HDPE | 150 |
| PS | 260 |
| LDPE | 420 |

[†] Trademark of The Dow Chemical Company.

resins are frequently coextruded with thick layers of PE or PS to provide gas barrier properties in films or sheets of these polymers.

The water vapor transmission rate through a 1-mil ply of the same polymers is shown in Table III. With the exception of Saran, the ranking for water vapor barrier properties changes significantly. Hydroscopic materials, such as nitriles, hydrolyzed ethylene vinyl acetate, and nylon, are poor barriers to water vapor. Polyethylene and PP are good barriers to water vapor. Saran is the only commercially available extrusion resin that has excellent barrier properties to both gases and water vapor.

## B.  Mechanical Properties of Multilayer Films

A polymer layer in a multilayer sheet or film can exhibit different mechanical behavior from a free film of the same material. Particularly striking effects can be observed in laminar composites containing a "hard" component and a "soft" component, or a "brittle" component and a "tough" component. Sword makers of early times hammered down alternate layers of hard and soft steel, obtaining blades that would take a fine cutting edge and yet were strong and tough. Similar effects can be observed with coextruded multilayer polymer systems.

Consider a laminar composite containing a layer of high-modulus, low-elongation polymer sandwiched (with good adhesion) between layers of a high-elongation material. When the composite is tested to failure in tension, the adhering high-elongation layers may act to prevent transverse crack

Table III

Approximate Water Vapor Transmission Rates
through Several Plastics

| Resin | Water vapor transmission rate (gm-mil/100 in.$^2$/24 hr) (100°F, 90%RH) |
|---|---|
| High-barrier Saran | 0.05 |
| Nylon | 0.6 |
| PP | 0.7 |
| HDPE | 0.7 |
| LDPE | 1.2 |
| Polyester | 1.8 |
| Rigid PVC | 3 |
| Nitrile | 5 |
| PS | 7 |
| Hydrolyzed ethylene–vinyl acetate | 45 |

propagation in the hard layer. With crack propagation so blocked, the hard layer may reach its ductile yield stress, and the entire composite stretch in a ductile manner to high elongation. This toughening effect may be termed "mutual interlayer reinforcement" and is indicated schematically in Fig. 18.

Interlayer interaction can sometimes result in the opposite result; a normally tough polymer may act in a brittle manner in a laminar composite.[†] In this case the high-elongation material is not able to block crack propagation in the brittle layer; a crack forms in the brittle layer, and continues into and through the tough layer, causing a localized failure at low overall elongation. For example, a skin of general-purpose polystyrene on a core of high-impact polystyrene can force the entire composite to undergo brittle failure on bending. This behavior has been called "mutual interlayer destruction" [25].

Such interlayer interactions, whether of a beneficial or a deleterious nature, require some degree of adhesion at the interfaces. However, the thinner the layers, the lower will be the degree of adhesion required in

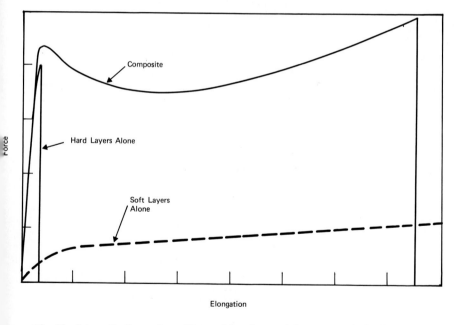

Fig. 18 Schematic illustration of "mutual interlayer reinforcement" obtained when "soft" layers block crack propagation in normally brittle "hard" layers.

[†] Djiauw and Fesko [26b] describe in detail the analogous behavior of elastomeric substrates coated with thin paint films. [—Editors]

order for one layer to influence the response of its neighbor. Thus, in micro-layer film, with hundreds of very thin layers, interlayer interactions can be anticipated even with very poorly adhering components.

The mechanical properties of a homogeneous polymer sheet or film depend strongly upon the pattern of molecular orientation. A unidirectionally oriented crystalline polymer, such as polyethylene or polypropylene, can exhibit a tensile strength (in the orientation direction) higher by an order of magnitude than that of the same polymer when unoriented. Strength in the transverse direction is low [27]. Polystyrene, a glassy amorphous polymer, can be made strong, tough, and craze resistant in one direction by uniaxial orientation, while being made extremely weak and brittle in the transverse direction [28]. Biaxial orientation of a sheet or film can impart desirable mechanical properties in all direction in the plane, and is commonly employed in the manufacture of plastic films of both the crystalline and the glassy amorphous types [29].

The mechanical properties of coextruded multilayer sheets and films also depend upon the degree of molecular orientation. An appropriate level of orientation can optimize the properties of the individual layers, and thereby the likelihood and the extent of interlayer reinforcement. For example,

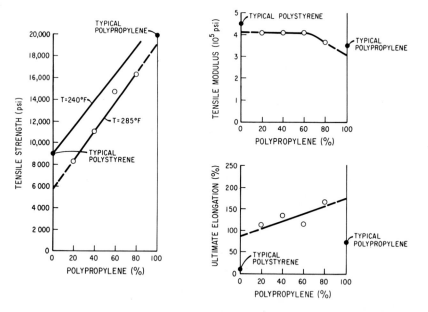

Fig. 19   Mechanical properties of biaxially oriented 125-layer films of PP and PS of different composition. Ultimate tensile strength can depend on the orientation temperature (240 versus 285°). (From Schrenk and Alfrey [25].)

biaxially oriented 125-layer films of polypropylene and polystyrene have been reported to exhibit high tensile strengths and ultimate elongations, over a composition range from 80/20 to 20/80 [25]. The reported tensile properties of these films are shown in Fig. 19.

## C. Optical Properties of Multilayer Films

A multilayer film made up of alternating layers of two transparent polymers of differing refractive indexes can exhibit vivid optical effects. Some wavelengths of light are strongly reflected; others are transmitted through the film [30a].

Figure 20 is a schematic drawing of a 125-layer film in which the odd layers have a refractive index of 1.6, and the even layers a refractive index of 1.5. When a beam of light of wavelength $\lambda$ shines upon the film, partial reflection occurs at each of the many interfaces. The reflections at the interfaces of increasing refractive index suffer a phase reversal. Since the layers differ only by 0.1 in refractive index, the individual reflections are weak. However, if reflections from the different interfaces all leave the film in phase with each other, the constructive interference yields a high-intensity reflection.

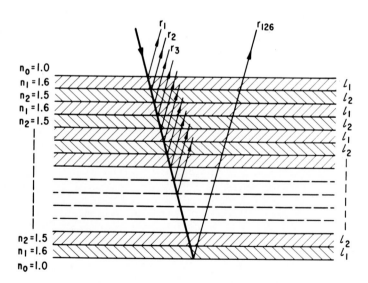

Fig. 20 Reflection of incident light from the many interfaces between polymers of different refractive index, which yields intense iridescent colors. The remaining light is transmitted through the film and is the complementary color to the reflected light. (From Schrenk and Alfrey [30a].)

The wavelength of the first-order reflection (for normal incidence) is given by

$$\lambda_I = 2(n_{odd}l_{odd} + n_{even}l_{even}) \tag{13}$$

For example, if the thicknesses of the odd and even layers are set at 700 and 746.5 Å, respectively,

$$n_{odd}l_{odd} = 1.6 \times 700 = 1120$$

$$n_{even}l_{even} = 1.5 \times 746.5 = 1120$$

$$\lambda_I = 2(1120 + 1120) = 4480 \quad Å$$

In addition to the first-order reflection $\lambda_I$, higher-order reflections occur at wavelengths $\lambda_{II}$, $\lambda_{III}$, etc.:

$$\lambda_{II} = (2/2)(n_{odd}l_{odd} + n_{even}l_{even}) \tag{14}$$

$$\lambda_{III} = (2/3)(n_{odd}l_{odd} + n_{even}l_{even}) \tag{15}$$

$$\lambda_M = (2/M)(n_{odd}l_{odd} + n_{even}l_{even}) \tag{16}$$

The relative intensities of these higher-order reflections depend upon the ratio of optical thicknesses of the two kinds of layer. For example, if the odd and even layers have equal optical thickness, the second-order and fourth-order reflections are suppressed. On the other hand, if $n_{odd}l_{odd} = 2n_{even}l_{even}$, the third-order reflection is suppressed.

Finally, the breadth of the reflection spectrum is related to the layer thickness gradient through the film. Uniform layers through the film yield a relatively narrow reflection band, while a layer thickness gradient reflects a wider portion of the spectrum.

An experimental study of iridescent coextruded multilayer films of polystyrene–PMMA has been reported [14]. Dimensions corresponding to

Table IV

Typical Reflecting Wavelengths and Layer Thicknesses for 231-Layer Films of Polystyrene and Poly(methyl methacrylate) (Equal-Thickness Layers)

| Light type | Reflecting wavelength | | Single-layer thickness | | | Total film thickness | |
|---|---|---|---|---|---|---|---|
| | Å | μm | Å | μm | mils | μm | mils |
| Ultraviolet | <3800 | <0.38 | <617 | <0.062 | <2.4 × 10⁻³ | <14.3 | <0.56 |
| Visible "pure" blue | 4730[a] | 0.473 | 768 | .077 | 3.0 × 10⁻³ | 17.7 | 0.70 |
| Visible "pure" green | 5130[a] | 0.513 | 833 | .083 | 3.3 × 10⁻³ | 19.2 | 0.76 |
| Visible "pure" yellow | 5770[a] | 0.577 | 937 | .094 | 3.7 × 10⁻³ | 21.6 | 0.85 |
| Infrared | >7600 | >0.76 | >1234 | >0.123 | >4.9 × 10⁻³ | >28.5 | >1.12 |

[a] From Burnham *et al.* [30b].

various colors are shown in Table IV. Observed reflectance spectra correlated well with measured layer thickness distributions.

In addition to reflecting wavelengths in the visible range of the spectrum to give iridescent colors, similar coextruded films can be designed that reflect in the near infrared or the ultraviolet portion of the spectrum. Films have been tested that can effectively reflect a high percentage of incident solar infrared radiation.

## VII. APPLICATIONS

Having reviewed the technology of coextrusion and some of the properties of coextruded film and sheet, a discussion of several commercial applications and development activities will illustrate how coextrusion is currently used to obtain desired end-use characteristics in products more economically than other laminating and coating processes.

One of the early coextrusions was a three-layer film of polyethylene–polypropylene–polyethylene (PE–PP–PE) for wrapping bread. Polyethylene provided heat sealability on high-speed packaging lines, while the polypropylene core added stiffness and toughness, and, because of its higher melting point, prevented burnthrough during heat sealing to broaden the effective heat-sealing range of the film. While the adhesion of PE to PP is generally poor, in thin films it was adequate for packaging low specific gravity products such as bread. Where better adhesion and stronger heat seals are desired, ethylene–vinyl acetate copolymers can be used.

Other early applications, using blown, coextruded films of low-density polyethylene–poly [ethylene-*co*-(vinyl acetate)] (LDPE–EVA), chlorinated polyethylene–low-density polyethylene (CPE–LDPE), and LDPE–HDPE, were for heavy-duty shipping bags for fertilizers, chemicals, salt pellets, etc. These combinations were designed to provide heat sealability, toughness, and chemical resistance. Also, a nonslip outer layer having a high coefficient of friction, such as CPE, could be used to improve stacking of bags on a pallet. Another early application in Europe was a two-color LDPE milk pouch film consisting of a white outer layer and a black inner layer to provide protection to the milk from light. Also, ionomer–LDPE coextrusions are used for sterilized medical product packaging, and applications are being found for EVA–HDPE films in snack food and cereal packaging.

Another pioneering application for coextrusion is a chill-roll-cast five-layer film of PE–EVA–Saran–EVA–PE (Saranex[†]) in which the Saran core

[†] Trademark of The Dow Chemical Company.

layer provides an excellent gas barrier, while the PE skin layers contribute good heat sealability [31]. The EVA layers bond the Saran and PE layers. These films find application in food packaging requiring excellent gas-barrier properties (meat, cheese, nuts and condiments, snack foods, fruit drinks, etc.) and certain medical applications.

Recently, The Mearl Corporation has commercialized the first microlayer film that consists of over 100 layers, each less than 1000 Å thick. The film is coextruded from two polymers having different refractive indexes, which gives the film striking iridescent color effects. The physical optics of these unique films have been discussed earlier. Applications for iridescent film include lamination to paper and other plastics for decorative uses [32].

Rasmussen [33] has developed a coextrusion process to manufacture bicomponent fibers by generating many multiple layers of three polymers in the sequence ABCABC..., where B adheres to A and C but A and C can delaminate at their interfaces. The multilayer stream is extruded through orifices, drawn into thin ribbons, and subsequently delaminated mechanically at a high rate of production to give a tow with a large number of very thin bicomponent fibers, each consisting of components A and C held together by the adhesive B.

Continued growth in coextrusion coating is forecast in which a multi-layer film is extruded through a flat die directly onto a substrate, such as paper, plastic foam, or cellophane. More recently coextrusion is being used to manufacture thick (greater than 0.010 in.) sheet. The major current use includes simple two- and three-layer structures for packaging. A thin layer of general purpose polystyrene (GPPS) is coextruded over a colored layer of high-impact polystyrene to impart high gloss for marketing appeal. It is desirable to keep the GPPS layer as thin as possible since it tends to embrittle the sheet by crack initiation and propagation through the rubber-modified layer (mutual interlayer destruction).

Two-color sheets, usually made as a three-layer sheet with the mixed colored scrap buried as a core layer, is also used for marketing appeal and improved economics since the scrap can be reused without color matching. In addition to packaging, a similar approach is being used for interior vinyl automobile trim. Before the advent of coextrusion, the scrap from each color of interior trim had to be carefully segregated and color matched for re-cycle. With coextrusion all of the mixed colored scrap can be collected at one place and coextruded with a surface layer of virgin vinyl of the desired color. The economic advantage of this approach is obvious.

In Europe the use of coextruded sheet for thermoformed packages is highly advanced, and multilayer sheet containing as many as five or six layers is being produced for applications requiring good thermoformability, heat sealability, stress crack resistance, barrier to gases and water vapor,

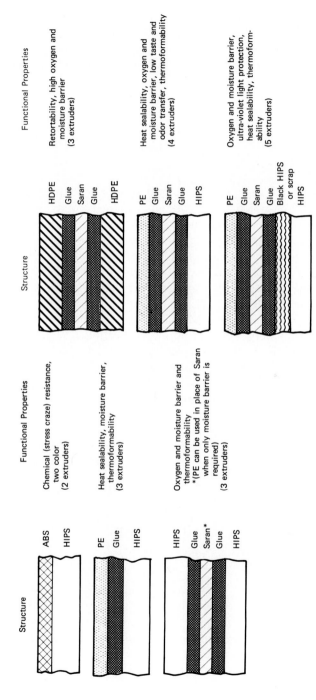

Fig. 21 Schematic illustration of several coextruded sheet structures used in packaging in which each layer provides a specific functional characteristic to the product.

and low taste and odor transfer. Some of the multilayer constructions and functional properties of these sophisticated structures are illustrated in Fig. 21. They are being used to package margarine, fruit drinks, cheese, puddings, sterilized milk, cakes, pickles, mayonnaise salads, and other items [22].

Developmental work has demonstrated the potential of manufacturing plastic cans that can withstand retorting at 250°F for 90 min [34]. A five-layer coextrusion of HDPE–EVA–Saran–EVA–HDPE can be formed into a can and hermetically sealed by double-seaming an aluminum lid to the plastic can body. The Saran layer provides a gas barrier, and HDPE layers give the can thermal stability during autoclaving.

While most of the early commercialization of coextrusion has been in film and sheet for packaging, its use for durable goods is increasing. Appliance manufacturers are developing refrigerator door liners and food boxes that may consist of a thin layer of ABS on HIPS to replace the all-ABS sheet currently used, at considerable cost saving in raw materials [35]. A modified polyethylene layer may also be coextruded on the other surface to act as a barrier to the polyurethane foam-fluorocarbon blowing gases, which tend to craze ABS and HIPS.

Coextrusion is being used in outdoor applications, such as in the fabrication of window shutters, house trim, and recreational vehicles. In these applications a thin layer of pigmented acrylic is coextruded over ABS or HIPS to provide weatherability by protecting the styrenics from ultraviolet exposure.

Other applications include coextrusions of a surface layer having a high coefficient of friction for shower mats. Also, decorative signs, luggage, boat hulls, and bath units are being made from coextruded sheets.

The ability to coextrude multilayer films and sheets economically from a wide variety of polymers, each adding a desired property to the end use characteristics of the product, has made coextrusion a major, growing fabrication process.

## VIII.  SCRAP RECYCLE

The reuse of scrap from coextrusion is an important economic consideration. The recycling of scrap from compatible polymer combinations is fairly straightforward as long as the amount of scrap recycle is controlled to maintain consistent physical properties. For example, in a coextruded sheet of GPPS and HIPS, or styrene–acrylonitrile (SAN) and acrylonitrile–butadiene–styrene copolymer (ABS), the scrap may be extruded back into the rubber-modified layer as a mixed polymer blend. The dilution of the rubber-modified phase with unmodified skin-layer polymer may cause some loss in impact strength, but is acceptable if the skins are thin and the amount of recycle remains consistent.

The recycling of dissimilar (incompatible) polymers poses a more difficult problem since the mixed blend can cause severe loss of mechanical properties in some cases. This is an important area of continuing research. Frequently the glue layers used in coextrusion of nonadhering polymers can act as compatibilizers in the mixed recycle blend.

The recycling of incompatible polymers, usually as a buried core layer, has been demonstrated for certain polymer combinations, such as ABS and HIPS; PS, EVA, and PE; PS, styrene-butadiene, and SARAN; and HDPE, EVA, and Saran. It is difficult, however, and must be approached carefully. The loss of properties due to a buried core layer of incompatible polymers is reduced when tough, virgin skin layers are used because of mutual interlayer reinforcement.

Approaches being taken to improve the compatibility of multilayer scrap recycle include:

1.  polymer modification to improve compatibility, including copolymers with a minor amount of a comonomer to achieve bonding;
2.  compatibilizers, such as CPE and other glue-layer polymers, as an intermediate phase to promote bonding; and
3.  mixing, as there is evidence that the loss of properties is reduced by intensive mixing of the blended scrap.

The technology developed for polymer blends will undoubtedly be used in coextrusion to solve the important economic problem of scrap recycling. Meanwhile, coextrusion can still show economic advantages over other processes for many film and sheet products even when total loss of scrap is considered.

## IX. NEW DEVELOPMENTS

Brief mention should be made regarding some new developments in coextrusion.

Foamed plastics are being coextruded with one or two solid polymer skins to obtain lightweight, rigid-sheet structures. With present flat-die coextrusion technology, foam layer densities of about $0.5 \text{ gm/cm}^3$ are possible. Limitations on foam density achieved in flat-die coextrusion include corrugations in the sheet due to lateral expansion of the foam, and collapse of the foam cells in thick-sheet coextrusion because of insufficient cooling rate. Tubular coextrusion of thinner, lower-density, multilayer solid-skin–foam composites is being developed. The tubular coextrusion is expanded circumferentially over a cooling mandrel to permit foam expansion in all three directions. The potential for savings of raw material by using a lightweight foam core in combination with solid skins to obtain rigidity will stimulate applications in packaging and durable goods. One problem is that some foam composites tend to be brittle. Improved reinforcement of the foam layer by the skin layers or internal reinforcement of the foam core by including thin ductile layers within the foamed core are possible solutions.

Multilayer injection molding of finished objects by simultaneous injection of two polymers into a mold has been demonstrated [36, 37]. This technology is still new, and applications are being developed.

Squeeze tubes and blow-molded bottles, using a tubular coextrusion die to produce a multilayer parison, have been made and are undergoing development for several applications.

Finally, a new solid-phase forming process is being commercialized that can form multilayer containers and similar objects without the trim scrap normally associated with conventional thermoforming [38]. The scrapless feature of this new process will undoubtedly expand the use of coextruded sheets.

# REFERENCES

1. U.S. Patent 3,008,696.
2. U.S. Patent 3,127,152.
3. W. J. Schrenk, K. J. Cleereman, and T. Alfrey, Jr., *SPE Trans.* **3**, 192 (July 1963).
4. W. J. Schrenk, D. S. Chisholm, and T. Alfrey, Jr., *Mod. Plast.* **46**, 152 (January 1969).
5. U.S. Patent 3,223,761.
6. J. E. Johnson, *Plast. Technol.* **22**, 45 (February 1976).
7. U.S. Patent 3,308,508.
8. W. J. Schrenk and T. Alfrey, Jr., *SPE J.* **29**, 38 (June 1973).
9. W. J. Schrenk and T. Alfrey, Jr., *SPE J.* **29**, 43 (July 1973).
10. U.S. Patent 3,677,873.
11. L. M. Thomka and W. J. Schrenk, *Mod. Plast.* **49**, 62 (April 1972).
12. U.S. Patent 3,557,265.
13. W. J. Schrenk, *Plast. Eng.* **30**, 65 (March 1974).
14. J. A. Radford, T. Alfrey, Jr., and W. J. Schrenk, *Polym. Eng. Sci.* **13**, 216 (1973).
15. U.S. Patent 3,565,985.
16. J. H. Southern and R. L. Ballman, *in U. S. Jpn. Seminar Polym. Process. Rheol.* (D. C. Bogue, M. Yamamoto, and J. L. White, eds.); *Appl. Polym. Symp.* No. 20.
17. J. L. White, R. C. Ufford, K. R. Dharod, and R. L. Price, *J. Appl. Polym. Sci.* **16**, 1313 (1972).
18. C. D. Han, *J. Appl. Polym. Sci.* **17**, 1289 (1973).
19. A. E. Everage, Jr., *Trans. Soc. Rheol.* **17**, 629 (1973).

20. S. Yamakawa, *Polym. Eng. Sci.* **16**, 411 (1976).
21. S. Y. Hobbs, *Polym. Eng. Sci.* **14**, 621 (1974).
22. H. Maack, Markets and structures for coextruded sheeting (Europe), *TAPPI 1975 Int. Coextrusion Seminar, Copenhagen, Denmark* (September 1975).
23. U.S. Patent 3,908,070.
24. French Patent 1,401,443.
25. W. J. Schrenk and T. Alfrey, Jr., *Polym. Eng. Sci.* **9**, 393 (1969).
26a. U.S. Patent 3,579,416.
26b. L. K. Djiauw and D. G. Fesko, *Rubber Chem. Technol.* **49**, 1111 (1976).
27. R. J. Samuels, "Structured Polymer Properties." Wiley, New York, 1974.
28. L. J. Broutman and F. J. McGarry, *J. Appl. Polym. Sci.* **9**, 609 (1965).
29. L. S. Thomas and K. J. Cleereman, *SPE J.* **28**(4), 2; (6), 9 (1972).
30a. T. Alfrey, Jr., E. F. Gurnee, and W. J. Schrenk, *Polym. Eng. Sci.* **9**, 400 (1969).
30b. R. W. Burnham, R. M. Hanes, and C. J. Bartileson, "Color: A Guide to Basic Facts and Concepts," Section 3.1.2a2. Wiley, New York, 1963.
31. J. Eichhorn, *Proc. Annu. Nat. Packaging Forum, 28th* **4**, 242 (1966).
32. W. J. Schrenk and J. Pinsky, Coextruded iridescent film, *TAPPI Paper Synthetics Conf., Atlanta, Georgia* (September 1976).
33. O. B. Rasmussen, *Amer. Chem. Soc. Div. Organ. Coatings Plast. Chem.* **32**(1), 264 (1972).
34. L. M. Thomka, *Package Eng.* **18**, 60 (February 1973).
35. C. R. Finch, Plastics in appliances, *SPE Nat. Tech. Conf., Louisville, Kentucky* (November 1975).
36. British Patent 1,156,217.
37. *Mod. Plast.* **53**, 40 (July 1976).
38. U.S. Patent 3,739,052.

# Chapter 16

# Fibers from Polymer Blends

## D. R. Paul

*Department of Chemical Engineering*
*The University of Texas*
*Austin, Texas*

# I. INTRODUCTION

In fiber technology, two or more polymers may be incorporated into a product to obtain combinations of individual material characteristics or new ones by either of two distinctly different routes, both of which are known as blending. The first is the older and more well-developed concept of fiber–fiber blending in which conventional synthetic or natural fibers are mixed [1, 2]. These are referred to here as *fiber blends*. Fiber blending can be done in many different ways. For example, two types of fibers can be blended into a yarn, or more than one yarn, each composed individually of a single fiber type, can be woven or knitted into a fabric. Very often these are the approaches used respectively for staple and continuous filaments. Fiber blends give combinations of such characteristics as comfort, ease of care, durability, and appearance in a product not available with a single fiber type. Well-known examples are polyester–cotton and nylon–wool. However, in this chapter only the second approach is discussed, in which polymer-polymer blends comprise the individual filaments. These are referred to as *blend fibers*. This concept is newer but rapidly developing as an approach for new fiber products and problem solving for existing products. It is almost exclusively limited to synthetic fibers since the polymer blending occurs prior to fiber formation rather than in later steps of textile processing. As a result, much of the technology for blend fibers lies in polymer processing, which shifts upstream the demand for expertise in the conversion of resources to products. In some cases, either method of blending may be a route to the same end; however, generally they are not interchangeable but are used to solve different types of problems. Basically, blend fibers are best suited when it is necessary that each fiber have the desired characteristics, whereas fiber blends are more appropriate when it is only necessary that the product as a whole have these characteristics.

A wide range of phase configurations are possible for fibers made from two or more polymers, and some of these are highly structured. A broad view of the blend concept in fibers is taken in this chapter and includes morphologies in which the two polymers are segregated into only two well-defined regions. Such low states of dispersion might be more properly thought of as composites in other technologies such as plastics or rubber. Because of the very small size of the fibers themselves, however, these phases are very small on an absolute basis, and it is appropriate to include them along with the other examples of more highly dispersed blend fibers. An analogous situation exists in film products with laminated structures (see Chapter 15).

The concepts and applications of polymer blending in fibers have grown rapidly in recent years, and as a result it is not practical to present an exhaustive treatment here. Instead, an attempt is made in this chapter to generalize to the extent possible by considering the basic types of blend fibers, the fundamental principles involved in their fabrication and properties, and then selected examples of problem solving by this approach. In most instances, blending is not the only approach to the problem to be solved. However, it is beyond the scope of this chapter to give equal space to all alternatives, so what follows is not a balanced treatment of fiber problem solving. Many of the fundamental principles relevant to blend fibers have been developed more fully in other chapters of this book and references to these areas are made where appropriate.

Despite the large industrial activity in blend fibers there is still only meager detailed scientific literature available. Many of the significant ideas and most of the details are proprietary and are only available through the patent literature. The bulk of these patents have been issued outside the United States in recent years, and only abstracts are readily available to most readers. To provide a data base of current activity for this chapter, an exhaustive survey of the Textiles Section of Chemical Abstracts was made for the years of 1974 and 1975, which produced in excess of 350 relevant patents. The ideas from many of these have been incorporated into this chapter, but for reasons of space not all are referenced.

## II. CLASSIFICATION OF BLEND FIBERS BY PHASE MORPHOLOGY

In this section the phase morphology of individual fibers made from two or more polymers is grouped into major categories. Techniques for generating these structures and their characteristics are covered in more detail in later sections. Typically these structural features can be observed through the conventional microscopy techniques normally used to prepare and examine fiber cross and longitudinal sections.

The terms "bicomponent" and "conjugate" fibers are frequently used in the literature in connection with blend fibers of the structured, heterogeneous type described in Section II.B; however, there appears to be no formal definition that would preclude using the term *bicomponent* to describe a homogeneous blend fiber although this is not done. Dictionaries of textile terms [3] and the U.S. Code of Federal Regulations define a fiber made by physically joining together two generically different polymers as a "biconstituent," whereas the term "bicomponent" is reserved for fibers from two generically similar but chemically or physically different polymers. This distinction in terminology is not always practiced in the literature.

## A.  Homogeneous Blends

In an ideal sense, this category is characterized by a single phase of miscible polymers. As discussed in Volume 1, Chapter 2, most polymer pairs are not miscible, but an important number of exceptions exists, particularly among polymers that are closely related members of a copolymer family. In a real sense, most fibers made from only one polymer are not homogeneous because of crystallinity, porosity, additives, etc. Consequently, a homogeneous-blend fiber does not have composition inhomogeneities because it is a blend; however, this does not preclude these other types of heterogeneities. Such a fiber would show none of the morphological features described in the next section, but their absence as determined by a particular microscopy technique does not assure homogeneity down to the polymer segmental level as implied in the above definition. Proof of this usually requires other techniques, which are described in Volume 1, Chapters 2, 4, and 5. Miscible mixtures may be expected to behave in some ways like a single polymer, for example, a single glass transition temperature ($T_g$), similar to a copolymer of the parent polymers. Its homogeneity would give it optical properties important to fibers, for example, luster, similar to a pure polymer. This type of blend fiber is not common first because of the lack of identified miscible polymer pairs suited for fibers and second because two phases are essential to achieve some of the effects sought by blending.

A homogeneous blend is probably desired when a second polymer is added to give dye sites; this approach has been employed in some commercial fibers [4]. Blending for dyeing purposes is considered more fully in Section VI.E. Homogeneity is probably acceptable when the objective of blending is to improve flame retardance, antistatic, or moisture sorption characteristics. For the usual system, however, homogeneity is not likely to occur. This property would be undesirable when a second polymer is added for the purpose of improving certain mechanical properties, as discussed in Sections VI.B and C. Homogeneous structures cannot be used to make self-crimping fibers since this characteristic requires spatial segregation of the two components. This does not rule out use of miscible pairs for this purpose, however, as is evident in later sections.

## B.  Heterogeneous Blends

In this more common category the two polymers are segregated into spatial regions that are composed essentially of one or the other pure component. Usually, the two polymers are immiscible but in principle they could be miscible but not mixed. Ordinarily, the phase morphologies of interest can be readily observed in fiber cross or longitudinal sections using standard optical microscopy techniques aided by differential staining or pigmentation. The possible morphologies could be divided further into *structured* and *random*, but there is not a clear boundary between these extremes. The number of potential phase arrangements is almost unlimited; thus, in the following only the major types that are generally recognized [5–12] are classified and possible variants suggested.

In heterogeneous structures, adhesion between the two phases is a critical concern (see Volume 1, Chapter 6). Adhesion is essential for the integrity of some structures but not others. Poor adhesion is a requirement for some applications discussed in Section VI.F, but usually good adhesion is required for many important blend fiber properties.

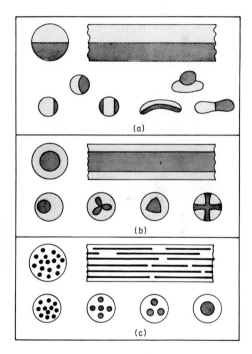

Fig. 1 Typical phase arrangements in heterogeneous-blend fibers: (a) side-by-side; (b) sheath–core; (c) matrix–fibril. Both cross-sectional and longitudinal views are shown for the three major categories, while some possible variations within each category are shown in cross section only. The cloverleaf and cross shapes in part (b) are the basis for the Japanese commercial products known as Siderea and Belima, respectively [6].

*1. Side-by-Side*

Here the two polymers lie side by side within the fiber as the name suggests and as shown schematically in Fig. 1a. The components must adhere or two fibers of different composition would result. This structure is generated to produce a self-crimping fiber that is bulky like wool or has a high extensibility, as discussed later. Some variations from the ideal shown in the upper part are shown in the lower part of Fig. 1a. These include cases in which there are unequal amounts of the two polymers, curved interfaces, two regions of one polymer, and the many possibilities when the fiber cross section is nonround, as in wet or dry spun fibers. One of the earliest commercial bicomponents, a member of the Orlon family [13–18], had shapes similar to some of the latter. Clearly there are many other possibilities, and this geometry has a significant influence on the ultimate fiber characteristics [13, 17, 18]. All of the individual fibers in a tow may have essentially the same morphology or variations from fiber to fiber may exist.

### 2. Sheath–Core

Here both polymers are not usually exposed to the fiber surface but rather one polymer is encased (core) by the other (sheath) as shown in Fig. 1b. Adhesion is not always essential for fiber integrity. This structure is employed when it is desirable for the surface to have the property of one of the polymers such as luster, dyeability, or stability, while the core may contribute to strength, reduced cost, and the like. A selection of variations is shown in the lower part of Fig. 1b. These include an eccentric core, which leads to self-crimping capability, and core shapes that may or may not extend to the surface at some points. A highly contorted interface between sheath and core with undercuts can lead to a mechanical interlocking that may be desirable in the absence of good adhesion. In some cases, the sheath may be used to protect the core from the environment. An assortment of uses for this morphology is discussed in subsequent sections.

### 3. Matrix–Fibril

The two classes just described are highly structured and require special fiber formation equipment; however, the more random matrix–fibril fibers can be made without these special provisions. The ideal situation is illustrated in the upper part of Fig. 1c, where many fine fibrils of one polymer are dispersed in a matrix of another. In a cross section, the fibrils appear random in size and location; however, because of inherent orientation processes in all fiber processes these fibrils are aligned in the axial direction. The simplicity of fabrication makes this an attractive structure for those cases in which such a configuration has the potential to achieve the objective. Adhesion is not essential to fiber integrity, and because of the finer dispersion and high interfacial area, a lack of adhesion may not be as apparent in this formation as in sheath–core structures. The size and number of fibrils depend on proportions and the rheological conditions of fiber formation. Clearly, there is a continuous spectrum of structures between this extreme and the sheath–core as the lower part of Fig. 1c suggests. A small but precise number of accurately placed fibrils in a matrix requires a structuring of the spinning fluid using special equipment. Such structures have been referred to as *islands in a sea* [5, 6, 19].

### 4. Others

Although there is no end to the numbers of possible structures, it is of some interest to point out a few miscellaneous ones because of their potential value and to illustrate the imaginative engineering displayed in this field. Into the latter category fall many combinations of the three types of structures described above, some examples of which appear in Fig. 2. A side-by-side

Fig. 2 Examples of combinations of the three morphology types shown in Fig. 1.

(a)        (b)        (c)

structure with a sheath is shown in Fig. 2a. One patent [20] claims that such a fiber has woollike characteristics, after some heat and chemical treatments, when the core contains nylon 6 and a nylon 6 copolymer and the sheath is poly(ethylene terephthalate). A sheath–core with a matrix–fibril structure for the sheath is shown in Fig. 2b. In one example of this [21], the fibrils and core were nylon 6, and the sheath matrix was polyethylene, which was subsequently dissolved away to leave a nylon 6 fiber surrounded by a number of highly entangled microfilaments of nylon 6. A similar concept is shown in Fig. 2c in a side-by-side arrangement with one or both sides having a matrix–fibril structure. In one example, it is reported [22] that the side-by-side spinning of polyesters with different molecular weights to give a crimpable composite fiber is improved by adding 5–20% nylon 6 to the side with the higher melt viscosity.

Two structures that have not been frequently discussed in the literature are illustrated in Fig. 3. The striated one in Fig. 3a can result under certain

(a)        (b)

Fig. 3 Schematic representation of two additional phase morphologies possible for blend fibers: (a) stratified; (b) interpenetrating network.

rheological situations (see Volume 1, Chapter 7) as shown by numerous photographs of rather large extrudates of blends [23–28], and it has been observed in fibers [29, 30]. Some characteristics of this structure would be particularly unique, for example, dye diffusion would take place through a series arrangement of the two phases. The structure in Fig. 3b is frequently referred to as an interpenetrating network or IPN (see Chapter 11) since both phases are simultaneously continuous. This structure has been reported to exist in a number of blends [23, 31–37]. Apparently, similar structures occur under certain rheological conditions when the two polymers are in nearly equal proportions; however, for miscible polymers with a phase boundary, this structure can be formed by the process known as spinodal decomposition [38–41] (see Volume 1, Chapter 4). It has many important possibilities, one of which is a relaxation of the need for adhesion between phases.

A familiar phase morphology in composites is a simple dispersion of spheres in a matrix. If both phases are mobile during fiber formation, such a structure will be converted into the matrix–fibril system discussed earlier. However, rigid spheres in a fluid matrix could be made into this structure. An example of this is provided by mixing a PVC latex with an aqueous poly(vinyl alcohol) spinning solution as discussed later in connection with flame retardance [42–44].

All of the morphologies described to this point have been concerned only with the distribution of the polymer in the fiber cross section, which in the ideal case does not vary with axial position along the fiber. There are reported instances in which the fiber composition or morphology has been intentionally varied along the fiber length to achieve unique effects [45].

## III.  PHASE MORPHOLOGY GENERATION AND CONTROL

Many of the possible phase organizations of two or more polymers in a blend fiber were described in the previous section. These structures are generated during fiber formation or spinning, and it is the purpose of this section to outline the basic principles needed to understand how these structures can be made. First a review of some of the standard techniques and elements commonly used in fiber spinning is presented. Then the special equipment features required to generate highly structured morphologies are discussed. This is followed by a discussion of how certain rheological factors may affect the operation of these systems. Finally, the spinning of fibers with random structures from dispersed mixtures of polymers, without the use of special equipment, is considered.

## A.   General Features of Fiber Spinning

To form any polymer into a fiber it is necessary to first make the polymer fluid. In many cases this can be done neat by merely heating the polymer to form a melt, for example, nylon or polyesters. However, some polymers do not "melt" readily, owing to very high melting points and/or lower degradation temperatures. These are dissolved in solvents and are solution spun, for example, acrylics, poly(vinyl alcohol), and most high-temperature polymers. The fluid melt or solution is pumped through a distribution system, usually with filtration, to a multiple-hole spinnerette. Fluid filaments emerging from these holes must be solidified, which is done by cooling in melt spinning. Solution spinning may be "dry," in which case the solvent is removed by evaporation, or "wet," whereby the filament is coagulated by diffusional exchange with a liquid bath, either of which solidifies the filament. Variations are possible such as "dry-jet" wet spinning in which the spinnerette is dry as in dry spinning but the filaments then enter a liquid bath as in wet spinning.

In wet spinning the individual spinnerette holes are very small (diameter $\sim 0.002$–$0.005$ in.) compared with melt spinning (diameter $\sim 0.007$–$0.012$ in.). In melt spinning the number of holes may be as small as ten, whereas in wet spinning one spinnerette may have more than 50,000 holes. Linear speeds in wet spinning are slow, of the order of 100 ft/min, while they may be more than two orders of magnitude faster in melt spinning. Dry spinning is usually intermediate to wet and melt spinning in these factors. The spinning process and the resulting fibers are greatly influenced by the many rheological, thermal, and physicochemical processes that operate during fiber formation [46–52].

After solidification, the filaments are taken up on a roll under stress and are usually stretched or drawn by a factor of 3–10 to impart a high degree of axial molecular orientation before arriving at the final take-up system. Further solvent removal occurs during these steps for solution spun fibers, with drying to remove the wash water. Owing to the mechanism of coagulation, wet-spun fibers are porous (often less than half the fiber volume is polymer) up to the point of drying, where most all porosity is usually collapsed [53–56]. The cross-sectional shape of fibers varies greatly as the examples in Fig. 4 show [11, 13, 46, 55]. Most melt-spun fibers will be round unless the spinnerette holes have other shapes used to achieve special fiber effects [1]. Most dry spinning is from round holes but the fibers are usually ribbon shaped with fat, round ends, and the descriptive term "dogbone" is often used. This shape results from fixing the perimeter of the fiber before most of the solvent is removed. Wet-spun fibers take on a variety of shapes ranging from the well-known "kidney bean" to round. This shape depends greatly on extrusion and coagulation conditions.

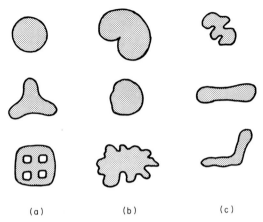

(a)                    (b)                    (c)

Fig. 4   Examples of possible fiber cross-sectional shapes obtainable from (a) melt, (b) wet, and (c) dry spinning. The nonround melt-spun examples are typical of the Antron nylon series [1, 10, 57] made from shaped spinnerette holes.

Frequently fibers are subjected to various thermal or mechanical treatments in the spinning line or afterwards. A very important response for subsequent sections is the significant shrinkage that occurs when the fiber is heated near or above its $T_g$ resulting from relaxation of the molecular orientation imported during drawing.

## B.   Mixed Stream Distribution and Spinnerette Systems

To make the highly structured side-by-side and sheath–core type morphologies requires specialized equipment beyond that described above for the distribution of the spinning fluids to the spinnerette, which is also specially designed in most cases. However, downstream of the spinnerette the process is usually the same as that for nonblend fibers. A few early references [14, 58–60] describe the basic aspects of mixed stream spinning, and a recent review [5] provides a more up-to-date summary of techniques.

The conceptual requirements for a single spinnerette hole are shown in Fig. 5. The early [13, 58, 59] conjugate jet for side-by-side spinning that keeps the two streams separated by a septum or knife edge until just above the hole where the two streams meet and flow through the hole together is illustrated in Fig. 5a. A similar arrangement for sheath–core spinning is shown in Fig. 5b that replaces the knife edge with a small tube through which the core is extruded, while the sheath flows in the annular space up to the hole. Clearly, distributing two fluids in this way to the very large number of holes required in commercial practice would be difficult. However, more

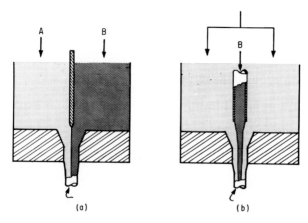

Fig. 5. Elementary stream distribution to individual spinnerette holes for producing (a) side-by-side and (b) sheath-core fibers.

advanced and simpler techniques for side-by-side spinning have been developed that are amenable to large-scale production. For sheath-core spinning the basic idea shown in Fig. 5b must be repeated for each hole although clever schemes to do this are available [5]. Basically sheath-core spinning is more difficult and expensive than side-by-side spinning. A part of this results from fabrication and placement of the tiny tubes for the core extrusion. It is not so difficult for melt spinning in which the hole sizes are large, but for the much smaller sizes involved in wet spinning it is often not practical. Eccentric sheath–core structures result from locating the core tube to one side. Core shapes other than round may be had by appropriately shaping the internal geometry of the core tube. Multiple cores can be produced by use of several tubes per hole. Such variations, of course, increase the complexity and cost beyond basic sheath–core spinning.

Advances in side-by-side spinning were made after it was realized that the knife edge in Fig. 5a was unnecessary since two highly viscous polymer fluids under proper circumstances may flow for considerable distances in a layered configuration without mixing or altering their structure. Even changes in stream diameter will not alter the relative stream configuration if done properly [60]. This makes it possible to use distribution systems such as those shown in Fig. 6, which generate either layers or concentric rings of the two fluids. It is easy to see that the pipe-in-pipe device in Fig. 6b could be used to make as many alternating concentric rings as desired. Once formed, these mixed streams may be pumped through the distribution system, with filtration, to the spinnerette. If all of the filaments are to have the same perfect side-by-side configuration, it is necessary to have a specially designed spinnerette having an outlay of holes that matches the stream configuration, as shown in Fig. 6. For example, the concentric

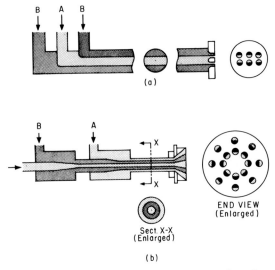

Fig. 6 Advanced systems for spinning pure side-by-side fibers through many holes simultaneously: (a) parallel streams; (b) concentric streams.

ring system requires the spinnerette to have the holes laid out on a circle with as many circles as there are fluid interfaces. The mixed stream will adjust itself until each interface coincides with a ring of holes, thus producing the desired flow situation without the septum of Fig. 5a. This process has remarkable self-adjusting features that relax the demand for precision matching of the spinnerette to the flow distribution system. There are certain problems of stability for mixed streams, however, which are treated later. An examination of Fig. 6b should explain the origin of the curved interface often seen in side-by-side bicomponent fibers (see Fig. 1a). The proportions of the two polymers are simply adjusted via the relative pumping rates of the two fluids.

When the spinnerette hole outlay does not match the mixed stream configuration [60], as shown schematically in Fig. 7, all of the fibers will not be the same. Some holes may not have an interface, and thus its fibers will be homofibers of one polymer or the other. Other holes may have an interface but the flow rate of the two polymers may not be in the same ratio as the average. As a result, a wide distribution of configuration is produced from the same spinnerette, and these are often called "mixed" or "random" bicomponents in contrast to "pure" bicomponents. For many cases such a mixture is not undesirable and may be decidedly advantageous, as described

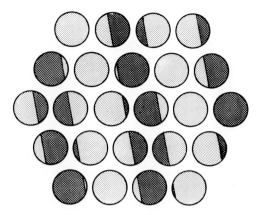

Fig. 7 Schematic illustration of "random" bicomponent fibers resulting from a lack of match between spinnerette hole outlay and mixed stream configuration.

later. For mixed bicomponents even simpler and cheaper methods of spinning can be used. A variety of simple devices, such as plate mixers, for structuring the two streams into lamellae have been described [5, 60]. The requirements for the spinnerette may also be relaxed considerably. However, attention has been devoted to matching the two to achieve a desired or optimum distribution.

## C. Rheological Considerations in Mixed Stream Spinning

Preparation of highly structured blend fibers requires the generation of appropriately structured configurations of the two spinning fluids, which must remain stable during transport up to and through the spinnerette hole. There are rheological processes that can act to alter this structuring (see Volume 1, Chapter 7) and spell the demise of the desired blend fiber morphology if proper precautions are not exercised. Similar processes are involved in the coextrusion of laminated film products (see Chapter 15). Basically, these processes stem from the desire of the flow system to seek its state of minimum energy dissipation [60–67], which for two immiscible Newtonian fluids is to have the lower viscosity fluid located in the regions of highest rate of shearing. For pipe or capillary flow the lower viscosity fluid would like to encapsulate the higher viscosity component in a configuration similar to a sheath–core arrangement. For viscoelastic fluids, the elastic character of the two components is expected to play some role [24]. As a practical matter, however, it appears that the viscosity difference dominates [62], but this may result in part from the fact that elasticity and viscosity are related to the same features of the polymer. Mixed streams with the components having identical rheological characteristics would not be expected to show instabilities of this type. However, rearrangements driven purely by interfacial tension forces would still be possible, although for highly viscous polymer fluids these appear to be much less important.

A sheath–core mixed stream with the more viscous component as the core is stable and will not tend to rearrange during flow, whereas, with a higher-viscosity component as the sheath it would be unstable. The side-by-side structure is always unstable if the two components have different rheological characteristics, and it will rearrange into a sheath–core with the lower viscosity component as the sheath [62]. The rearrangement is a kinetic process the driving force of which is the degree of mismatch in viscosities and the resistance of which depends on the absolute level of the viscosities [67]. As a result, rearrangement requires time to occur, and one may expect the flow geometry to evolve into the equilibrium configuration with axial position along the capillary [68]. Most spinnerette holes are relatively short. In wet spinning the length-to-diameter ratio $L/D$ is generally one or less, whereas somewhat larger values are used in melt spinning. Consequently, there is little time for rearrangement within the spinnerette hole itself for most commercial spinning systems. It is usually not a problem in wet spinning, but it may be a concern in melt spinning when the viscosity difference is large [62, 63].

The more likely place to encounter this problem is in the structured stream that is used to feed a multiple hole spinnerette for side-by-side filaments. There may be considerable distances over which this feed stream has an opportunity to rearrange. These problems have usually been solved by efforts to match viscosities and to minimize residence time. It is also important to eliminate flow disturbances that would create problems from yet other sources.

For non-Newtonian fluids, viscosity matching is not simple owing to the fact that the viscosity depends on shear rate. This may result in matched viscosities under one set of extrusion conditions but not others. Southern and Ballman [62, 63] and others [69, 70] have constructed clever experiments that vividly demonstrate these effects. The essence of these observations is summarized in Fig. 8. Polymers A and B were polystyrene melts with appropriately selected molecular weight distributions so that a viscosity crossover occurred within the shear rate range of interest. A relatively perfect side-by-side arrangement of Polymers A and B was introduced to the entrance of a capillary. The component configuration of these extrudates are shown schematically in Fig. 8. At low shear rates where B was more viscous, the interface between components was curved convexly for B, indicating that A was beginning to engulf B. At higher shear rates where B was less viscous, the interface has the opposite curvature since B now was

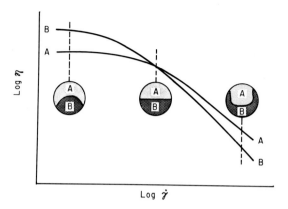

Fig. 8   Schematic illustration of the flow instability of the side-by-side configuration owing to viscosity $\eta$ differences. For non-Newtonian fluids the degree of imbalance depends on the shear rate $\dot{\gamma}$. (After Southern and Ballman [62, 63].)

trying to encapsulate A. The capillary was sufficiently short so that these rearrangements did not have time to go to completion. At the crossover point the interface was not curved owing to the stability of such a balanced system. However, it would be desirable to have the viscosities of the two components the same for all shear rates since otherwise the extrusion would only be stable for one extrusion rate. Shear rates in the spinnerette are generally much larger than in the flow distribution system, so for the situation shown in Fig. 8, the flow could not be stable in both places simultaneously.

### D.   Blend Fibers from Randomly Mixed Immiscible Polymers

Blend fibers may be prepared from spinning fluids composed of two polymers that have been mixed into an intimate dispersion. Since the process can be similar to that used for making single polymer fibers, it is more attractive economically than those described above. Matrix–fibril structures usually result although others, including those shown in Fig. 3, are possible. The special case of totally miscible polymer blends is of no interest here. The phase structure is a random one, the details of which are affected by many of the processing steps and how they are executed. The purpose of this section is to elaborate on many of the factors that may be important and to summarize the relevant findings reported in the literature.

#### 1.   Preparation of Spinning Fluid

The first step in making such blend fibers is the preparation of the spinning fluid from the two polymers. For solution processes, there are at least three distinct ways to go about this. Most common is to dry blend the two polymers in the desired proportions followed by dissolution in a solvent. A less common route is to dissolve the two polymers separately followed by mixing the two solutions by a procedure to disperse the two intimately. Finally, one polymer may be dissolved and mixed with the monomer of the second polymer, which is subsequently polymerized in the presence of the first polymer. This approach is substantially different from the other two since it opens the possibility for some grafting of the second polymer to the first, as occurs in making high-impact polystyrene (see Chapter 11–14). This could have profound effects on the subsequent structure and properties of the final fiber. For melt spinning, the two polymers are usually dry blended in pellet form, melted, and then intensively mixed. In some cases the polymers are mixed in an extruder and pelletized. The pellets are reextruded through the melt spinning machine to obtain a finer scale of dispersion [71]. In all cases the objective is usually to obtain a fine dispersion; therefore, high-intensity mixing is necessary in the spinning-fluid preparation step.

The stability of the blend fluid is an important consideration. In solution processes, long holding times may be necessary if the mixing is done continuously or in batches. Mixing may be necessary during this time to prevent gross phase separation by coalescence. In the melt spinning of condensation polymers, certain chemical reactions may occur. For example, Hayes [8] points out that terminal amine groups in polyamides may react with ester linkages in the following manner:

$$\sim\!\!\text{NH}_2 + \sim\!\!\text{O}\!-\!\overset{\overset{\textstyle O}{\|}}{\text{C}}\sim \rightarrow \sim\!\!\text{NH}\!-\!\overset{\overset{\textstyle O}{\|}}{\text{C}}\sim + \sim\!\!\text{OH}$$

This produces an undesirable random polyesteramid copolymer. One solution is to remove the terminal amines by end-capping with monofunctional carboxylic acids. However, it is possible that the small quantities of block copolymer resulting from a controlled amount of this reaction might be beneficial to interfacial adhesion (see Chapter 12). In any event, reduced residence times for the mixed spinning fluid appear desirable.

### 2.  Thermodynamic Considerations

Even though two polymers may be largely immiscible in the solid or molten state, their miscibility may be enhanced in the presence of large quantities of a common solvent as in solution spinning (see Volume 1, Chapter 2). A schematic ternary phase diagram that might apply to this situation is illustrated in Fig. 9. Here the two polymers are immiscible with one another but each is completely miscible with the solvent. The question is whether the spinning solution is in the one-phase (point $X$) or the two-phase (point $Y$) zone.

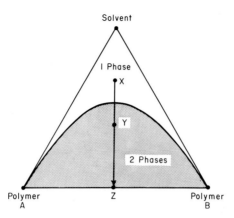

Fig. 9   A typical isothermal phase diagram for a ternary solvent–polymer–polymer system showing regions of one and two phases.

Solvent removal during fiber spinning will take the mixture to point $Z$. A process starting at $X$ is likely to be influenced also by the thermodynamics of the phase transformation as this boundary is crossed (see Volume 1, Chapter 4). It is possible that the kinetics of the phase change do not allow equilibrium structures to be formed in all cases. This transformation could proceed by nucleation and growth or spinodal decomposition, depending on the nature and severity of the driving force. Rheological processes will occur simultaneously with the formation of the phases to affect their morphology. Many interesting possibilities exist here for controlling blend fiber structure; however, too little information is available presently to warrant further speculation.

Phase boundaries are also possible in melt-spun systems when no solvent is present. In principle one might expect an analogous situation to that described above in which the polymers are miscible at high melt temperatures but two amorphous phases develop at lower temperatures, but there appear to be no documented cases of such an upper critical solution temperature for polymer blends reported to date. However, several systems (see Volume 1, Chapter 4) are known to have lower critical solution temperatures. Such systems might allow the generation of a controlled phase separation at high temperatures that can be preserved by quenching. This might be interesting because of the possible diffuseness of the interface that would result in good adhesion and stress transfer. Another interesting possibility is the growing evidence for interdiffusion between two "immiscible" polymers in the melt state [72, 73]. There is no evidence from the literature that such thermodynamic factors have ever been considered or exploited in blend fibers. However, it does appear to be an important area for future attention that might provide interesting leads to solve problems like adhesion through control of thermodynamic history during spinning.

### 3.   *Rheological Considerations and Fiber Formation*

The gross rheological characteristics of the blend spinning fluid are not important for the discussion here although there is a great deal of relevant literature [69, 74–81] (see Volume 1, Chapter 7, and Chapter 22). It is the microrheology of the two individual phases controlling stability and morphology that is of more interest. Roughly, three stages of fluid operations are involved: spinning fluid preparation as noted above, flow to the spinnerette, and fiber formation in the vicinity of the spinnerette hole.

The state of dispersion achieved in a mixing step depends on the type and magnitude of the fluid motions induced, the rheological characteristics of the two phases, and the interfacial tension (see Volume 1, Chapter 6) between the two phases. Imposed stresses cause dispersed domains to deform and break into smaller ones; coalescence is a competing process. The generation of new interfacial area requires the expenditure of mechanical energy with especially large requirements for highly viscous fluids. Because of the interest to a number of industries, these processes have been studied fundamentally [24, 31, 32] and empirically [82] (see Volume 1, Chapter 7, and Chapters 19 and 21). It is generally agreed that when the viscosity and elasticity of the minor component are greater than that of the major component, the minor component will be dispersed coarsely, whereas when these characteristics are reversed, the minor phase will be dispersed finely. When both phases have the same characteristics, the minor component will be finely dispersed. If both phases are present in roughly equal amounts, there may be no true dispersed and continuous phases. Reducing the interfacial tension would increase the fineness of dispersion and opportunities for achieving this exist through judicious use of graft or block copolymers (see Chapter 12).

After the mixing step, the blend polymer spinning fluid must be delivered by laminar flow under relatively small stresses to the spinnerette. During this transport, opportunities exist for deleterious changes in the state of dispersion. Small particles may coalesce into larger ones without compensating breakup owing to the relatively low stress conditions. Even though complex flows in the mixing step may disperse one phase in another, it was shown in Volume 1, Chapter 7 that this may not be the stable configuration for certain mixtures during simple shearing flows such as those in the transfer lines to the spinnerette. The dispersed state may evolve into a stratified or laminated structure similar to that shown in Fig. 3a, depending on the rheological characteristics of the two phases. Van Oene [24] has formulated simple rules for molecular weight distributions that may lead to or avoid this possibility. In general such changes are undesirable, and one way to avoid them is to add some form of in-line mixing upstream of the spinnerette, perhaps of the static mixer type as has been reported on by several authors [33, 83].

At the hole entrance, a rapid acceleration begins, which produces an elongational flow that deforms dispersed particles into fibrils in the manner depicted in Fig. 10. While shearing also occurs, the elongational mode is

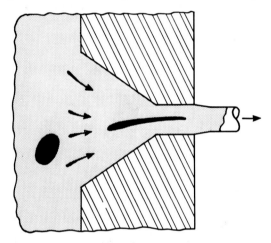

Fig. 10   Conversion of a nearly spherical particle into a fibril shape by elongational flow at the spinnerette hole entrance.

usually the most important. The extent of the particle elongation depends on the relative rheological characteristics of the two phases, the interfacial tension, and hole geometry. As a basis for comparison, a single-phase spinning fluid may be imagined to have a phantom particle of identical characteristics. This imaginary particle would deform to an extent defined by the characteristics of the fluid and the details of the flow. Quantification of this change in shape is a difficult problem in rheology; however, the qualitative changes from it for real situations are easy to predict. If the dispersed phase had a higher viscosity or rigidity than the matrix fluid, the extent of deformation would be less. A relatively rigid element, such as a latex particle, would not deform at all. If the interfacial tension representing a retractive force were large, the deformation would likewise be smaller even if rheological properties of the two phases were the same. In general, to produce fibrils with large $L/D$ ratios by this route, it is important that interfacial tensions be small and the relative viscosity or elasticity of the dispersed phase not be large compared to the matrix. It is also important to realize that fibrillar elements of fluid are unstable [84, 85] and may tend to break up into small droplets. This would be favored if the dispersed phase were low in viscosity and the interfacial tension large. Elastic fracture of an overly elastic dispersed phase is yet another possibility [86–90]. Clearly, all of these factors plus details of spinnerette hole design are important considerations to achieve the desired results.

After emerging from the spinnerette hole other factors are important to the preservation of the fibrils that have been formed. Without any restraints imposed by a fiber take-up device, the elasticity of the spinning fluid would cause some recovery of this deformation, that is, die swell [47, 50], and the fibrils would return partially to their original shape. However, the restraints of extrudate take-up will tend to retard this [50] although in the absence of rapid solidification, stress relaxation or partial recovery may occur. One might expect dry jet–wet spinning to produce shorter fibrils than wet spinning. If the disperse phase were much more elastic than the matrix phase, it could relax preferentially more than the matrix. Generally, polymer solidification after leaving the spinnerette preserves much of this structure. The draw down or jet stretch of the fluid filament prior to solidification may also elongate the dispersed phase into fibrils [91]. In melt spinning, the amount of jet stretch is so great in certain cases that it may have a larger effect on fibril geometry than the extensional flow into the spinnerette hole. In wet spinning, however, the jet stretch ratio is often less than one [92], and consequently elongational flow in the spinnerette hole must be the dominant mechanism for fibril formation.

Further changes in the fiber structure occur during drawing operations, and these are discussed in a subsequent section.

## 4.   Fiber Phase Structure

The phase structure of fibers prepared in the manner described above have been reported in detail by a number of authors for a wide range of polymer combinations and process techniques. One of the more extensively studied systems [7, 8] is a mixture of nylon 6 and poly(ethylene terephthalate) introduced in the late 1960s as a commercial product tradenamed Source for tire cords and subsequently carpets. This fiber constitutes a nearly ideal matrix–fibril system with as many as 3500 fibrils in the cross section of a single 4-denier filament. The fibrils are 100–200 $\mu$m in length with diameters of the order of 0.1 $\mu$m [8] in fibers in which one component is clearly in excess of the other. Compositions of nearly equal parts of each component are complex with a more or less binodal distribution of fibrils [7], the larger ones being nonround with dimensions 1–2 $\mu$m and smaller ones more nearly round with a diameter of 0.06 $\mu$m. Clever infrared analyses based on a reflectance technique [93] have shown that the surface consists of a larger proportion of nylon 6 than the interior; this is also evident from photomicrographs of cross sections [7, 94], and has important implications for dyeing and adhesion in tires. No doubt this factor is controllable by the relative rheological characteristics of the two polymers. In other studies [23, 26, 27, 94] it is suggested that there is a gradation of fibrillar nature across the fiber diameter, which probably arises from the inhomogeneity of the flow field that generates these fibrils. Apparently, migration of blend components in shear fields is common [95].

Kitao *et al.* [35] have described similarly prepared melt-spun blends of nylon 6 with a range of polyamides and a polyester. The structures generated are similar to those described above. The patent literature also describes similar structures [91, 96–100] for a wide range of polymers and conditions, some of which claim to give especially small fibril sizes.

Cates and White [36] give one of the most extensive descriptions of solution-spun blend fibers to be found in the open literature. They showed that fibers from mixtures of polyacrylonitrile and cellulose acetate also have structures similar to those described above. Many subsequent reports have appeared in the Soviet Union literature on this [101–103] and other systems. Issues of the translation *Fibre Chemistry* contain many papers on this subject but most do not allow general conclusions to be made.

All of the work described above indicates that the matrix–fibril structure idealized in Fig. 1c is the one usually found for blend fibers made in this manner, but there are important deviations from the ideal picture. These plus other sources [32–34] suggest that structures approaching an inter-penetrating network of phases (see the diagram in Fig. 3b) are possible. This seems to be most prevalent in mixtures of nearly equal proportions of the two components in which neither has a mandate to play the role of either continuous or dispersed phase. No other simple guidelines have emerged, to help one obtain this structure, but there is evidence that it can be manipulated by rheological means [33]. Further, no ways to characterize the extent of interpenetration seem to be available other than microscopic examination, which is limited in concept and technique. Such structures, however, have many advantages and may be desirable for many situations.

## IV.  FUNDAMENTAL CONSIDERATIONS OF BLEND FIBERS FROM IMMISCIBLE POLYMERS

In previous sections a variety of possible blend fibers and some of the details of the methods available to generate them have been dealt with in a conceptual way. How they have been or could be useful is discussed in subsequent sections. There are many fundamental factors that have not yet been considered that are important to the viability of the process and the adequacy of basic fiber properties in these examples. A few of these factors are introduced in this section. The literature does not contain in-depth studies on any of them, and as a result the following discussion must be brief.

### A.   Processing Differences and Interactions between the Two Polymers

The viscoelastic response of the spinning fluid to extrusion and subsequent filament attenuation plays an important role in process performance and fiber properties [46–56, 92, 104–108]. For blend fibers, spinning fluids of quite different rheological character may be combined into any of the phase arrangements described previously, and at the present there is a small but growing literature on two-phase viscoelastic fluids to guide process development [70]. For example, die swell has been shown to be one important response [47, 50, 92]. Some die swell results have been reported for dispersed polymer blends [26] that show complex, nonadditive responses to blend composition. Apparently no studies are available for other phase arrangements. Of more direct concern are productivity limitations such as melt or solution fracture [86–88] and the maximum take-up rate for the fiber [47]. Limited studies [91, 109] suggest that the latter limitation sometimes may be less severe in dispersed blends than for the pure components. Han [90] has reported a few results for melt fracture of dispersed blends but no unifying conclusions have yet evolved. Shaw [110] has made observations that may be relevant to sheath–core spinning. He coextruded a core of molten polyethylene surrounded by a sheath of lower-viscosity silicone grease and found that the polyethylene fractured at lower rates in this situation than without the grease. He concluded that the lower-viscosity grease reduces the shearing component of deformation in the polyethylene thus causing more extensional deformation that promotes fracture [89].

It is also recognized that the rate of polymer solidification (crystallization, coagulation, etc.) during fiber formation affects spinning behavior and fiber characteristics [89]. Almost no attention has been paid to the possibilities that may arise in blend spinning of two polymers that solidify at different rates. Some effects this has on the interface is considered in the next section. Several studies on melt-spun blends have suggested that the crystallization of one of the polymers is affected by the presence of the other polymer [7, 35, 111]. This is supported by recent fundamental studies, which show that crystalline polymers may serve as nucleation substrates for crystallizing polymers and thereby affect crystallization kinetics, morphology, and level of final crystallinity [112]. Even amorphous polymers may affect crystallization in blends by constraints they impose [113]. The partial miscibility of the two polymers also affects crystallization [114–116], the usual result being to retard the rate and reduce the amount of crystallinity (see Volume 1, Chapter 5).

In fiber drawing, under given conditions, each polymer system has a maximum draw ratio that can be achieved without breaks which is related to its $T_g$, molecular weight, and so forth [51, 92, 112]. In the absence of interactions between the two polymers, damage will occur to one of the phases of a blend with a parallel-phase arrangement if the draw ratio used exceeds the limit for this polymer although the limit for the other polymer is not exceeded. However, interactions evidently do occur since it has been observed that the maximum obtainable draw ratio may exceed that of either pure polymer [35]. The opposite situation exists as well. It is frequently observed in matrix–fibril systems [7, 35, 36] that drawing induces voids in the blend fiber because one component draws more readily than the other and there is not adequate interfacial adhesion. This has important consequences in dyeing and optical properties.

### B.   Adhesion between Phases

Adhesion between phases in blend fibers has an important effect on the final product just as in all composites. It is essential in side-by-side types, which are similar to laminated films (see Chapter 15). Some of the fundamental issues that affect polymer–polymer adhesion are examined in Volume 1, Chapter 6. If the two polymers are miscible or partially miscible, interdiffusion of the two at the interface will provide the needed adhesion between phases. However, in theory it is only necessary that the two materials wet each other for good adhesion. In many blend fibers of the side-by-side type, the two polymers are sufficiently similar, perhaps differing only slightly in comonomer contents, so that adhesion can occur by either of the mechanisms mentioned above. In some instances polymers that are entirely different chemically may adhere well to each other because of complementary polarity, which provides good wetting. However, poor adhesion between any two polymers is not uncommon, and this may limit their use in certain blends. If adhesion is marginal, properly selected molecular weights and processing conditions may be found that prove adequate. For many systems there are no conditions that are adequate. Addition of graft or block copolymers may be a solution, as discussed in Chapter 12, and this route has been used in blend fibers [102, 118]. There are indications that blending of the type shown in Fig. 2c helps, but it is not a complete solution nor is it generally applicable. Mechanical interlocking of phases such as interpenetrating networks (Fig. 3) or irregularity of the interface lessens the requirement for good adhesion.

The integrity of the interface may be influenced by differential volume or shape changes between the two phases. Sheath–core structures from nylon and polyesters provide an interesting example to illustrate the possibilities. For an as-spun fiber the nylon will probably be crystalline and the polyester still amorphous because of the large differences in their crystallization rates. Subsequent heat treatment will cause the polyester to crystallize, resulting in a significant reduction in its volume. If the polyester is the core, this differential volume change may cause loss of nylon–polyester contact at the interface. In the reverse situation of a polyester sheath, a significant increase in effective adhesion could result from the normal forces created by the polyester shrinking down on the nylon. Simple thermal expansion would produce similar but smaller effects. In wet spinning, differences in gel density of the two polymers on coagulation [48, 53–56] and their subsequent collapse on drying may lead to similar interfacial effects. Relaxation steps after spinning cause fiber shrinkage. Differential shrinkage between the two polymers leads to helix formation in side-by-side blends; however, in most sheath–core or matrix–fibril systems this mechanism for accommodating to the stresses produced is not available and extreme cases may lead to actual movement between phases. Very little information is available to assess to what extent these possibilities affect blend fibers.

## C.  Blend Fiber Mechanical Properties

The mechanical properties of fibers are important to their fabrication into useful products and their subsequent uses. The individual fiber is subjected to a complex array of deformations including pulling, twisting, bending, etc. For simplicity, the present discussion is divided into the limiting fundamental modes that contribute to the overall response in real situations. A part of the huge amount of experimental and theoretical effort that has been devoted to analyzing composite materials [12] is applicable to the mechanical behavior of two phase blend fibers. Composite theory suggests that the reversible, small deformation behavior of two-phase systems is a function of (a) the mechanical properties of the component phases, (b) the phase organization or morphology, and (c) the interfacial adhesion as shown in Volume 1, Chapter 8. However, ultimate mechanical behavior is sensitive to other factors such as defects and is less amenable to analysis. Through simple models, Nielsen [119] has provided some insight into this more difficult problem. In Chapters 13 and 14 some of the factors, in addition to the three listed above, are dealt with that contribute to the failure behavior of plastics. There are some similarities in mechanical behavior between blend fibers and blend film (see Chapters 15 and 21).

In trying to rationalize any response of blend fibers it is important to remember that the behavior of the phases may not be the same as those of fibers made from each polymer individually. These phases may be different in composition, crystallinity, molecular orientation, etc. than the corresponding fiber of the pure polymer [35, 72, 73, 111]. This fact will frustrate the use of notions from composite theory.

### 1. Longitudinal Properties

It is easier to test fibers in simple tension than in any other way, and as a result this mode accounts for the majority of the mechanical characterization that is done on fibers. The side-by-side and sheath–core types of blend fibers form essentially perfect "parallel" systems. As a first approximation, interfacial adhesion is of no consequence, and the uncrimped blend fiber will behave as a fiber blend of unconnected fibers from Components A and B. In this limit, the entire strain–strain relation for the blend fiber is the weighted average of those for the components. If the components follow the relations of stress to strain, $\sigma_A(\epsilon)$ up to an ultimate value $\epsilon_A$ and $\sigma_B(\epsilon)$ up to $\epsilon_B$ where $\epsilon_A < \epsilon_B$, then the following will describe the blend:

$$\sigma(\epsilon) = \sigma_A(\epsilon)\,\phi_A + \sigma_B(\epsilon)\,\phi_B \qquad \text{for} \quad \epsilon < \epsilon_A$$
$$= \sigma_B(\epsilon)\,\phi_B \qquad \text{for} \quad \epsilon_A < \epsilon < \epsilon_B \qquad (1)$$

Total failure will occur at $\epsilon_B$. The fact that the two are connected with some degree of adhesion may influence the behavior at yielding and beyond because of inhomogeneous deformations or because one phase imposes a different mechanism of yielding or failure on the other. While data on mechanical behavior of commercial products are available, the extensive experiments required to determine the extent of such effects have not been published. In all cases, the modulus at small deformations may be expected to follow the simple parallel model (see Volume 1, Chapter 8)

$$E = E_A\phi_A + E_B\phi_B \qquad (2)$$

Matrix–fibril systems are somewhat more complex, depending on the length of the fibrils. For infinitely long fibrils, the picture described above applies. In the other extreme, spherical inclusions, a completely different picture results, which has been described extensively in the composite literature (see Volume 1, Chapter 8). The actual situation is usually fibrils with $L/D$ values of 100–1000, which more nearly approaches the infinite case. However, there is some chance for interfacial slippage if adhesion is

very poor, and certainly the fibril ends must be regarded as stress concentrators or defects. Small-deformation behavior should still be guided by
Eq. (2). While there is no sound quantitative basis for the ultimate properties
in such systems, one can consult limiting cases for guidance. Some data on
real systems of this type are worthy of mention here. The mechanical
properties of one solution-spun system [36] of blends fall more or less
between the limits of the pure components as one would expect, but this
comparison is frustrated by a lack of suitable controls in certain instances.
The once commercial system Source seems to be very much in line with
these expectations [7]. Another series of melt-spun blends reported by
Kitao et al. [35] show the same behavior except in one very interesting
instance. This exception is the nylon 6–nylon 6, 10 system in which the blends
are all stronger than either pure fiber. The most highly drawn 50/50 blend
has a tenacity of roughly 12 gm/denier, which is about twice as large as
that obtained with either pure fiber. This has interesting commercial
ramifications but also raises the fundamental question: Why does it occur?
No doubt there is some interaction between components that leads to phase
properties not obtainable individually. There are numerous distinctions
that Kitao [35] found about this system, compared with the others studied,
but two stand out most prominently. First, the blends have a lower specific
volume than was calculated on the basis of volume additivity (no other blend
showed this), which suggests that crystallinity is higher in the blend than
in the pure components. The undrawn 50/50 blend had the same tenacity
as the undrawn pure fibers, and the higher tenacity only developed on
drawing. Perhaps this supertenacity is a result of a more nearly perfect
molecular orientation caused by and/or aided by the increased crystallinity.

## 2.   Transverse Properties

The mechanical properties transverse to the fiber axis are almost never
measured directly because of the inherent difficulty to do so. However, this
property is very important to many practical characteristics of fibers since
they are also stressed in this manner in their end use. Poor transverse
properties are responsible for fibrillation, splitting, low knot strength, poor
retention of strength on twisting as in tire cords, and other failure modes
that detract from performance and appearance. All fibers can be expected to
have poorer transverse than longitudinal strength because of preferential
molecular orientation inherent to fibers. The interface between the two
polymers is another potential source of lateral weakness for blend fibers.

Axial deformations in most blend fibers do not stress the interface as much
as transverse deformations do. The simple view taken earlier in which the
axial characteristics were treated in terms of a "parallel"-phase model would

relegate the transverse characteristics to be described by a "series"-phase model. In the latter case, if there is no adhesion, the system fails. While this is true for a side-by-side structure, the model is too extreme for all others. The general conclusion remains, though, that axial properties are relatively insensitive to adhesion compared to lateral properties.

### D. Optical Properties

The visual appearance of a fiber product depends on a number of complex optical characteristics. Some applications require an opaque fiber, which is easily made in practice by pigments, while others require transparent fibers. A more subtle characteristic is "luster." It is determined by the relative proportion of specular-to-diffuse scattering of incident light, which is influenced by both surface roughness and internal homogeneity. The basic principles of optical behavior of blends discussed in Volume 1, Chapter 9 are of direct concern to blend fiber products.

Because of scattered light, polymer blends usually have low optical clarity unless one of the following applies: (a) the two polymers are miscible, (b) the domain sizes are much smaller than the wavelength of light, or (c) the refractive indexes of the two phases are the same. When optical clarity is necessary, the latter two are the most likely routes to achieve this property for matrix–fibril systems since the first is not usually found. Frequently, the fibrils can be made small enough to meet the second condition. The size required depends on the difference in refractive index between the two phases (see Volume 1, Chapter 9). An alternate route is to match refractive indexes through chemical modification of one or both polymers. This approach is commonly practiced in the plastics industry to give clear impact-modified polymer blends of poly(vinyl chloride), acrylics, and ABS [120]. Simultaneous manipulation of both factors is a powerful way to tailor optical properties.

Interfacial voids can lead to serious losses in clarity for blend fibers since this space has a refractive index of unity compared with the 1.45–1.65 range for polymers. A few voids can render a blend fiber opaque that otherwise would be quite transparent. As a result, the proper control and design of process steps, such as drawing, are crucial to the appearance of blend fibers.

## V. MECHANICS OF SELF-CRIMPING CONJUGATE FIBERS

Textile processing into spun yarns requires fibers to have crimp to provide the fibrous mass with structural integrity during yarn manufacture, reduce fiber slippage, and increase yarn strength. This crimp or texture also gives the end product the desirable aesthetic and functional features of stretch and bulk (or cover). Numerous thermomechanical techniques of texturing have been developed [121, 122], but these are not suitable for all fibers or situations. The self-crimping possibilities of conjugate-spun fibers have been exploited for this purpose. This concept stems from the pioneering work of Horio and Kondo [123], who showed that wool has a bilateral structure composed of the orthocortex and paracortex, which are responsible for the reversible crimp triggered by moisture conditions, a property that this fiber is well known for. These authors also showed that certain self-crimping rayon fibers have a bilateral structure induced by nonsymmetrical coagulation conditions during fiber formation [124]. Sisson and Morehead [58] later showed that different viscose solutions could be coextruded through the same spinnerette hole to produce a conjugate filament. This idea was extended to coextrusion of two acrylic-type polymers, which formed the basis of the new fiber Orlon Sayelle [15]. One side of this fiber shrinks more than the other on heating, which produces a high degree of three-dimensional spiral crimp. On immersion in water, this side swells more than the other, which reduces the crimp. Recrimping occurs on drying. From here, the commercial interest in side-by-side and eccentric sheath–core type structures has been enormous. During 1974–1975, *Chemical Abstracts* indexed over 125 patents in this area. Polyamides are easily textured by thermochemical methods; however, a survey has shown at least 183 patents were issued between 1962 and 1973 involving polyamides in bicomponent fibers. Most fiber manufacturers today offer bicomponent fibers to provide texturing for fibers difficult to texture by other means, for example, acrylics, or to achieve special types of texturing [5, 6, 125]. The principles and applications of this concept are developed more fully in the following sections.

### A.   Principles of Crimp Development

Helical crimp develops in conjugate fibers by the same principle that produces lateral movement in bimetallic strips. Figure 11a is an illustration of such a bilateral element, which bends to a certain radius of curvature $\rho_0$ in response to a differential length change in the components induced by any number of mechanisms. A long strip will curl into a helix. The original theory was developed by Timoshenko [126] to describe the coiling of bimetallic strips composed of materials of different thermal expansion coefficients on heating or cooling. This basic theory has been extended and applied to conjugate fibers [5, 13, 17, 18, 60, 121, 122, 127–130] with considerable success. The radius of curvature depends on the size and shape of the

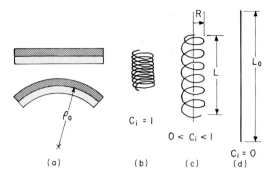

Fig. 11 (a) The bending of a conjugate material caused by differential changes in length of the two components. (b, c, d) The different helical states described in the text.

fiber, the distribution of components and their moduli, and the differential length change achieved [121, 122]. A generalization of the theory may be cast into the following form [5, 121, 122]:

$$1/\rho_0 = (K/d)\,C_p\,\Delta S \tag{3}$$

where $K$ is a constant that depends weakly on the ratio of component moduli and the overall fiber shape, $d$ a characteristic measure of fiber size (the diameter for round cross-sections), $\Delta S$ the differential change in length (or shrinkage) between the two components on a fractional basis, and $C_p$ the crimp potential, which depends on the distribution of components [121, 122]. The crimp potential is greatest for simple side-by-side types with equal proportions of the two polymers and is zero for ones with axes of symmetry such as the middle structures in Fig. 1a. For sheath–core types it depends on the degree of eccentricity.

Other parameters have been defined [13] to further describe the state of crimp in a fiber. These are visualized with the aid of the diagrams in Fig. 11b. A fiber whose straightened or contour length of $L_0$ will have $C_n$ loops:

$$C_n = L_0/2\pi\rho_0 \tag{4}$$

when crimp with a radii of curvature $\rho_0$ is introduced. The crimp frequency $C_f$ per unit of uncoiled length is

$$C_f = C_n/L_0 = 1/2\pi\rho_0 \tag{5}$$

These loops may be tightly or loosely packed, as shown in Fig. 11b, depending on the restraints imposed during crimp development. This may be characterized by the crimp index $C_i$, defined as

$$C_i = (L_0 - L)/L_0 \tag{6}$$

In turn, $C_i$ is related to the helix radius $R$ by the following equations

$$R/\rho_0 = C_i(2 - C_i) \tag{7}$$

or

$$R = C_i(2 - C_i)/2\pi C_f \tag{8}$$

which are easily derived from simple geometrical considerations. The crimp period is given by $(1 - C_i)/C_f$. Computer generated visualization techniques for fibers with complex crimp characteristics have been developed [11, 131, 132].

The differential length change required for crimp development may be reversible or irreversible [11]. The former generally refers to differential changes in swelling of the two components by water as humidity conditions are cycled. The latter usually involve differential shrinkage induced by heating the fiber. This may be made somewhat reversible by crimping with heat and pulling out the crimp by further drawing. A second heating will reinstate the crimp. A wide variety of techniques may be used to develop the differential shrinkage illustrated in Fig. 12, and numerous examples

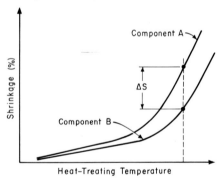

Fig. 12   Schematic diagram showing the different shrinkage responses of two fibers as a function of heat-treatment temperature.

follow. Obviously two different generic polymers [133, 134] such as polypropylene–nylon 11 or poly(ethylene terephthalate)–nylon 6 would produce the situation in Fig. 12. This approach is often frustrated by a lack of adhesion between widely different polymers, and hence it is common to use polymers more closely related. Random copolymers are the most common for this purpose. Most acrylic fibers are copolymers of acrylonitrile with vinyl acetate or methyl acrylate, and to make bicomponent products, two polymers that differ slightly in comonomer content [135] are used. For polyesters, one polymer may have some of its terephthalic acid replaced by isophthalic acid [136]. Copolymers are also used in polyamide systems [13, 137]. One patent describes the cospinning of polypropylene with a block copolymer of propylene and ethylene [138], while another [139] describes use of a cellulose ether grafted with acrylonitrile for cospinning with polyacrylonitrile. Others describe the use of branching [140] or cross-linking [141]. An attractive route is to cospin chemically identical polymers that differ only in molecular weight [142]. A variant on this is to spin only one polymer but to expose part of it to conditions that cause molecular weight degradation and recombine the two streams for coextrusion [143]. Self-crimping fibers may also be made from one polymer by proper use of additives on one side. Silica often reduces fiber shrinkage [144, 145], whereas compounds like tris(2,3-dibromopropyl) phosphate increase shrinkage [145]. The spinning of matrix–fibril blends as one of the components has also received attention [22, 146–148]. Any of these methods that will develop the needed difference in shrinkage is potentially viable because adhesion should be quite good.

## B. Bicomponent Fibers for High Bulk

A "bulky" fabric or carpet consists of kinky fibers that occupy a lot of volume. Because of their contour, they will not pack together efficiently. As a result, it takes a smaller amount of fiber to "cover" the same area or fill the same volume compared with fibers that are straight. Companion benefits are good insulation qualities, a stretchy behavior, and a nice feel. Wool has this characteristic naturally. Helical crimp from bicomponent fibers is more energetic in resisting and recovering from deformation than most monocomponent fibers that have been mechanically crimped. As a result, there are many commercial products based on this concept. The following illustrate some examples from the United States [5]: acrylic carpet fibers (Acrilan 71 and 94), acrylic knitted fabrics (Acrilan 57, Orlon 21 and 24), polyolefin knitted fabrics (Herculon 404), nylon for stockings and pantyhose (Cantrell).

Ideally, the volume occupied by a fiber with helical crimp is determined by the helix radius $R$ shown in Fig. 11a and Eq. (8). However, systems of identical, perfect helixes will tend to interpenetrate severely and thus reduce the bulkiness. Real fibers naturally develop crimp reversals that tend to retard their intermeshing, as illustrated in Fig. 13 [5, 60]. The perfect helix will develop only when each fiber is free to rotate and twist as needed. When large numbers of fibers are crimped together in a tow, restrictions prevent this movement, and the helix reverses itself in order to accommodate to the inability of the fiber to rotate freely. Crimping conditions can influence the frequency of crimp reversals, and this phenomenon can be used to combat the intermeshing problem. Another very interesting way to solve this problem is to

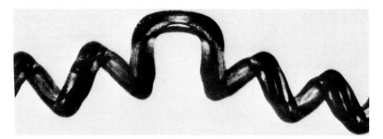

Fig. 13  Photomicrograph of a crimp reversal in an acrylic bicomponent fiber. (Courtesy of Monsanto Textiles Company.)

introduce other random features to the crimp by varying $C_i$ from fiber to fiber. Mixed-stream spinning can produce "random bicomponent" fibers as described earlier (see Fig. 7) in which there is a distribution of the proportions of the two components in the fibers. This creates a distribution in the crimp potential $C_p$ from fiber to fiber (see Fig. 14). This results in a distribution of crimp parameters (frequency, amplitude, and period) among the fibers that greatly reduces the problem of helix intermeshing. At least one commercial product line [149, 150] employs this principle.

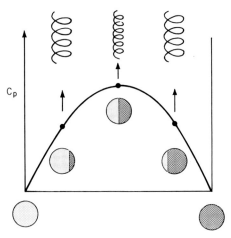

Fig. 14  Effect of component proportions on the crimp potential parameter and resulting crimp structure [60].

## C.  Biconstituent Fibers for High Extensibility

All crimped fibers have a certain amount of stretchiness, but there are applications that require extensibilities far greater than that normally found in textured fibers. These needs are often filled by elastomeric fibers of the spandex (polyurethane) type usually in conjunction with another fiber in a complex, composite yarn [151, 152]. In recent years, interest has focused on achieving these effects by combining the spandex material into each fiber through conjugate spinning. A side-by-side biconstituent fiber from polyurethane and nylon 6 known as Monvelle, which has certain unique characteristics [151–153], is now available. Numerous patents [154–156] describe similar combinations of polyurethanes with acrylics, polyesters, polyolefins, and the like.

For these structures, the very high shrinkage of the polyurethane component causes a very tight crimp with closely packed coils, as shown in Fig. 15. This side view of a crimped Monvelle fiber shows the diameter of the

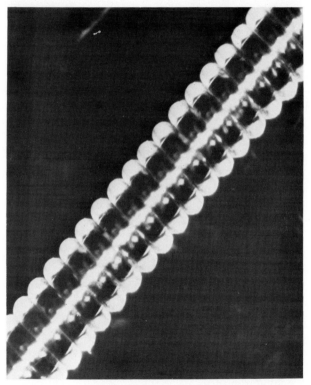

Fig. 15   Side view of a tightly coiled helix of Monvelle side-by-side biconstituent fiber of nylon 6 and a segmented polyurethane [151–153]. (Courtesy of Monsanto Textiles Company.)

cylinder of coils is only three to four times the diameter of the fiber, which is unlike most of the cases described earlier in which the coil radius $R$ was large compared with the fiber radius $r$. The finite volume of the fiber prohibits developing a perfect crimp index of unity; instead, the fiber shown has a value of $C_i \sim 0.91$. Such a structure may be stretched like a coil spring. Figure 16 is an illustration of the coils of Fig. 15 after application of a modest load. This fiber would have to be stretched by a factor of ten to uncoil it completely. After uncoiling, the slope of the force–elongation curve would be given directly by the elongational modulus of the fiber, whereas the initial slope governing small extensions such as that shown in Fig. 16 would be much less. The following formula describes the load ($F$) deformation ($\epsilon$) characteristics of simple, monocomponent coil springs [157]:

$$F = (r^5 G / 2R)\,\epsilon \qquad\qquad (9)$$

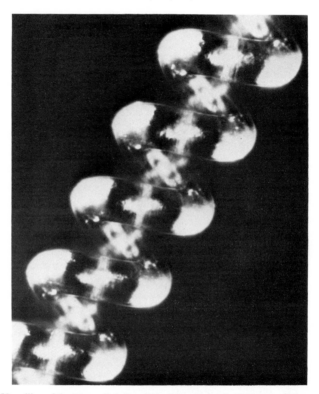

Fig. 16    Uncoiling of the Monvelle helix of Fig. 15 under load. (Courtesy of Monsanto Textiles Company.)

which in the limit that $E = 3G$ predicts the ratio of these slopes (fully extended/initial) to be $6\pi(R/r)^3$. For the fiber in Figs. 15 and 16, $R/r \sim 3.5$, and this ratio is of the order of 800. The load–extension curve for such a coil is concave upward, starting from a small slope and gradually building to a much higher value at very high deformations as described by more complex analyses than given here [13, 17, 158, 159]. This stretch power has proved valuable in hosiery, especially of the support type, making it possible to fit most legs with relatively few sizes [152].

## VI. PROBLEM SOLVING AND NEW PRODUCTS VIA POLYMER BLENDING

In this section several examples of significant commercial interest have been selected to illustrate how polymer blending in fibers offers a way to solve problems and generate new products. These examples do not comprise an exhaustive list, nor in some cases is blending the only route to achieve the desired result, as will be pointed out. Often, one polymer simply serves as a convenient fiber additive that has advantages over other potential additives because its high molecular weight makes it less prone to loss by evaporation or leaching and thereby more permanent.

### A. Permanent Antistatic Synthetic Fibers

The buildup of static electrical charges has been one of the severe deficiencies of synthetic fibers that is well known to everyone who has walked over a nylon carpet when the humidity is low. Many fundamental approaches to this problem have been considered [57, 160]. While many factors are involved [161], it is clear that the low electrical conductivity of most synthetic polymers, which prevents the rapid decay of electrical charges induced by triboelectrification, is the main problem. Early attempts to alter this centered on the application of conductive finishes to the fiber surface [161]; however, this solution was not lasting since the finish is rapidly removed by use and cleaning. A novel approach described in two key patents [162, 163] based on polymer blending offers a more lasting solution. These are based on the fact that polyethers like poly(ethylene oxide) have an innate electrical conductivity orders of magnitude higher than most other polymers. Binks and Sharples [164] have shown that this stems from an inherent ionic process rather than from impurities such as water. Poly(ethylene oxide) typically has a specific conductivity of the order of $10^{-10}$–$10^{-9}\Omega^{-1}$ cm$^{-1}$ compared with the $10^{-15}$–$10^{-14}$ range for nylon. This fact has been exploited by making a matrix–fibril structure from a poly-(ethylene oxide)-related polymer dispersed in a matrix of nylon or polyester. This significantly increases the electrical conductivity of the fiber and reduces the extent of the undesirable discharge. Variations on this approach have been the basis for over 60 patents abstracted in the 1974–1975 period. Since the olefin oxide polymers are usually water soluble, some attention has been devoted to reducing their extraction during washing. Only a few fundamental studies have been reported [6, 165].

This concept has resulted in at least one commercial product (22N antistatic nylon 66 from Monsanto, later called Cadon and Ultron) [166].

According to a patent [163], the polymeric additive is a castor oil derivative with the general structure

$$CH_2—A_x$$
$$|$$
$$CH—A_y \quad\quad —A_x = \quad —O—\overset{\overset{O}{\|}}{C}—(CH_2)_{10}—CH—O—(CH_2CH_2O)_x—H$$
$$| \quad\quad\quad\quad\quad\quad\quad\quad\quad\quad\quad\quad\quad\quad\quad |$$
$$CH_2—A_z \quad\quad\quad\quad\quad\quad\quad\quad\quad\quad\quad\quad\quad C_6H_{13}$$

**(I)**

Cross-sectional and longitudinal surface views of this commercial fiber are shown in Fig. 17. The structure is clearly of the matrix–fibril type. One

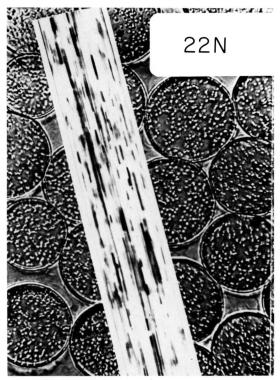

Fig. 17   Superimposed cross-sectional and longitudinal surface photomicrographs of commercial 22N antistatic fiber. (Courtesy of Monsanto Textiles Company.)

test to determine the antistatic qualities of a fiber is to induce a static charge on a fabric or carpet by a rotating drum and observe the rate of charge decay. The above-mentioned patent [163] gives some data from this type of test, which are shown in Fig. 18 as the half-life for static charge decay. Only 10% of the above additive reduces this decay time by more than

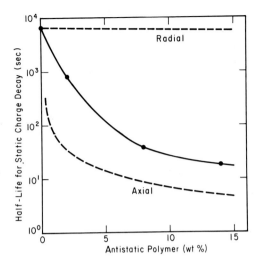

Fig. 18  Effect of the amount of antistatic polymer additive on half-life for charge decay. Dotted lines were computed assuming limiting resistance was radial or axial conduction in a perfect matrix-fibril composite in which the fibrils are $10^4$ more conductive than the matrix [163].

two decades. One way to understand these results is to view the fiber as the resistor in an RC electrical circuit although this is surely an over simplified model. The dotted lines in Fig. 18 were computed from this model [167], assuming that the limiting resistance was either axial (lower curve) or radial conduction (upper curve) in a perfect matrix–fibril structure in which the fibrils are $10^4$ times more conductive than the matrix. These limiting cases nicely bracket the experimental data suggesting that combinations of axial and radial conduction control the charge decay. In reality, the $L/D$ for these fibrils is not infinite (see Fig. 17) as the axial curve assumes. In addition, the limitations of the basic model cannot be ignored.

Other blend approaches to making antistatic fibers have been described [168].

### B.  Improved Flat-Spotting Behavior of Nylon Tire Cord

The introduction of high-tenacity tire cord in the early 1950s made a significant impact on the tire industry, and in time nylon essentially displaced rayon from this market. However, nylon has the annoying characteristic of "flat spotting" familiar to most automobile drivers. Several analyses [169, 170] showed that the source of this problem is the low value of the $T_g$ of nylon and its effect on the modulus of the fiber. Frictional processes heat the ordinary tire to the vicinity of 70–80°C, which is above the nylon $T_g$, and as a result the tire cord modulus is significantly lower at operating temperatures than at ambient temperature. On cooling at rest, the tire cord drops below its $T_g$ and stiffens, which gives the foot of the tire a temporary flat spot until, in use, the tire warms up again. The effect of moisture, sorbed before tire construction, on the nylon $T_g$ is also a factor [170]. Clearly a solution would be a fiber with a higher and moisture-insensitive $T_g$ such as poly(ethylene terephthalate) (PET). However, this fiber was not available in this market at that time because the technology did not exist for bonding it to the rubber. This technology now does exist, and poly(ethylene terephthalate) has largely displaced nylon from the passenger tire market.

Before the advent of adhesive systems for polyesters, many approaches were considered for modifying nylon to remove the flat-spotting deficiency. These included nylons with higher $T_g$ [170] and block copolymers of nylon [171]. Blends played a role, however, and ultimately some commercial products were offered [7–9, 169, 171]. The idea was this: a parallel array of nylon with a polymer with a less temperature-sensitive modulus would produce a composite with a less temperature-dependent modulus according to Eq. (2). However, if the second polymer were poly(ethylene terephthalate), the exterior of the fiber should be mostly nylon in order to obtain adhesion of the tire cord to the rubber. As described earlier, this condition can be built into matrix–fibril systems of the nylon–polyester type once offered for this application [7, 169]. An interesting comparison of nylon–polyester composites in different configurations has been reported by Fontijn [9] in which he considered the many different factors important to a tire cord. On balance, sheath–core and properly constructed matrix–fibril structures performed similarly. A side-by-side structure would be expected to develop poorer adhesion because of the exposed portion of polyester on the surface.

The development of adhesion systems for polyesters and the advent of radial tires that permit the use of other materials such as glass and steel have greatly reduced the need for the ingenious blend products of the 1960s produced in response to this challenge.

## C. Improved Hot–Wet Mechanical Properties

Piece dyeing of some fibers poses a problem similar to the flat-spotting situation in the previous section. Dyeing usually requires the immersion of the fabric or carpet into boiling water for a period of time to achieve dye penetration into the fibers. These hot–wet conditions carry the fiber above its $T_g$ (a necessity to achieve reasonable dye diffusion rates), and in polymers with low or poorly developed crystallinity such as acrylics this causes a severe reduction in modulus. The limp fiber under these hot–wet conditions becomes more rigid again after removal from the dye bath, leaving the article with undesirable deformations that detract from its appearance. More crystalline fibers do not suffer this problem so severely because their crystalline phase preserves mechanical stability under the dyeing conditions (note that dye diffusion is through the amorphous phase, which is above $T_g$).

One solution to this problem is to blend with a less heat–moisture sensitive polymer and make a fiber with a matrix–fibril structure. Very dramatic improvements can be contemplated. For example, if the original hot–wet modulus is a very low (0.5 gm/denier), Eq. (2) predicts that blending 25 % by volume of a polymer with a modulus of 18 gm/denier will increase the hot–wet modulus by one order of magnitude. However, this second polymer must be capable of being spun, drawn, etc. simultaneously with the first polymer.

### D.  Flame Retardance

The introduction of stringent safety laws has stimulated the development of fibers that resist combustion on exposure to high temperatures and flames [6]. Some polymers used in fibers such as highly halogenated modacrylics, vinyls, and most high-temperature-resistant polymers have an acceptable resistance as a natural part of their chemical makeup. However, these fibers by no means fill all the requirements of the market, and methods to endow existing fiber types with some level of flame retardance are being sought. Flame-retardant finishes for the natural fibers have been used for some time, but these treatments are rarely permanent enough. Clearly it is more desirable to build in the needed flame-retardant properties, and this is possible for synthetic fibers. In some cases this may be done through co-polymerization, but incorporation of additives that accomplish this is a more versatile approach [172]. Many of these additives are polymers, and this is the part of this overall approach of interest here. Polymer blending for this purpose has been the subject of much activity, as indicated by over 40 patents abstracted in this area during 1974–1975. Fiber burning is a complex phenomenon, and numerous factors such as tendencies to drip, char, or retract away from the flame play an important role in the overall safety considerations. Additives may alter burning characteristics chemically by producing incombustible gases, cooling the flame by endothermic processes, or trapping radicals [173]. The physical effects of additives may also be important. In many cases the physical state of dispersion or solubilization of these additives is not important for flame resistance because the flame-retardant reactions occur in the gas phase [173]. However, considerable research on degradation of blends [174–181] has shown that the radical

*D. R. Paul*

fragments from additives may or may not interact with the polymer they are mixed with to alter its decomposition kinetics. In general, very little fundamental information is available to guide the selection and use of additives for achieving the desired results.

A major consideration in using polymeric flame retardants in fibers is their compatibility with the process used to make the fiber. Solution-spun fibers from a water-based solvent system may employ a latex of the flame retardant. For example, poly(vinyl alcohol) is spun from water solutions, which readily accept lattices of highly halogenated polymers such as those based on vinyl chloride and/or vinylidene chloride [42]. This is the basis for some commercial fibers that apparently have excellent flame retardant characteristics [6]. Acrylic fiber processes that use an aqueous salt solution as solvent are also amenable to the latex route [182]. Processes that use organic solvents will not accept lattices usually because either the water from the latex will precipitate the base polymer or the organic solvent will dissolve the latex particles.

While acrylic polymers may be made flame retardant by incorporation of halogenated monomers such as vinyl chloride, vinyl bromide, or vinylidene chloride, into the chain, there are many cases in which it is undesirable to do so. As an alternative, a highly halogenated polymer may be blended with the base polymer in the spinning solution prior to spinning [183]. This gives certain advantages in physical characteristics of the fiber over copoly-merization and simplifies polymerization and inventory logistics in a plant that makes many products, since the same base polymer may also be used for other products. In this approach, the flame-retardant modifier must be similarly soluble and processible as the base polymer.

Halogeniated vinyl-type polymers, while effective flame-retardant additives, cannot be cospun with polymers such as polyesters because they lack the thermal stability needed to endure melt spinning temperatures of 250–300°C. As a result, other classes of polymers must be used. Some halogenated additives that have been considered are poly(tribromoneopentyl acrylate) [184], poly(vinylidene fluoride) [185], and copolyesters with brominated terephthalate residues [186]. However, most attempts to make polyesters flame retardant have resorted to the flame-retarding characteristics of phosphorus [187]. While a large number of phosphorus compounds have been considered [172], many of the patents on polymer blends have used members of the generalized family shown below [188–195]:

$$X = \text{—}\langle\bigcirc\rangle\text{—} , \text{—}\langle\bigcirc\rangle\text{—}\langle\bigcirc\rangle\text{—} , \text{—}\langle\bigcirc\rangle\text{—}Z\text{—}\langle\bigcirc\rangle\text{—}$$

$$Z = \text{—O—}, \text{—S—}, \text{—SO}_2\text{—}, \text{—}\overset{\displaystyle CH_3}{\underset{\displaystyle CH_3}{C}}\text{—}, \text{etc.}$$

Very little information is available on these polymers regarding their physical characteristics, efficacy as flame retardants, processibility, or the effect they have on other fiber properties.

Less attention has been paid to blending for flame-retarding nylon fibers since they are not as naturally deficient in this regard as polyesters or acrylics.

In most instances, it is not likely that the flame-retardant polymeric additives used in fibers are miscible with the base polymer. The phase morphology of blends in which one polymer is in latex form will be of the spherical dispersion type, whereas most others will probably be of the matrix–fibril kind. Minor attention has been paid to achieving flame retardance by blending into other structures such as the sheath–core type [196]. It would be interesting to see a detailed comparison of the influence of blend structure on overall performance.

## E. Dyeability

Most fibers are colored to achieve the pleasing effect in clothing, carpets, home furnishings, etc. that makes everyone's surroundings more aesthetically appealing. Coloration would be relatively simple if this were done during fiber manufacturing, but this would create an enormous scheduling and inventory situation. As a result, it is more logical to have the fabric or carpet manufacturer color his products as needed, and dyeing is the only practical means. In some instances, fibers are producer dyed or pigmented, but for the most part fibers must be capable of being dyed by the processor. This involves diffusing large dye molecules throughout the fiber interior to achieve this color. To attract enough dye into the fiber to develop the depth of shades desired and to hold it there during laundering, many synthetic fibers today have ionic dye sites to attract oppositely charged dyes. Other mechanisms such as reactive and disperse dyeing are also used. To accept a cationic or basic dye the fiber must have strong acid groups in its structure, whereas anionic or acid dyes require basic groups. Condensation polymers such as nylon have unreacted end groups that can serve this purpose. Addition polymers like the acrylics may have sulfonic acid groups at the chain ends resulting from redox initiators. Alternatively, ionic monomers may be copolymerized into the polymer. Apart from this, dyeing also involves the kinetic process of diffusion, which may be the limiting factor in commercial dyeing operations. Two aspects of dyeing are of interest in this chapter on blends. First, blending is an additional route to incorporate dye sites into a fiber and to achieve certain special effects. Second, products made by blending for another purpose such as those considered above have to be dyed, and, therefore, transport of dye in composite structures is of concern (see also Volume 1, Chapter 10).

To illustrate the latter, a matrix–fibril fiber made for any of the reasons given earlier is considered here. It is likely that the fibrillar phase will not accept any appreciable dye, whereas the matrix phase is readily dyeable. This would be true of an acrylic with poly(vinyl chloride) dispersed in it as a flame retardant or for nylon with polyester fibrils [8]. The equilibrium dye uptake for the blend fiber will be lower than that of the unblended matrix by a factor $(1 - \phi)$, where $\phi$ = volume fraction of fibrils. The kinetic situation would be dye diffusion through a matrix with dispersed impermeable cylinders, normal to the diffusion flux, that act as obstructions. Barrer [167] has considered this problem experimentally and his results show that blending lowers the effective diffusion coefficient approximately by the factor $(1 - \phi)/(1 + \phi)$. For $\phi = 0.35$, the initial dye sorption rate and equilibrium sorption will be about 45 and 65%, respectively, of that for the fiber without the obstructing fibrils. However, as a practical matter the blend fiber may have internal voids associated with the interfacial zones as pointed out earlier, which may make it dye even more readily than the fiber without fibrils [7]. Several authors [13, 197] have discussed the dyeing characteristics of self-crimping conjugate fibers.

Next, it is of interest to consider the different ways in which blending can be used to achieve dyeability (over 45 patents for this purpose were abstracted during 1974–1975). The most common situation is to add a second polymer with dye sites to a base polymer that has none.

Before going into detail on this, some other novel approaches will be mentioned. One example is a nylon sheath–polyester core structure [118]. Polyesters are much more difficult to dye than nylon; however, polyesters are cheaper and have some physical property advantages. Through this combination one can manufacture a fiber with improved properties and low cost while retaining the dyeability of the nylon, since for many applications dyeing of the nylon sheath will provide adequate coloration. Switching the location of the two components would yield a fiber that dyes as poorly as polyester without affecting greatly the other properties or cost. Special dyeing effects have been sought by mixing polymers that dye differently, such as a side-by-side structure in which one component accepts only acid dyes while the other only accepts basic dyes [198]. These fibers can be dyed with either acid or basic dyes or both to give a two-color effect.

It is difficult to build dye sites into some polymers, for example, polypropylene, directly through polymerization. In other cases, it is merely advantageous in terms of the process not to do so. In these situations dye sites are often added by blending in a small amount of a second polymer that has a high content of dye sites. However, simply adding dye sites is not enough since dyeing is a kinetic process. Some general requirements for this second polymer are that it should be readily processible with the base polymer, resistant to leaching, and capable of being highly dispersed in the base polymer [199]. In fact it is preferable that the two polymers be miscible. Poor dyeing rates may be expected if the polymer with the dye sites exists as well-defined discrete phases in a matrix of the undyeable base polymer since this would require the dye to diffuse through the matrix (where dye "solubility" is low) to reach the dye sites. In the extreme case, the kinetics for this would be very slow. A molecular dispersion of the two would be equivalent to having the dye sites as part of the base polymer molecule. As a result, some degree of "compatibility" of the two polymers is essential [199]. An additional advantage of blending in dye sites rather than having them part of the molecule of the fiber-forming polymer is that the number of dye sites can be varied without altering the molecular weight or composition of the main polymer as the latter requires. Some comments on specific systems follow.

Numerous routes have been explored to make a dyeable polypropylene fiber [200, 201]; however, at least one commercial product has been announced that achieves this by polymer blending [199]. A very interesting paper [199] describes the tradeoffs involved in selecting an additive polymer for this purpose. Basic dyes have poor lightfastness in polypropylene, so it usually is desirable to add cationic dye sites. Many polymers have been considered for this, but those containing vinyl pyridine appear most popular [199].

Poly(ethylene terephthalate) fibers are notoriously difficult to dye because of their hydrophobicity and lack of strong dye sites [201]. Strong dye sites may be added through copolymerization with monomers such as [202, 203]

$$NaSO_3 - \bigcirc \begin{matrix} -COOH \\ | \\ COOH \end{matrix} \qquad \begin{matrix} CH_3-CH-CH_3 \\ | \\ N \\ \diagup \quad \diagdown \\ HOCH_2CH_2CH_2 \qquad CH_2CH_2CH_2OH \end{matrix}$$

(III)                                                      (IV)

but this route dramatically deteriorates important thermal properties of the fiber [201]. An alternate approach is to incorporate rather significant quantities of such monomers into a polyester, which is then blended in small quantities into the fiber. At the same final dye site content, under ideal conditions, the effect on fiber thermal behavior by blending should be quite negligible compared with that occurring on copolymerization.

Several commercial acrylic fibers have been made dyeable through the blend route [4]. The additive polymer usually contains acrylonitrile with appropriate amounts of ionic comonomer selected to give among other things the desired type of dye sites. Some frequently mentioned monomers for this purpose are [204]

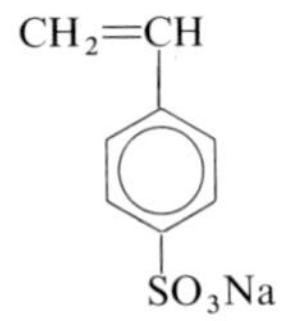

$$CH_2{=}CH$$

$SO_3Na$

**(V)**

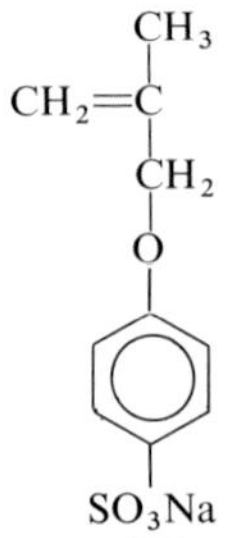

$$CH_2{=}C{-}CH_3$$
$$CH_2$$
$$O$$

$SO_3Na$

**(VI)**

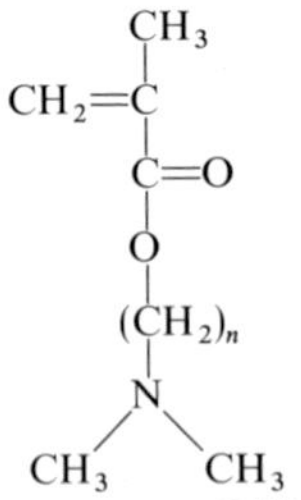

$$CH_2{=}C{-}CH_3$$
$$C{=}O$$
$$O$$
$$(CH_2)_n$$
$$N$$
$$CH_3 \quad CH_3$$

$n = 2{-}4$

**(VII)**

$$CH_2{=}C{-}CH_3$$
$$C{=}O$$
$$O$$
$$(CH_2)_n$$
$$(CH_3)_3N^+SCN^-$$

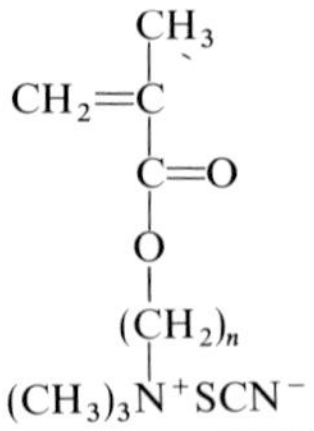

$n = 2{-}4$

**(VIII)**

## F.  Ultralow-Denier Fibers, Synthetic Fibrils, and Other Fibrillated Products

In this section brief consideration is given to some special cases in which the two polymers in a blend fiber are ultimately split apart to achieve a different kind of product than those sought previously. As a result, cospinning is simply a means to achieve this product. Previous discussions emphasized that good adhesion between the phases is desirable although not always attainable. Poor adhesion is used to advantage here.

There is a need for fibers or fibrils much finer than can be spun by ordinary processes. One example is synthetic polymer pulp for paper making [91]. Numerous patents [96–100] describe ways to make such fibers by cospinning incompatible polymers into a matrix–fibril structure and then dissolving away the matrix to leave very fine fibrils with diameters often less than 0.1 $\mu$m. An alternate route for synthetic paper fibrils is to fibrillate the blend mechanically [96], which is easy to do if adhesion is poor [205] and has the advantage of utilizing both polymers.

A somewhat different fibrillated product may be made from sheath–core structures in which the core has points that extend to the fiber surface. For example, the core may be a star [206] or a cross, as illustrated on the extreme right in Fig. 1b. Fibrillation of these structures splits the two components apart to make a fiber–fiber blend. One commercial product, named Belima, is based on the shape in Fig. 1b [6]. Its cross is nylon and the bays are polyester. On separation, a fiber–fiber blend results in which each composite fiber contributes one fiber with a cross shape and four fibers shaped like a triangle with curved sides. The final product is said to be similar to silk in some characteristics [6].

## G.  Miscellaneous

This chapter is concluded with the brief mention of an assortment of patents, which illustrate that the possible effects and products from the blend fiber concept is almost limitless.

Most synthetic fibers have low moisture sorption, which is a contributing factor to their less comfortable feel than natural fibers. To alter this, hygroscopic polymers have been blended into the base fiber [207, 208]. However, appreciable swelling of the fibrils in a matrix–fibril structure could lead to delustering or possibly splitting. Some commercial fibers now achieve a "soil-hiding" characteristic (e.g., Ultron, Cadon, and Antron III) by internal light reflections from voids left in the fiber after fibrils from a matrix–fibril structure are leached out [168]. Other related optical effects have also been claimed [209]. One patent describes incorporation of a fungicide, poly(tributyl itaconate), with polypropylene into a matrix–fibrillar fiber for fishing nets [210].

Sheath–core structures offer many possibilities for clever product applications. An important development in nonwoven fabrics has been the so-called heterofil fibers, which have high-melting polymers as cores and lower-melting ones as the sheath to aid bonding [1, 211]. Basically the core is chosen for strength while the sheath is selected to provide an easier means to obtain fiber-to-fiber connections either by heat or solvent bonding. Improved fiber optics has been claimed by the appropriate selection of polymers in a sheath–core arrangement [212]. In one case a fragrance was added to the core polymer, and the sheath acted as a release-rate-controlling membrane [213]. In another case, an ultraviolet stabilizer was added to the sheath to protect the sensitive core polymer [214]. Numerous patents describe porous hollow fibers overlaid with a thin sheath for membrane applications [215] (see also Volume 1, Chapter 10). And on the list could and probably will go in the future.

ACKNOWLEDGMENTS

The author gratefully acknowledges the information and understanding provided by the following individuals: P. H. Hobson, J. H. Saunders, Y. G. Bryant, N. A. Ednie, and P. V. Papero.

# REFERENCES

1.  R. W. Moncrieff, "Man-Made Fibers," 6th ed. Wiley, New York, 1975.
2.  J. W. S. Hearle, P. Grosberg, and S. Backer, "Structural Mechanics of Fibers, Yarns, and Fabrics," Vol. I. Wiley (Interscience), New York, 1969.
3.  "Encyclopedia of Textiles" (American Fabrics Magazine Editors), 2nd ed. Prentice-Hall, Englewood Cliffs, New Jersey, 1972.
4.  J. W. S. Hearle, *Skinner's Silk and Rayon Record* **32**, 46 (January 1958).
5.  *Ciba-Geigy Rev.* No. 1 (1974).
6.  R. Meredith, *Textile Progr.* **7**(4), (1975).
7.  P. V. Papero, E. Kubu, and L. Roldan, *Textile Res. J.* **37**, 823 (1967).
8.  B. T. Hayes, *Chem. Eng. Progr.* **65** (10), 50 (1969).
9.  W. J. Fontijn, Properties of poly(amide)-poly(ester) composite yarns in dependence on the polymer distribution, presented at *Nat. Amer. Chem. Soc. Meeting, 159th, Houston, Texas, February 22–27* (1970).
10.  E. E. Magat and R. E. Morrison, *J. Polym. Sci. Part C* **51**, 203 (1975).
11.  O. Heuberger and A. J. Ultee, Bicomponent fibers, presented at the *Int. Man-Made Fiber Conf., 13th, Dornbirm, Austria, September 10–12* (1974).
12.  J. A. Manson and L. H. Sperling, "Polymer Blends and Composites." Plenum, New York, 1976.
13.  E. M. Hicks, E. A. Tippetts, J. V. Hewett, and R. H. Brand, *in* "Man-Made Fibers: Science and Technology" (H. F. Mark, S. M. Atlas, and E. Cernia, eds.), Vol. I, p. 375. Wiley (Interscience), New York, 1967.
14.  R. A. Buckley and R. J. Phillips, *Chem. Eng. Progr.* **65** (10), 41 (1969).
15   E. M. Hicks, J. F. Ryan, R. B. Taylor, and R. L. Tichenor, *Textile Res. J.* **30**, 675 (1960).
16.  E. A. Tippetts and J. Zimmerman, *J. Appl. Polym. Sci.* **8**, 2465 (1964).
17.  R. H. Brand and S. Backer, *Textile Res. J.* **32**, 39 (1962).
18.  R. H. Brand, *Textile Res. J.* **41**, 70 (1971).
19.  M. Okamoto, British Patent 1,325,776 (to Toray) (1973) [*Chem. Abstr.* **80**, 16297 (1974)].
20.  T. Mashiko and I. Sasaki, Japanese Patent, 38,454 (to Teijin) (1972) [*Chem. Abstr.* **80**, 84582 (1974)].
21.  T. Yamane, Japanese Patent, 20,424 (to Kuraray) (1974) [*Chem. Abstr.* **81**, 79244 (1974)].
22.  K. Hirose, Japanese Patent, 06,818 (to Unitika) (1975) [*Chem. Abstr.* **83**, 12039 (1975)].
23.  C. D. Han and T. C. Yu, *J. Appl. Polym. Sci.* **15**, 1163 (1971).
24.  H. Van Oene, *J. Colloid Interface Sci.* **40**, 448 (1972).
25.  J. L. White, R. C. Ufford, K. Dharod, and R. L. Price, *Amer. Chem. Soc. Organ. Coatings Plast. Chem. Preprints* **32**(1), 256 (1972).
26.  C. D. Han and T. C. Yu, *Polym. Eng. Sci.* **12**, 81 (1972).
27.  Z. K. Walczak, *J. Appl. Polym. Sci.* **17**, 169 (1973).
28.  T. C. Yu and C. D. Han, *J. Appl. Polym. Sci.* **17**, 1203 (1973).
29.  T. Endo, German Patent, 2,430,533 (to Toray) (1975) [*Chem Abstr.* **83**, 29695 (1975)].

30. M. Masao, S. Tokura, and H. Takahashi, Japanese Patent, 50,006 (to Kanebo) (1972) [*Chem. Abstr.* **80**, 122229 (1974)].

31. J. M. Starita, Ph.D. Dissertation, Princeton Univ. (1970).

32. J. M. Starita, *Trans. Soc. Rheol.* **16**, 339 (1972).

33. C. D. Han, Y. W. Kim, and S. J. Chen, *J. Appl. Polym. Sci.* **19**, 2831 (1975).

34. N. K. Baramboim and V. F. Rakityanskii, *Colloid J. USSR* **36**, 108 (1974).

35. T. Kitao, H. Kobayashi, S. Ikegami, and S. Ohya, *J. Polym. Sci. Polym. Chem. Ed.* **11**, 2633 (1973).

36. D. M. Cates and H. J. White, *J. Polym. Sci.* **20**, 155, 181 (1956).

37. S. I. Slepakova and B. E. Geller, *Khim. Volokna* No. 1, p. 44 (1975) [*Chem Abstr.* **82**, 126438 (1975)].

38. J. W. Cahn, *J. Chem. Phys.* **42**, 93 (1965).

39. J. W. Cahn, *Trans. Metall. Soc. AIME* **242**, 166 (1968).

40. L. P. McMaster, in "Copolymers, Polyblends, and Composites" (N. A. J. Platzer, ed.), Adv. in Chem. Ser., Vol. 142, p. 43. Amer. Chem. Soc., Washington, D.C., 1975.

41. T. Nishi, T. T. Wang, and T. K. Kwei, *Macromolecules* **8**, 227 (1975).

42. M. Washimi, Japanese Patent, 01,971 (to Unitika) (1973) [*Chem. Abstr.* **80**, 28272 (1974)].

43. A. Ohmori and M. Ande, Japanese Patent, 62,729 (to Kuraray) (1974) [*Chem. Abstr.* **81**, 171178 (1974)].

44. S. Kobayashi and A. Ohmori, Japanese Patent, 13,625 (to Kuraray) (1975) [*Chem. Abstr.* **83**, 29720 (1975)].

45. F. Higuchi and T. Kikuchi, Japanese Patent, 08,523 (to Toray) (1973) [*Chem. Abstr.* **80**, 134682 (1974)].

46. A. Ziabicki, "Fizka Procesów Formowania Wlókien." Wydawnictwa Nankowo-Techniczne, Warsaw, 1970.

47. D. R. Paul, *J. Appl. Polym. Sci.* **12**, 2273 (1968).

48. D. R. Paul, *J. Appl. Polym. Sci.* **12**, 383 (1968).

49. D. R. Paul, *J. Appl. Polym. Sci.* **13**, 817 (1969).

50. D. R. Paul and A. A. Armstrong, *J. Appl. Polym. Sci.* **17**, 1269 (1973).

51. A. L. McPeters and D. R. Paul, *Appl. Polym. Symp.* **25**, 159 (1974).

52. W. E. Fitzgerald and J. P. Craig, *Appl. Polym. Symp.* **6**, 67 (1967).

53. J. P. Craig, J. P. Knudsen, and V. F. Holland, *Textile Res. J.* **32**, 435 (1962).

54. J. P. Knudsen, *Textile Res. J.* **33**, 13 (1963).

55. M. Takahashi, Y. Nukushina, and S. Kosugi, *Textile Res. J.* **34**, 87 (1964).

56. J. P. Bell and J. H. Dumbleton, *Textile Res. J.* **41**, 196 (1971).

57. M. C. Geohegan, C. P. Malone, and E. Rivet, ANTRON III new nylon antistatic fiber, presented at *Textile Res. Inst., New York, March 19* (1975).

58. W. A. Sisson and F. F. Morehead, *Textile Res. J.* **23**, 152 (1953).

59. W. A. Sisson, *Textile Res. J.* **30**, 153 (1960).

60. W. E. Fitzgerald and J. P. Knudsen, *Textile Res. J.* **37**, 447 (1967).

61. R. S. Schechter, *AIChE J.* **7**, 445 (1961).

62. J. H. Southern and R. L. Ballman, *Appl. Polym. Sci.* **20**, 175 (1973).

63. J. H. Southern and R. L. Ballman, *J. Polym. Sci. Polym. Phys. Ed.* **13**, 863 (1975).

64. D. L. MacLean, *Trans. Soc. Rheol.* **17**, 385 (1973).

65. A. E. Everage, *Trans. Soc. Rheol.* **17**, 629 (1973).

66. J. L. White and B. L. Lee, *Trans. Soc. Rheol.* **19**, 457 (1975).

67. M. C. Williams, *AIChE J.* **21**, 1204 (1975).

68. N. Minagawa and J. L. White, *Polym. Eng. Sci.* **15**, 825 (1975).

69. C. D. Han, *J. Appl. Polym. Sci.* **17**, 1289 (1973); **19**, 1875 (1975).

70. C. D. Han, "Rheology in Polymer Processing." Academic Press, New York, 1976.

71.   I. Hamana, Japanese Patent, 40,892 (to Teijin) (1972) [*Chem. Abstr.* **80**, 4785 (1974)].

72.   J. Letz, *J. Polym. Sci. Part A-2* **7**, 1987 (1969); **8**, 1915 (1970).

73.   M. Kryszewski, A. Galeski, T. Pakula, and J. Grebowicz, *J. Colloid Interface Sci.* **44**, 85 (1973).

74.   L. B. Kandyrin and V. N. Kuleznev, *Colloid J. USSR* **36**, 431 (1974).

75.   C. D. Han, K. U. Kim, J. Parker, N. Siskovic, and C. R. Huang, *Appl. Polym. Symp.* **20**, 191 (1973).

76.   C. D. Han, *J. Appl. Polym. Sci.* **15**, 2579 (1971).

77.   K. Maciejewski and R. G. Griskey, *SPE Tech. Papers* **32**, 23 (1974).

78.   A. S. Hill and B. Maxwell, *Polym. Eng. Sci.* **10**, 289 (1970).

79.   D. Feldman and M. Rusu, *Eur. Polym. J.* **7**, 215 (1971).

80.   W. M. Prest and R. S. Porter, *J. Polym. Sci. Part A-2* **10**, 1639 (1972).

81.   I. Z. Zakirov, V. K. Pshedetskaya, and B. E. Geller, *Fibre Chem.* **2**, 250 (1971).

82.   M. H. Walters and D. N. Keyte, *Trans. Inst. Rubber Ind.* **38**, 140 (1962).

83.   Y. Oyanagi, *Kobunshi Ronbunshu Eng. Ed.* **4**, 503 (1975).

84.   S. Tomotika, *Proc. Roy. Soc. London* **A150**, 322 (1935).

85.   D. C. Chappelear, *Amer. Chem. Soc. Polym. Preprints* **5**, 363 (1964).

86.   J. P. Tordella, *in* "Rheology: Theory and Applications" (F. R. Eirich, ed.), Vol. 5, Chapter 2. Academic Press, New York, 1969.

87.   J. H. Southern and D. R. Paul, *Polym. Eng. Sci.* **14**, 561 (1974).

88.   D. R. Paul and J. H. Southern, *J. Appl. Polym. Sci.* **19**, 3375 (1975).

89.   A. E. Everage and R. L. Ballman, *J. Appl. Polym. Sci.* **18**, 933 (1974).

90.   C. D. Han and R. R. Lamonte, *Polym. Eng. Sci.* **12**, 77 (1972).

91.   J. R. Stell, D. R. Paul, and J. W. Barlow, *Polym. Eng. Sci.* **16**, 496 (1976).

92.   D. R. Paul and A. L. McPeters, *J. Appl. Polym. Sci.* **21**, 1699 (1977).

93.   J. P. Sibilia, *J. Appl. Polym. Sci.* **17**, 2911 (1973).

94.   T. I. Ablazova, M. B. Tsebrenko, A. B. V. Yudin, G. V. Vinogradov, and B. V. Yarlykov, *J. Appl. Polym. Sci.* **19**, 1781 (1975).

95.   H. P. Schreiber, *J. Appl. Polym. Sci.* **18**, 2501 (1974).

96.   W. A. Miller and C. N. Meriam, U.S. Patent, 3,097,991 (to Union Carbide) (1963).

97.   C. N. Meriam and W. A. Miller, U.S. Patent, 3,099,067 (to Union Carbide) (1963).

98.   E. Sommer, K. Gerlach, and H. Werner, U.S. Patent, 3,223,581 (to Vereinigte Glanzstoff-Fabriken) (1965).

99.   A. L. Breen, U.S. Patent, 3,382, 305 (to DuPont) (1968).

100.   H. D. Anspon, B. H. Clampitt, D. G. Ashburn, and F. E. Brown, Canadian Patent, 897,923 (to Gulf) (1972).

101.   N. G. Scherbakova, T. A. Strizhakova, V. E. Lozhkin, G. N. Kukin, and E. N. Chernov, *Fibre Chem.* **1**, 139 (1970).

102.   I. Z. Zakirov, A. A. Geller, Yu. B. Monakov, S. I. Slepakova, and B. E. Geller, *Fibre Chem.* **1**, 623 (1970).

103.   E. A. Rassolova, M. A. Zharkova, G. I. Kudryavtsev, V. G. Kulichikhin, and V. S. Klimenkov, *Fibre Chem.* **4**, 458 (1973).

104.   A. Ziabicki and R. Takserman-Krozer, *Kolloid-Z.* **198**, 60 (1964).

105.   J. Ferguson and K. M. Ibrahim, *Polymer* **10**, 135 (1969).

106.   C. D. Han and L. Segal, *J. Appl. Polym. Sci.* **14**, 2973 and 2999 (1970).

107.   C. D. Han and R. R. Lamonte, *Trans. Soc. Rheol.* **16**, 447 (1972).

108.   P. D. Griswold and J. A. Cucolo, *J. Appl. Polym. Sci.* **18**, 2887 (1974).

109.   C. D. Han and Y. W. Kim, *J. Appl. Polym. Sci.* **18**, 2589 (1974).

110.   M. T. Shaw, *J. Appl. Polym. Sci.* **19**, 2811 (1975).

111.   H. Mitomo and H. Tonami, *Sen-i Gakkaishi* **30**, T338 (1974).

112. A. M. Chatterjee, F. P. Price, and S. Newman, *J. Polym. Sci. Polym. Phys. Ed.* **13**, 2369, 2385, 2391 (1975).
113. Z. Mencik, H. K. Plummer, and H. Van Oene, *J. Polym. Sci. Part A-2* **10**, 507 (1972).
114. D. R. Paul and J. O. Altamirano, *in* "Copolymers, Polyblends, and Composites" (N. A. J. Platzer, ed.), Adv. in Chem. Ser., Vol. 142, p. 371. Amer. Chem. Soc., Washington, D.C., 1975.
115. R. L. Imken, D. R. Paul, and J. W. Barlow, *Polym. Eng. Sci.* **16**, 593 (1976).
116. D. C. Wahrmund, M.S. thesis, Univ. of Texas (1975).
117. M. Miyanoki and Y. Joh, *J. Appl. Polym. Sci.* **20**, 715 (1976).
118. A. Nakagawa, Japanese Patent, 17,611 (to Toray) (1973) [*Chem. Abstr.* **80**, 84602 (1974)].
119. L. E. Nielsen, *J. Appl. Polym. Sci.* **10**, 97 (1966).
120. B. F. Conaghan and S. L. Rosen, *Polym. Eng. Sci.* **12**, 134 (1972).
121. W. E. Fitzgerald and J. P. Knudsen, *Annu. Conf. Textile Inst.* **51**, 134 (1966).
122. W. E. Fitzgerald and G. B. Hughey, *Amer. Dyestuff Reporter* **55**(6), 37 (1966).
123. M. Horio and T. Kondo, *Textile Res. J.* **23**, 373 (1953) (see also E. H. Mercier, *ibid.* **23**, 387 (1953)).
124. M. Horio and T. Kondo, *Textile Res. J.* **23**, 137 (1953).
125. Yarns and Fibers Catalog-75. Monsanto Textiles Co. (1975).
126. S. Timoshenko, *J. Opt. Soc. Amer.* **11**, 233 (1925).
127. S. Backer and S. Batra, *Textile Res. J.* **41**, 1008 (1971).
128. A. El-Shiekh, J. F. Bogdan, and R. K. Gupta, *Textile Res. J.* **41**, 281, 916 (1971).
129. W. L. Yang and S. K. Yang, *Textile Res. J.* **42**, 298 (1972).
130. B. S. Gupta and W. George, *Textile Res. J.* **45**, 338 (1975).
131. R. H. Brand and P. Kende, *Textile Res. J.* **40**, 169 (1970).
132. R. H. Brand and R. E. Scruby, *Textile Res. J.* **43**, 544 (1973).
133. F. Ogata and K. Nagamine, U.S. Patent, 3,788,940 (to Kanegafuchi) (1974) [*Chem. Abstr.* **81**, 50917 (1974)].
134. T. Nichida, Japanese Patent, 36,954 (to Kanebo) [*Chem. Abstr.* **81**, 79253 (1974)].
135. S. Satake and M. Nishioka, Japanese Patent, 46,846 (to Mitsubishi Rayon) (1972) [*Chem. Abstr.* **80**, 60911 (1974)].
136. V. P. Krapotkin, G. I. Fantina, V E. Geller, and E. M. Aizenshtein, *Khim. Volokna* **15**(5), 12 (1973) [*Chem. Abstr.* **80**, 49022 (1974)].
137. T. Ohzeki, Japanese Patent, 00,289 (to Mitsubishi Rayon) (1973) [*Chem. Abstr.* **80**, 84569 (1974)].
138. I. Aijima, Japanese Patent, 17,567 (to Asahi) (1969) [*Chem. Abstr.* **72**, 13750 (1970)].
139. K. Nakao, Japanese Patent, 46,848 (to Kanebo) (1972) [*Chem. Abstr.* **80**, 60910 (1974)].
140. I. Takeda, Japanese Patent, 11,767 (to Teijin) (1973) [*Chem. Abstr.* **80**, 134680 (1974)].
141. Y. Sugaya, Japanese Patent, 85,398 (to Asahi) (1974) [*Chem. Abstr.* **82**, 44900 (1975)].
142. H. Nakajima and M. Fujiwara, Japanese Patent, 63,212 (to Daiwa) (1975) [*Chem. Abstr.* **83**, 148951 (1975)].
143. F. Geleji and G. Druzsbaczky, *J. Polym. Sci. Part C* **42**, 713 (1973).
144. T. Aribara, Japanese Patent, 29,810 (to Toyobo) (1973) [*Chem. Abstr.* **81**, 38759 (1974)].
145. J. K. Thomas, U.S. Patent, 3,798,296 (to American Cyanamid) (1974) [*Chem. Abstr.* **81**, 38786 (1974)].
146. E. Wakita, Japanese Patent, 42,069 (to Asahi) (1972) [*Chem. Abstr.* **80**, 71959 (1974)].
147. H. Kato, Japanese Patent, 99,419 (to Asahi) (1973) [*Chem. Abstr.* **81**, 14590 (1974)].
148. S. Ejima, Japanese Patent, 00,409 (to Chisso) (1974) [*Chem. Abstr.* **81**, 50956 (1974)].
149. G. D. Wilkinson, F. C. Flindt, and S. R. Lowy, Carpet Technology: A-71. Monsanto Co. (1969).
150. G. D. Wilkinson and F. C. Flindt, Sweater Technology: A-57. Monsanto Co. (1969).

151. F. McNeitney, *Mod. Knitting Manage.* **51**(8), 1 (1973).
152. A. H. Bruner, N. W. Boe, and P. Byrne, *J. Elastoplastics* **5**, 201 (1973).
153. J. H. Saunders, J. A. Burroughs, L. P. Williams, D. H. Martin, J. H. Southern, R. L. Ballman, and K. R. Lea, *J. Appl. Polym. Sci.* **19**, 1387 (1975).
154. S. Ishiguro and S. Takenishi, Japanese Patent, 39,729 (to Nisshin) (1972) [*Chem. Abstr.* **80**, 4812 (1974)].
155. J. M. Chamberlin, U.S. Patent, 3,761,348 (to Monsanto) (1973) [*Chem. Abstr.* **80**, 16280 (1974)].
156. A. Ichikawa, Japanese Patent, 29,130 (to Mitsubishi) (1974) [*Chem. Abstr.* **82**, 87558 (1975)].
157. F. L. Singer, "Strength of Materials," Chapter 3. Harper, New York, 1951.
158. J. C. Guthrie, D. H. Morton, and P. H. Oliver, *J. Textile Inst.* **45**, T912 (1954).
159. H. W. Holdaway, *J. Textile Inst.* **47**, T586 (1956).
160. M. C. Geoghegan, E. Rivet, and C. P. Malone, Static control needs for carpets, presented at *Textile Res. Inst., New York, March 19* (1975).
161. D. A. Seanor, *Polym. Plast. Technol. Eng.* **3**, 69 (1974).
162. E. E. Magat and D. Tanner, U.S. Patent, 3,329,557 (to Du Pont) (1967).
163. L. W. Crovatt, U.S. Patent, 3,388,104 (to Monsanto) (1968).
164. A. E. Binks and A. Sharples, *J. Polym. Sci. Part A-2* **6**, 407 (1968).
165. G. R. Bhat, Ph.D. dissertation, North Carolina State Univ. (1974) [*Diss. Abstr. Int. B* **35**, No. 11, 5443 (1975)].
166. O. A. Pickett and L. W. Crovatt, Permanent antistatic nylon, given at *AATCC Meeting, March 26* (1970).
167. R. M. Barrer, *in* "Diffusion in Polymers" (J. Crank and G. S. Park, eds.), Chapter 6. Academic Press, New York, 1968.
168. E. E. Magat and R. E. Morrison, *J. Polym. Sci. Part C* **51**, 203 (1975).
169. P. V. Papero, R. C. Winckhofer, and H. J. Oswald, *Rubber Chem. Tech.* **38**, 999 (1965).
170. E. A. Tippets, *Textile Res. J.* **37**, 524 (1967).
171. J. Zimmerman, E. M. Pearce, I. K. Miller, J. A. Muzzio, I. G. Epstein, and E. A. Hosegood, *J. Appl. Polym. Sci.* **17**, 849 (1973).
172. M. Lewin, S. M. Atlas, and E. M. Pearce (eds.), "Flame-Retardant Polymeric Materials." Plenum, New York, 1975.
173. L. Mascia, "The Role of Additives in Plastics." Edward Arnold, London, 1974.
174. I. C. McNeill and D. Neil, *Eur. Polym. J.* **6**, 143, 569 (1970).
175. D. L. Gardner and I. C. McNeill, *Eur. Polym. J.* **7**, 1 (1971).
176. N. Grassie, I. C. McNeill, and I. Cooke, *J. Appl. Polym. Sci.* **12**, 831 (1968).
177. C. Z. Carrol-Porczynski, *Composites* **4**, 9 (1973).
178. L. P. Blanchard, V. Hornof, H. H. Lam, and S. L. Malhotra, *Eur. Polym. J.* **10**, 1057 (1974).
179. N. Grassie and W. B. H. Leeming, *Eur. Polym. J.* **11**, 819 (1975).
180. N. Hurduc, C. N. Cascaval, I. A. Schneider, and G. Riess, *Eur. Polym. J.* **11**, 429 (1975).
181. I. A. Schneider and G. Riess, *Rev. Roum. Chim.* **18**, 1685 (1973).
182. K. Takeya, Japanese Patent, 68,820 (to Japan Exlan) (1973) [*Chem. Abstr.* **80**, 4780 (1974)].
183. S. Ito, Japanese Patent, 64,224 (to Mitsubishi) (1973) [*Chem. Abstr.* **80**, 16253 (1974)].
184. W. C. Dickason, D. E. van Sickle, and J. M. McIntire, U.S. Patent, 3,755,498 (to Eastman Kodak) (1973) [*Chem. Abstr.* **80**, 16261 (1974)].
185. I. Kobayashi and H. Ishii, Japanese Patent, 43, 376 (to Kanebo) (1974) [*Chem. Abstr.* **83**, 12012 (1975)].
186. Z. Orito, Japanese Patent, 81,939 (to Mitsubishi) (1973) [*Chem. Abstr.* **80**, 97181 (1974)].

187. A. Granzow and J. F. Cannelongo, *J. Appl. Polym. Sci.* **20**, 689 (1976).
188. K. Kazuya, Japanese Patent, 10,242 (to Mitsubishi) (1974) [*Chem. Abstr.* **81**, 79236 (1974)].
189. M. Mizuno, Japanese Patent, 48,056 (to Kuraray) (1975) [*Chem. Abstr.* **83**, 116757 (1975)].
190. N. Takeuchi and T. Tsuji, Japanese Patent, 56,441 (to Kuraray) (1975) [*Chem. Abstr.* **83**, 116875 (1975)].
191. M. Mizuno and K. Igi, Japanese Patent, 36,728 (to Kuraray) (1975) [*Chem. Abstr.* **83**, 81078 (1975)].
192. M. Mizuno and K. Ohno, Japanese Patent, 34,649 (to Kuraray) (1975) [*Chem. Abstr.* **83**, 116727 (1975)].
193. M. Mizuno, Japanese Patent, 48,057 (to Kuraray) (1975) [*Chem. Abstr.* **83**, 133310 (1975)].
194. K. Hirakawa, Japanese Patent, 40,659 (to Kuraray) (1975) [*Chem. Abstr.* **83**, 148964 (1975)].
195. Japanese Patent, 53,240 (to Mitsubishi) (1974) [*Chem. Abstr.* **83**, 133184 (1975)].
196. Z. Orito, Japanese Patent, 91,318 (to Mitsubishi) (1973) [*Chem. Abstr.* **81**, 14593 (1974)].
197. R. P. Beal, M. E. Dullaghan, M. W. Goodell, and A. Lulay, *Textile Chem. Colorist* **6**, No. 1, 30 (1974).
198. K. Fottiguchi and T. Kikuchi, Japanese Patent, 10,488 (to Toray) (1973) [*Chem. Abstr.* **80**, 60950 (1974)].
199. M. Farber, *SPE J.* **24**, 82 (1968).
200. A. V. Galanti and C. L. Mantell, "Polypropylene Fibers and Films." Plenum, New York, 1965.
201. M. E. Carter, "Essential Fiber Chemistry." Dekker, New York, 1971.
202. E. C. Thomm and B. R. Knowlton, U.S. Patent, 3,846,507 (to Union Carbide) (1974) [*Chem. Abstr.* **82**, 172485 (1975)].
203. E. Marcus, German Patent, 2,452,363 (to Union Carbide) (1975) [*Chem. Abstr.* **83**, 99051 (1975)].
204. R. H. Yocum and E. B. Nyquist, "Functional Monomers," Vols. I and II. Dekker, New York, 1974.
205. G. Schmack, W. Buger, and H. W. Kammer, *Faserforsch. Textiltech./Z. Polymerforsch.* **26**, 284 (1975).
206. H. Yasuda and R. Masuda, Japanese Patent, 09,015 (to Toyobo) (1973) [*Chem. Abstr.* **80**, 4803 (1974)].
207. T. Okada, Japanese Patent, 46,849 (to Mitsubishi) (1972) [*Chem. Abstr.* **80**, 28285 (1974)].
208. T. Mizoguchi, Japanese Patent, 13,337 (to Unitka) (1973) [*Chem. Abstr.* **80**, 28298 (1974)].
209. W. A. Smithey, U.S. Patent, 3,846,226 (to American Cyanamid) (1974).
210. H. Miyamoto, Japanese Patent, 04,858 (to Mitsubishi) (1973) [*Chem. Abstr.* **80**, 84574 (1974)].
211. S. Shimauchi, Japanese Patent, 04,381 (to Teijin) (1975) [*Chem. Abstr.* **83**, 29831 (1975)].
212. J. Kawamura, Japanese Patent, 20,480 (to Teijin) (1973) [*Chem. Abstr.* **80**, 84599 (1974)].
213. K. Kifune and K. Sasaki, Japanese Patent, 93,714 (to Unitika) (1973) [*Chem. Abstr.* **80**, 122255 (1974)].
214. T. Kawakita, Japanese Patent, 35,610 (to Chisso) (1973) [*Chem. Abstr.* **81**, 27080 (1974)].
215. Y. Hanada, Japanese Patent, 62,380 (to Asahi) (1974) [*Chem. Abstr.* **81**, 154435 (1974)].

# Chapter 17

# Polymeric Plasticizers

## C. F. Hammer

*Plastic Products and Resins Department*
*E. I. du Pont de Nemours & Company*
*Wilmington, Delaware*

## I. INTRODUCTION

### A. Plasticizers

#### *1. Definition*

A plasticizer is a chemical used to reduce the stiffness of an amorphous (glassy) thermoplastic resin. As an added benefit, it may also improve the processibility of the product. As is evident in subsequent sections, the softening effect of plasticizers is a highly nonlinear function of the amount added since the plasticizer acts mainly to lower the glass transition temperature. The use of the term "plasticizer" to denote a toughener is too great a misuse of the term for consideration in this chapter. In practical cases, a toughener may provide a small amount of flexibilization, but that effect is usually undesirable and is the fault of our technical ability. It should not confuse our terminology.

Conventional low molecular weight plasticizers are often called "monomeric," which is incorrect terminology because they are not polymerizable. These plasticizers improve the flexibility and workability of a resin but usually exhibit an undesirable lack of permanence (volatility, exudation, migration, extractability, etc.). A polymeric plasticizer provides flexibility to a resin, may improve the workability, and provides a marked increase in permanence over its monomeric counterpart.

This chapter also includes some information on low molecular weight "polymeric" plasticizers, which are usually liquids, as well as the true polymeric plasticizers, which are high molecular weight solids.

Many of the commercial polymeric plasticizers have complex molecular structures, which usually are not disclosed for proprietary reasons. Consequently, these products must be referred to in this chapter by their trade designations. In addition, it is common practice to use combinations of monomeric and polymeric plasticizers in commercial applications. For this reason, it is impossible to obtain from the technical literature some of the comparative data that would be of interest here.

## 2. Molecular Function of Plasticizers

A description of the behavior of polymers plasticized with low molecular weight plasticizers occupies an important portion of the literature on physical behavior of high polymers. An excellent review of the subject has been given by Darby and Sears [1], but such a thorough discussion is out of place here. Data on the use of low molecular weight plasticizers is included when suitable for comparison with the behavior of the polymeric plasticizers.

For comparison of low molecular weight plasticizers with polymeric plasticizers it is important to understand the mechanism of plasticization. The fundamental molecular effect of a monomeric plasticizer is to interact on a molecular scale with the segments of the polymer to speed up the viscoelastic response of the polymer. A *polymeric* plasticizer produces exactly the same result. Thus, a plasticizer increases the molecular mobility of the polymer chains. This increase in mobility produces a lowering of the glass transition temperature ($T_g$). This change may be observed by a variety of techniques. Curves of dynamic (low strain, low frequency) modulus and mechanical loss versus temperature are particularly suitable for understanding plasticization. Most of the physical changes observed, stiffness, hardness, low temperature stiffness, and the like, are a result of changes in viscoelastic responses.

The change in the $T_g$ as measured by the peak in the dynamic loss curve (see Volume 1, Chapters 5 and 8) is often used to characterize the behavior of polymeric plasticizers and to compare their behavior with that of low

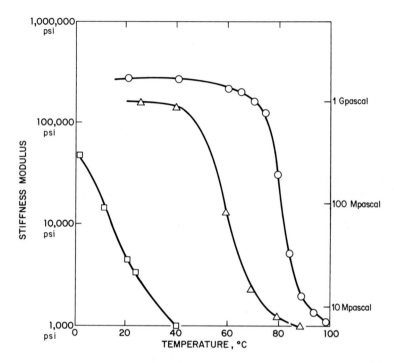

Fig. 1 Stiffness modulus versus temperature of plasticized poly(vinyl chloride) polymerized at 40°C: (○) no plasticizer; (△) 10% dioctyl phthalate (DOP); (□) 30% DOP. $T_g$ are assigned, respectively, as (○) 80°C; (△) 59°C; (□) 16°C. (From Reding *et al.* [3].)

molecular weight plasticizers. It is well known that a plot of modulus versus temperature has an inflection point at approximately the same temperature as the peak in the plot of dynamic loss factor versus temperature [2]. The highest-temperature major mechanical loss peak for an amorphous polymer is commonly designated the $T_g$. In Fig. 1 are shown modulus versus temperature curves for poly(vinyl chloride)(PVC) containing no plasticizer, 10% dioctyl phthalate (DOP) (a typical low molecular weight plasticizer), and 30% DOP [3]. This shows the characteristic effect of interaction of a plasticizer with a rigid plastic. The inflection point in the modulus curve is generally chosen as the $T_g$ of the blend. The dynamic loss curve for each blend (not shown) has a peak at a temperature corresponding roughly to this inflection point. Either measurement is suitable for obtaining the $T_g$. The plasticizer lowers the temperature at which this inflection point occurs. It also spreads the loss peak over a wider range of temperatures. Thus, the practical result of incorporating a plasticizer is a product that has a lower degree of stiffness at a given temperature than the parent thermoplastic resin. Addition of plasticizer results in a change in $T_g$; however, the stiffness does not change uniformly with concentration.

Other performance factors also change with the increase in flexibility. The tensile strength drops, and the elongation increases. The increase in molecular mobility also produces a change in the electrical properties. The dielectric loss increases while the resistivity drops.

The curves in Fig. 1 show that the $T_g$ for PVC drops about 2.1°C for each percentage point of DOP added. This change is reasonably uniform up to about 50% DOP. On the other hand, it is clearly seen that the modulus at 23°C does not change uniformly. The first 10% DOP drops the modulus in half, but 30% DOP decreases the modulus to about 1% of the initial value. This behavior also shows the importance of the $T_g$ of the rigid resin being plasticized. A resin with a $T_g$ only slightly above room temperature may be reduced in flexibility at room temperature say to 1% of its initial value by blending with only 10% plasticizer. Poly(vinyl acetate) is such a resin. On the other hand, a resin having a higher $T_g$, for example, over 100°C, may show only a small drop in room temperature stiffness with mixtures of 30% plasticizer, but then show a large drop in stiffness with the addition of more plasticizer.

### 3. Reasons for Use

Low molecular weight plasticizers may be used to flexibilize a rigid resin or to improve its processibility. If the rigid resin has a low $T_g$, both effects are obtained. On the other hand, if the rigid resin has a high $T_g$, and a high melt viscosity, the plasticizer may produce a dramatic improvement in the processibility of the melt with a relatively minor decrease in the rigidity of the resin.

### 4. Drawbacks and Compromises

Low molecular weight plasticizers are generally efficient in reducing stiffness and in improving processibility. On the other hand, such plasticizers always provide a degree of impermanence to the blend. Since there are several thousand metric tons of plasticizer used each year in the United States alone, this impermanence is evidently something we have learned to accept. Nevertheless, the impermanence limits the use of the plasticized resins and reduces the service life of many commercial products, for example, film and wire coatings. As films used in upholstery, shoes, pond liners, baby pants, shower curtains, etc. lose flexibility, they crack and must be discarded. Coatings on wire may lose flexibility and fail when bent.

Migration of the plasticizer to another site is also a problem. Plasticizer may migrate from a limp plastic used as a gasket to a rigid plastic and ruin the utility of both; for example, windshields in automobiles become coated and pick up dirt.

In the uses of a plasticized resin and in the absence of a polymeric plasticizer a less than satisfactory compromise may be necessary. Enough plasticizer is added to obtain a satisfactory flexibility of processibility, but this may also make the impermanence an important factor. As a result, the desirability of a polymeric plasticizer is apparent.

### 5. *Importance of Diffusion to Permanence*

Most plasticized resins are used at temperatures above the $T_g$ for the blend, and consequently there is considerable molecular motion. Thus, diffusion of plasticizer molecules is relatively rapid.

In a plasticized film there is a surface layer of segments of the polymer combined with molecules of the plasticizer. There is an equilibrium such that as many plasticizer molecules diffuse into the mass of the film as diffuse to the surface. If a sink at the surface into which the plasticizer molecules can go is provided (for example, air or another plastic), then the central mass of the film loses plasticizer molecules to the surface through the diffusion process. Impermanence, then, depends both upon the rate at which plasticizer molecules can leave the surface, and the rate at which they diffuse to the surface. It may be quite easy for low molecular weight plasticizers to leave the film surface. For polymeric plasticizers, however, only segments rather than the entire molecule are usually at the surface. Thus, a polymeric plasticizer may have a fairly high segmental diffusion rate inside the film, but the entire molecule is usually unable to leave the surface of the film.

### B. Polymer Blends

Whenever two polymers are blended in the melt, or from solution, there is always a trace amount of Polymer A dissolved in Polymer B and vice versa. For most polymer pairs this solubility limit is very low, and the polymers are classified as *incompatible*. When there is an infinite solubility of one in the other, the pair is classified as *compatible* (see Volume 1, Chapter 2). Compatible blends will have a single $T_g$ intermediate between the two values for the separate polymers. In a strict sense the polymer with the lower $T_g$ is a polymeric plasticizer for the polymer with the higher $T_g$. This point might be missed with some polymer pairs because experimenters tend, for practical reasons, to focus attention on room-temperature behavior. Such situations occur if the polymer pairs have both $T_g$'s well above or well below room temperature, but such blend phenomena are relatively unimportant

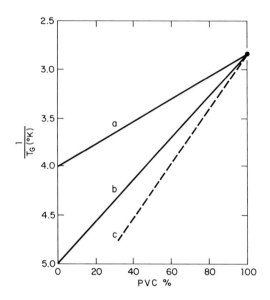

Fig. 2   $1/T_g$ versus concentration for blends with PVC and various plasticizers: (a) ethylene copolymers (Hammer [4]; Hickman and Ikeda [5]); (b) PCL 700 (polycaprolactone) [6,7]; (c) PVC–DOP.

for the purposes of this chapter since it is the change in modulus at room temperature that is of primary significance here.

The conventional way to represent the effect of a plasticizer on the $T_g$ of the polymer is in terms of a plot of $1/T_g$ versus plasticizer concentration. A comparison of three polymeric plasticizers with a monomeric plasticizer, di-2-ethylhexyl phthalate (DOP), is shown in Fig. 2. The most efficient plasticizer (greatest reduction in $T_g$ per unit of plasticizer) in this case is the monomeric DOP. Polycaprolactone (see Chapter 22), PCL-700, is next, and the two ethylene copolymers are less efficient. Not shown, but still less efficient, is a highly chlorinated polyethylene. This behavior is directly related to the value of $T_g$ for each plasticizer.

Poly(vinyl chloride) is the most widely plasticized polymer. The properties of this resin in the plasticized state are greatly affected by the presence of a small amount of crystallinity, as shown by Reding *et al.* [2]. For semi-crystalline polymers, a plasticizer will depress the melting point and reduce the degree of crystallinity. The compatible plasticizer, in this case, forms a molecular blend with the polymer in the amorphous phase, and the crystal-line phase contains very little if any of the plasticizer. Thus, one starts out with a two-phase system, crystalline and amorphous, in which both phases are chemically identical. With the addition of a "compatible" plasticizer, one obtains a two-phase system in which one phase is pure crystalline polymer and the other phase is a compatible amorphous blend.

The justification for calling such blends[†] "compatible" lies in the fact that the molten blend, at temperatures above the crystalline melting point, is a single homogeneous phase.

For PVC, however, a high degree of plasticization reduces the degree of crystallinity to a very low level. At this level, the crystallinity probably has an insignificant effect on the properties and will be ignored. For a polymer with a higher degree of crystallinity, the latter effect will probably be important and should be recognized.

## C.   General Aspects of Polymeric Plasticizers

### 1.   Molecular Weights

The commercial needs to overcome the deficiencies of monomeric plasticizers led to the development and commercial use of very low molecular weight polymers as plasticizers. Nevertheless, these plasticizers are referred to as "polymeric" rather than the more realistic term "oligimeric." These medium molecular weight or oligimeric plasticizers are included here.

The tremendous range of viscosities and molecular weights of plasticizer products termed "polymeric" in the literature is illustrated in Table I. The viscosities of the low molecular weight products are so low that measurements are generally made at room temperature. The viscosities of the very high molecular weight products are usually measured at about 190°C. A very rough comparison has been made by using values provided by the supplier and then estimating a viscosity value at 100°C. Molecular weights of plasticizers are seldom given by the suppliers. Those shown here may be of interest but should be used with caution.

### 2.   Reasons for Use; Drawbacks, Compromises

Polymeric plasticizers exhibit performance characteristics that are sufficiently valuable to the consumer to overcome their disadvantages. The trade-off depends upon the particular use. The reader may use the values given for PVC (below) as an example of the trade-offs that one should expect in comparing a monomeric with a polymeric plasticizer.

---

[†] It is common to use the term *compatible* in this situation; however, this term is used in Chapter 2 of Volume 1 in a much more restricted way and would not concur with the present usage.   [—EDITORS]

Table I

Plasticizers Classified According to Viscosity

| Plasticizer | Type and manufacturer | Viscosity (pascal-sec)[a] 25°C | 100°C | Molecular weight |
|---|---|---|---|---|
| Monomeric: | | | | |
| dioctyl phthalate | — | 0.081 | 0.008 | 374 ($\overline{M}_n$) |
| Low molecular weight polymeric | | | | |
| Paraplex G-54 | Polyester (Rohm & Haas) | 5.3 | 0.19 | — |
| Plastolein 9776 | — (Emery Industries) | — | 0.165 | — |
| Admex 761 | Polyester (Ashland Chemicals) | 5.3 | 0.19 | — |
| Flexol R-2H | — (Union Carbide) | 16.52 | 1.6 | — |
| Plastolein 9778 | — (Emery Industries) | — | 1.3 | — |
| Paraplex G-56 | Polyester (Rohm & Haas) | 11.0 | 1.0 | — |
| Medium molecular weight polymeric | | | | |
| Truflex 471 | Polyester (Teknor Apex) | 35 | — | — |
| Admex 760 | Polyester (Ashland Chemicals) | 118 | — | — |
| Paraplex G-25 | Polyester (Rohm & Haas) | 220 | 6.0 | 8000 ($\overline{M}_w$) |
| PCL 300 | Polycaprolactone (Union Carbide) | — | 10 | 15,000 ($\overline{M}_w$) |
| High molecular weight polymeric | | | | |
| PCL 700 | Polycaprolactone (Union Carbide) | — | 300 | 40,000 ($\overline{M}_w$) |
| Elvaloy 741 | Ethylene Copolymer (E. I. du Pont de Nemours & Co.) | — | 10,000 | 250,000 ($\overline{M}_w$) |

[a] Multiply values by 1000 to obtain viscosity in centipoise.

## D. Structural Features of Polymeric Plasticizers

### 1. Compatibility

Polymeric plasticizers must (a) be compatible on a molecular scale with the polymer that is to be plasticized; (b) have a sufficiently low $T_g$ so that it will efficiently lower the $T_g$ of the polymer to be plasticized; and (c) have sufficiently high molecular weight to justify the term "polymeric" (versus oligimeric). A reasonable molecular weight breakpoint is probably about $\overline{M}_n = 5000$.

General criteria for polymer–polymer compatibility are given in Volume 1, Chapters 1–4. It is interesting to note the specific cases of compatibility with PVC and to speculate on the specific structural features that are responsible. Some of the polymers known to be compatible with PVC are certain poly(methyl methacrylate)s, ethylene–vinyl acetate (EVA) copolymers, ethylene–carbon monoxide (ECO), copolymers [8, 9], ethylene–sulfur dioxide (ESO$_2$) copolymers [10], butadiene–acrylonitrile copolymers, certain terpolymer combinations, and polyesters such as polycaprolactone [6].

The structure of polycaprolactone consists of five CH$_2$ groups followed by an ester group in the chain. In the ethylene–vinyl acetate series of copolymers [4] one finds that copolymers containing up to about 60% vinyl acetate (VA) (by weight) are not compatible with PVC, while copolymers containing 65–70% VA are compatible. Those with greater than about 80% VA are again not compatible. We can think of inclusion of ester groups this way as a means of adjusting the polarity to give the best match with PVC. It is interesting to note that, on the average, the compatible structure range among the EVA polymers also corresponds to about five CH$_2$ groups per ester group.

The ethylene–carbon monoxide copolymers are not compatible with PVC below about 14% CO, while at 15% and above, the ECO copolymers are compatible with PVC [8]. With this system, however, we find that on the average only one ketone group separated by as many as 10–12 CH$_2$ groups is sufficient to provide a structure compatible with PVC. In terpolymer systems, CO can be substituted for vinyl acetate, and it is about four times as effective (on a weight basis) as vinyl acetate in producing a PVC-compatible resin [9].

Sulfur dioxide can be randomly copolymerized at moderate levels with ethylene to obtain resins compatible with PVC [10]. Compatibility is achieved with about 13% SO$_2$. For this copolymer, on the average, one sulfone group separated by about 30 CH$_2$ groups provides a resin with PVC compatibility. On a weight basis, SO$_2$ is about five times as effective as vinyl acetate in promoting compatibility.

It is important to recognize that compatibility in these cases is judged by shifts in $T_g$ determined by the peak of the mechanical loss curve. The polymer molecules are so intimately mixed that chain segmental relaxations consist of the motion of at least a portion of a PVC chain and a portion of the ethylene–SO$_2$ chain. It is interesting that the effect of one SO$_2$ group is sufficiently strong to involve 30 CH$_2$ groups in this motion. Molecular compatibility of these ethylene copolymers with PVC might be questioned in this light. However, Hickman and Ikeda [5] found that their blends with PVC have a significantly higher density than would be expected from a calculation based on additivity of volume. This suggests a negative $\Delta H$ for this particular system and thus true molecular compatibility.

## 2. Glassy Transitions of Ethylene Copolymers

Ethylene copolymers have two glass transitions that involve segmental motions of the main chain atoms of the polymer. The behavior of ethylene copolymers as polymeric plasticizers is directly related to the behavior of these two transitions. The lower-temperature (gamma) transition of polyethylene occurs at about $-120°C$. This involves a main-chain motion of a few adjacent $CH_2$ groups and probably does not involve motions of neighboring chains. For this reason it does not require much of an increase in free volume and is very difficult to detect by dilatometric techniques. This motion carries over into the ethylene copolymers such as EVA, ECO, and chlorinated polyethylene. But these copolymers also undergo a glass transition in the neighborhood of $-30°C$. This is an interchain motion that requires the cooperative motion of neighboring chains, and does result in an increase in free volume to accommodate the new mobility [4].

Blending two compatible polymers produces an intimate molecular mixing that increases the interchain molecular mobility of the higher-temperature softening resin and decreases the mobility of the resin with the lower softening temperature. In a 50/50 blend the molecular motion as reflected in $T_g$ of each resin is decreased or increased by more or less equal amounts. When an ethylene copolymer is used as the plasticizer, an interesting phenomenon is observed; the intrachain mobility of the $CH_2$ groups remains unaffected by the neighboring chains and continues to occur down to about $-120°C$; the interchain segmental mobility that does require cooperative chain motion now depends on the mobility of the polymer being plasticized. Therefore, the temperature of this transition is shifted upward to accommodate to the behavior of the higher $T_g$ polymer. The blend of an ethylene copolymer has physical properties that depend on both of these transitions. The blend complies with small deformations (stiffness) according to the interchain behavior. The low-temperature toughness of the blend, however, is enhanced remarkably by the intrachain deformability of the $CH_2$ groups in the ethylene copolymer.

It should also be recognized that the polymer to be plasticized must have a reasonable $T_g$ ($\sim 80$–$150°C$) so that plasticization is practical. It is simply not efficient to plasticize a resin from a $T_g$ of $200°C$ down to a $T_g$ of about $0°C$, with products having $T_g$ values of $-100$ to $-50°C$.

### 3. Polycaprolactone

The use of polycaprolactone (PCL-700) as a permanent plasticizer is unexpected because it is a crystalline polymer. This limits its use as a plasticizer. Molten blends of PVC with polycaprolactone are compatible. At high levels of PVC, the $T_g$ of the blend is near that of PVC and is too near the melting point of the polycaprolactone ($\sim 60°C$) to allow the polycaprolactone to crystallize. The freezing point of polycaprolactone is depressed slightly by the PVC. At these levels, polycaprolactone is a plasticizer. When more PCL is added to the blend, the $T_g$ falls below the freezing point of polycaprolactone, and slow crystallization does occur.

Table II

Plasticizer Efficiency in PVC: Concentration and Hardness at Equivalent Stiffness

| Type | Type and manufacturer | In PVC (%) | Stiffness (100% secant modulus) Mpascal | Stiffness (100% secant modulus) psi | Hardness (Shore A, 10 sec) |
|---|---|---|---|---|---|
| Elvaloy[a] 742 | Ethylene copolymer (E. I. du Pont de Nemours & Co.) | 41 | 7.2 | 1040 | 88 |
| PCL 780 | Polycaprolactone-*g*-(polystyrene) (Union Carbide) | 46 | 12.4 | 1800 | 80 |
| Hycar 1042[b] | Poly(butadiene-*co*-acrylonitrile) (B. F. Goodrich) | — | 7.6 | 1100 | 93 |
| CPE 4213[b] | Chlorinated polyethylene (Dow Chemical Company) | 50 | 8.3 | 1200 | 85 |
| Rucothane P-53[b] | Polyurethane (Hooker Chemical Company) | 50 | 9.9 | 1440 | 86 |
| Nuoplaz 6187[a] | Polyester (Tenneco Chemicals) | 38 | 9.0 | 1300 | 84 |
| Paraplex G-25 | Polyester (Rohm & Haas) | 40 | 10 | 1450 | 79 |
| Admex 761 | Polyester (Ashland Chemicals) | 38 | 13 | 1900 | 87 |
| Paraplex G-54 | Polyester (Rohm & Haas) | 38 | 11 | 1570 | 88 |
| Dioctyl phthalate | — | 32 | 8.6 | 1250 | 86 |

[a] Du Pont trademark for its resin modifiers.
[b] From Roesler and McBroom [11].

Unfortunately for commercial use of pure polycaprolactone, blends that are soft enough for use may crystallize (and stiffen) upon standing. To resolve this problem, one may add a low molecular weight plasticizer. Such a blend does have a low $T_g$, the freezing point of the PCL-700 having been sufficiently depressed to give a permanently flexible product, but some of the permanence is lost.

One useful method of circumventing the crystallization problem has been developed in which the polycaprolactone chains have been grafted onto polystyrene. This reduces the crystallization mobility of the polycaprolactone but does not affect the chain segmental mobility that provides flexibility. This product is coded PCL-780. (See Tables II, III, and IV for evaluation data, and Chapter 22 for further details on polycaprolactone.)

Table III

Low-Temperature Performance[a] of Plasticizers in PVC

| Type[b] | Low-temperature stiffness (°C) Clash–Berg (690 Mpascal, 100 kpsi) | Low-temperature brittleness (°C) ASTM D-746 |
|---|---|---|
| Elvaloy[c] 742 | −11 | −70 |
| PCL 780 | −18 | −25 |
| Hycar 1042[d] | −16 | −35 |
| CPE 4213[d] | −7 | −70 |
| Rucothane P-53[d] | −6 | −70 |
| Nuoplaz 6187[d] | −18 | −18 |
| Paraplex G-25 | −11 | −10 |
| Admex 761 | −5 | |
| Paraplex G-54 | | |
| Dioctyl phthalate | −34 | −27 |

[a] Concentrations adjusted to give equivalent stiffness values at room temperature (in the range of 1–2 Mpsi) (7–14 Mpascal).
[b] See Table II for manufacturers.
[c] Du Pont trademark for its resin modifiers.
[d] From Roesler and McBroom [11].

Table IV

Mechanical Properties of PVC Blends Containing Various Plasticizers

| Plasticizer[a] | Stiffness (100% secant modulus) | | Tensile strength | | Elongation (%) | Tear strength Elmendorf (gm) |
|---|---|---|---|---|---|---|
| | Mpascal | psi | Mpascal | psi | | |
| Elvaloy[b] 742 | 7.2 | 1040 | 17 | 2530 | 330 | 460 |
| PCL 780 | 12.4 | 1800 | 21 | 3000 | 350 | 300 |
| Hycar 1042[c] | 7.6 | 1100 | 13 | 1900 | 450 | — |
| CPE 4213[c] | 8.3 | 1200 | 14 | 2060 | 395 | — |
| Rucothane P-53[c] | 9.9 | 1440 | 23 | 3350 | 445 | — |
| Nuoplaz 6187[c] | 9.0 | 1300 | 22 | 3150 | 350 | — |
| Paraplex G-25 | 10 | 1450 | 16 | 2300 | 300 | — |
| Admex 761 | 13 | 1900 | 21 | 3100 | 285 | — |
| Paraplex G-54 | 11 | 1570 | — | — | — | — |
| Dioctyl phthalate | 8.6 | 1250 | 12 | 1800 | 210 | 220 |

[a] See Table II for manufacturers.
[b] Du Pont trademark for its resin modifiers.
[c] From Roesler and McBroom [11].

## II. PLASTICIZED SYSTEMS

### A. Poly(vinyl chloride)

#### 1. General

The behavior of polymeric plasticizers for amorphous, glassy polymers is quite similar when one system is compared with another. The same physical principles apply to each case, and the performance of the blends is qualitatively similar. The case of PVC is used as an example, because there are more data available for this system owing to its unique commercial importance. In most aspects, however, the behavior is general, and from the results with PVC one may predict qualitative behavior and sometimes quantitative behavior of other systems.

The use of polymeric plasticizers in PVC is not characteristic in one sense. PVC is a very viscous polymer and is difficult to process in commercial equipment. At temperatures high enough to give good flow, PVC undergoes rapid decomposition. Consequently, monomeric plasticizers were eagerly adopted by PVC processors. When the user demanded greater permanence, however, the polymeric plasticizer did not provide as great an improvement in flow as the monomeric plasticizer. By meeting the needs of the user (more permanence), the commercial difficulties were shifted back to the processor. Generally a compromise was reached. Such a compromise would not be necessary if PVC were chemically stable up to 275°C.

Many of the data supplied by producers of polymeric plasticizers for PVC are on blends, which contain about 10% monomeric plasticizer. This is true for chlorinated polyethylene, butadiene–acrylonitrile copolymers, and the graft copolymer of polycaprolactone on polystyrene [12]. These polymeric products are often termed "secondary plasticizers." Because data on such blends are all that is available, direct comparisons of permanence properties with other polymeric plasticizers is not generally possible.

### 2. Glass Transition Temperature Versus Plasticizer Concentration

The effect of various plasticizers on lowering the $T_g$ of PVC was shown in Fig. 2. The most rapid drop in $T_g$ is produced by a low molecular weight

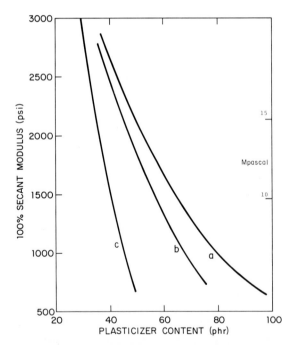

Fig. 3  Effect of plasticizer types on stiffness of vinyl compounds (c): (a) Elvaloy (du Pont trademark for its resin modifiers) 742 high molecular weight polymeric plasticizer; (b) Paraplex G-54; (c) DOP, plasticizers compounded with PVC of inherent viscosity 1.03.

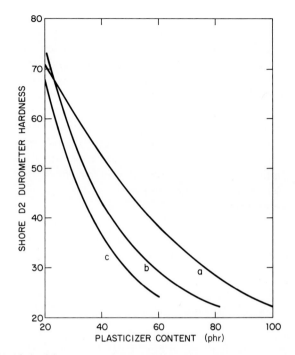

Fig. 4 Effect of plasticizer types on hardness (Shore D2) of vinyl compounds (c): (a) Elvaloy 742 high molecular weight polymeric plasticizer; (b) Paraplex G-54; (c) DOP, plasticizers compounded with PVC of inherent viscosity 1.03.

plasticizer such as DOP. The most effective polymeric product is the linear polyester. (See below for limitations for this type of product.) The ethylene copolymers are next in effectiveness. Not shown, but much less effective, are the chlorinated polyethylenes.

One may wish to speculate about the role of the structure of the polymeric plasticizer on the effectiveness of lowering the $T_g$ of PVC; that is, what is the cause for the difference in $T_g$ of the polymeric plasticizers? At this stage of understanding, one can only qualitatively relate the $T_g$ to the ease with which chains can slip past another. The polyester has only an ester carbonyl protruding beyond the H atoms along the chain, whereas the ethylene copolymers have bulkier groups such as acetate radicals or chlorine atoms.

### 3. Efficiency of Plasticizers: Stiffness and Hardness

As discussed in Section I.A.2, the nearly linear drop in $T_g$ causes a non-linear decrease in the stiffness and hardness of the polymer. Such curves are illustrated in Figs. 3 and 4. For practical purposes one wishes to compare plasticizers for PVC in blends that are essentially equivalent in stiffness or hardness—the reason for which plasticizers are added. A stiffness of hardness level of commercial importance has been selected and the amounts of various plasticizers needed to achieve this have been determined. This comparison is shown in Table II. The higher molecular weight plasticizers are almost always at a disadvantage on this basis.

For plasticized PVC systems there is a good, but not precise, correlation between stiffness and hardness. Thus, one may use concentrations to give equivalent hardness, or concentrations to give equivalent stiffness for comparing the performance of polymeric plasticizers with monomeric plasticizers. When precise efficiencies are needed for commercial purposes, however, the user must decide the main commercial reason for plasticizing the PVC and then make a selection of type and concentration accordingly.

The deviation from a precise correlation of stiffness with hardness is apparent in Table II. For example, PCL 780 gives a relatively soft blend with higher than normal stiffness, whereas HYCAR 1042 gives a relatively limp blend with a higher than normal hardness.

## 4.  Low-Temperature Performance: Stiffness, Brittleness

On cooling a limp polymer, one generally finds that the material becomes brittle as it becomes rigid. Plasticized PVC does not follow this expected behavior. On cooling, a blend containing DOP becomes brittle before it becomes stiff. On the other hand, blends containing high molecular weight polymeric plasticizers become rigid, but retain toughness behavior at much lower temperatures [13] (see Table III).

The low-temperature stiffness value is based on a torsion measurement (Clash–Berg) that is closely related to the dynamic modulus test. Thus, it is not surprising that the performance in this test closely parallels the $T_g$ blending behavior illustrated in Fig. 2. Low-temperature brittleness, on the other hand, is a much more complex phenomenon, involving factors other than stiffness.

The blend containing DOP becomes brittle at a higher temperature than the temperature at which it becomes stiff. This effect is caused by the large percentage of low molecular weight material present. Blends containing oligimeric materials become brittle as they stiffen. Three of the higher molecular weight polymeric plasticizers, on the other hand, retain toughness at much lower temperatures than the stiffness temperature. This phenomenon is related to the very low-temperature, intrachain segmental mobility that is unaffected by the glassy behavior of the PVC chains [4] (also see Section I.D.2).

Table V

Electrical Properties of Plasticized PVC Compounds[a]

| Dielectric constant and dissipation factor (ASTM D 150) | Elvaloy[b] 742 (80 phr) | | DOP (50 phr) | |
|---|---|---|---|---|
| | K | D | K | D |
| 100 Hz | 6.5 | 0.13 | 6.8 | 0.09 |
| 1 KHz | 5.4 | 0.11 | 5.9 | 0.11 |
| 10 KHz | 4.7 | 0.08 | 4.9 | 0.13 |
| 100 KHz | 4.2 | 0.07 | 4.0 | 0.13 |
| 1 MHz | 3.9 | 0.05 | 3.4 | 0.09 |
| Volume resistivity (ASTM D 257) $(\Omega \times 10^{12})$ | 36 | | 5 | |
| Dielectric breakdown, short time (ASTM D 149) (V/mil) | 1100 | | 1120 | |
| V/m | $43 \times 10^{6}$ | | $44 \times 10^{6}$ | |

[a] Plasticizers in quantities indicated compounded with poly(vinyl chloride) (inherent viscosity 1.1) and 5 parts "Dythal" lead stabilizer (NL Industries, Inc.). Concentrations are adjusted to provide approximately equal stiffness to the blends.
[b] Du Pont trademark for its resin modifiers.

Table VI

Permanence of Plasticized PVC Compounds

| | Elvaloy[a] 742 (70 phr)[b] | Paraplex G-54 (60 phr)[b] | DOP (45 phr)[b] |
|---|---|---|---|
| Extraction, % wt loss | | | |
| Ivory Soap, 1% soln., 7 days, 194°F (90°C) | 0 | 18 | 11 |
| Hexane, 24 hr, room temp. | 0 | 2.9 | 21 |
| Perchloroethylene, 1 hr, 140°F (60°C) | 0 | 14 | 18 |
| Volatiles, % wt loss, 158°F (70°C) | | | |
| 1 day | 0.4 | 0.3 | 0.6 |
| 3 days | 0.8 | 1.5 | 6 |
| 7 days | 0.7 | 1.6 | 15 |
| Migration | | | |
| To nitrocellulose, 48 hr, 158°F (70°C) 1 lb/in² (6.9 pascals) | No effect | Increased surface gloss | Dulled surface |
| To polystyrene, 20 hr, 185°F (85°C), 1 lb/in² (6.9 pascals) | No effect | Softened surface | Softened surface |

[a] Du Pont trademark for its resin modifiers.
[b] Phr.

## 5. Mechanical Properties at Room Temperature

The good low-temperature toughness, described above for blends containing polymeric plasticizers, is probably related to the absence of large quantities of low molecular weight molecules. It is expected and found to carry over into the physical properties of blends determined at room temperature. The tensile strength, the elongation, and the tear strength are all better for blends with polymeric plasticizers than for blends containing DOP (see Table IV).

## 6. Electrical Properties

Certain electrical measurements show some superiority for blends containing high molecular weight polymeric plasticizer (Table V). There is probably little or no difference at low frequencies, but the difference in loss factor at 1 MHz is probably a real one. The volume resistivity is about seven-fold higher (better) for the blend with the polymeric plasticizer.

## 7. Permanence: Extractability, Volatility, Migration

The primary disadvantage of the monomeric plasticizers is lack of permanence. This may be exhibited in use in a variety of ways. Volatilization of the plasticizer can cause undesirable fogging of neighboring surfaces such as windshields, along with the loss of stiffness and embrittlement of the product. Migration of plasticizer from PVC jacketing to polyethylene primary insulation can cause a serious reduction in the electrical properties of the primary insulation and render the cable unsuitable for its intended use. Migration of plasticizer from PVC electrical tape into the adhesive can markedly reduce the strength of the adhesive. Extraction of plasticizer by water can lead to embrittlement and failure of such home products as shower curtains, table cloths, and baby pants. Extraction by water or common solvents such as hexane can make it unfeasible to use PVC for commercial applications such as pond liners. An extraction of plasticizer by aqueous systems is highly undesirable for use of PVC in medical tubing or blood bags.

An increase in permanence can be achieved with a low molecular weight oligimeric plasticizer such as that given in Table VI. The ultimate in permanence is achieved, however, with the use of the high molecular weight polymeric resin also shown in Table VI.

The permanence of high molecular weight plasticizers is such that the average molecular weight of the polymeric plasticizer is no longer a factor. It is more likely that any slight impermanence is due to (a) additives in the blend such as stabilizers and soaps used in the PVC synthesis;

(b) stabilizers and catalyst fragments from the polymeric plasticizer; (c) stabilizer and lubricants used in making the blend; and (d) the often unavoidable low molecular weight tail of the molecular weight distribution for both the PVC and the polymeric plasticizer. (If such factors cause small amounts of impermanence, this should result in a leveling off with time in the various tests for permanence.)

The resistance to extraction of the high molecular weight polymeric plasticizers by soapy water, hexane, or perchloroethylene is illustrated by the comparison of Elvaloy[†] 742 with the oligimeric Paraplex G-54 and DOP. In this test, the oligimeric material is not so good as the high molecular weight plasticizer. In the test for volatiles or migration tests, the oligimeric product seems to perform almost as well as high molecular weight products.

### 8. *Melt Workability: Blendability, Viscosity*

The processibility of PVC has been a difficult problem since the early days of the product. The use of monomeric plasticizers has made possible the fabrication of PVC at reduced temperatures. The introduction of a new plasticizer is always approached with caution, lest this new aspect reintroduce unwanted chemical stability problems. The evaluation of such systems is conducted in a variety of ways.

The melt behavior of blends containing polymeric plasticizers involves two aspects that are considered below: the mechanics of making the blend and the melt performance of the finished blend.

*Blending.*   Blending with PVC is difficult. Commercial PVC is a high-viscosity resin at suitable processing temperatures, 160–200°C. At 200°C and above, decomposition occurs rapidly; hence, melt temperatures of say 210°C can be tolerated for only about 1 min. In addition, the processes used to make commercial PVC tend to produce a small fraction of particulate PVC of extremely high molecular weight. These problems are not necessarily encountered in other polymer–polymer blend systems.

The blending of a monomeric plasticizer into PVC tends to be relatively easy because of its rapid diffusion into the PVC. As the molecular weight of the plasticizer increases, this process becomes slower, and the problem then becomes one of obtaining intimate mixing of two fluids of vastly differing viscosity (see Volume 1, Chapter 7).

Compatible polymer blends, by definition, have a thermodynamic driving force to promote blending on a molecular scale. However, the diffusion rate across the interface is only of the order of 100 $\mu$m/hr [14]. To obtain suitable

[†] This is du Pont trademark for its resin modifiers.

blends in a commercially satisfactory time, for example, 2–4 min, the problem then becomes one of producing sufficient shear in an unblended mixture to create large areas of interface and to break down the unmixed regions into particles about 1 $\mu$m in size. Electron photomicrographs of blends of PVC with *incompatible* modifiers, for example, impact modifiers, show that breakdown into 1-$\mu$m particles can be achieved in 5 min or less. This may be done in any of several types of commercial melt mixers, including an extruder.

For compatible blends, the diffusion process begins as soon as the polymers are molten. This early-stage transfacial blending helps the equipment to apply shear. Therefore, the problem of shearing two liquids of different viscosity begins to be alleviated as the mixing proceeds. In spite of these factors, the blending of a polymeric plasticizer in PVC is more difficult than the blending of a monomeric one with PVC. A satisfactory degree of blending takes longer times and higher temperatures. As a guideline, one may assume that both a 20°C higher temperature and a two-fold increase in time is needed.

From the practical standpoint, such a penalty is a cost that must be included in considering a shift from a monomeric plasticizer to obtain the permanence of a polymeric plasticizer. There may be an unnecessary cost, however, in using optical transparency as a criterion for completeness of blending. The very high molecular weight PVC particles mentioned above comprise a very small fraction of the blend, but take extremely long times to work out. They do not seem to interfere with physical properties, however, and may readily be considered as a trace of a particulate filler. Particulate fillers such as $TiO_2$ tends to act as grinding aids and reduce the time needed to achieve suitable blending.

The use of a small amount of a low-viscosity "polymeric" plasticizer or a monomeric plasticizer is the most generally used method for alleviating the blending problem. About 10% of the total blend is often a liquid plasticizer. Such formulations are often recommended by suppliers of polymeric plasticizers for PVC. The general use of such dual-plasticizer systems in the technical literature has made it extremely difficult to obtain comparative data for single-plasticizer systems for this chapter.

The melt behavior of compatible blends qualitatively follow what one expects from the values for the components. A monomeric plasticizer gives a lower blend viscosity than a polymeric plasticizer at equivalent concentrations. For commercial use, however, less monomeric plasticizer is used to obtain equivalent flexibility of the blends. A comparison of melt viscosities of blends, based on roughly equivalent flexibility at room temperature, is given in Table VII. From this we see that the practical viscosity of a blend containing a polymeric plasticizer may not be vastly higher (less than two-fold) than that of the blend containing the monomeric plasticizer.

Table VII

Melt Viscosities of PVC Blends[a] Containing Various Plasticizers

| Plasticizer | Type and manufacturer | Plasticizer | Brabender torque at 190°C (m-g) |
|---|---|---|---|
| Elvaloy[b] 742 | Ethylene copolymer (E. I. du Pont de Nemours & Co.) | 50 | 1060 |
| Rucothane P-53[c] | Polyurethane (Hooker Chemical Company) | 50 | 710 |
| Nuoplaz 6187[c] | Polyester (Tenneco Chemicals) | 38 | 690 |
| DOP | | 32 | 590 |

[a] Blends contain given percentages of plasticizer to have essentially equivalent flexibility.
[b] Du Pont trademark for its resin modifiers.
[c] From Roesler and McBroom [11].

Table VIII

Properties of Free Films from Blends of Nitrocellulose[a] with Elvax ® 40 Resin[b,c]

| Elvax 40 in nitrocellulose[a] (%) | Tensile strength | | Elongation (%) | Young's modulus | |
|---|---|---|---|---|---|
| | Mpascal | psi | | Gpascal | psi |
| 20 | 79 | 11,500 | 4 | 3.1 | 451,000 |
| 25 | 68 | 9,800 | 4 | 2.4 | 347,000 |
| 33 | 48 | 6,900 | 9 | 1.6 | 226,000 |
| 50 | 38 | 5,500 | 41 | 1.0 | 150,000 |
| 67 | 27 | 3,900 | 185 | — | — |

[a] RS[TM] $\frac{1}{2}$-sec nitrocellulose.
[b] Ethylene–vinyl acetate copolymer containing 40% vinyl acetate.
[c] From Tech. Data [16].

This simple viscosity value is obviously an oversimplification. Such factors as the effect of shear rate on viscosity, melt fracture, melt drawability, and "nerve" may be important, depending upon the type of fabrication equipment being used. A determination of these aspects depends upon the PVC being used, the formulation, and to some extent the perfection of the blending. For commercial purposes, researchers must make comparisons of each specific formulation, using laboratory equipment that has been shown to correlate well with commercial equipment.

Table IX

Physical Properties of Films from Blends
of Nitrocellulose with Polycaprolactone

| PCL in nitrocellulose blend (%) | Tensile strength | | Elongation (%) |
| --- | --- | --- | --- |
| | Mpascal | psi | |
| 10 | 4.3 | 6300 | 17 |
| 20 | 4.1 | 6000 | 16 |
| 30 | 2.7 | 3900 | 16 |
| 40 | 1.7 | 2500 | 43 |
| 50 | 0.9 | 1300 | 83 |

[a] From New Product Bulletin [7].

## B.  Nitrocellulose

Nitrocellulose plasticized with a high molecular weight ethylene–vinyl acetate copolymer represents one of the first commercial uses of a polymeric plasticizer [15]. Its use as a lacquer is not that of a conventional thermoplastic. As a result, there are very few data related to the information normally obtained on molded plastics. Nevertheless, from the data shown in Table VIII one can see that the tensile strength and elongation behave as predicted for the plasticization of a polymer with a moderately high $T_g$. There is very little elongation until one uses 50% of the plasticizing resin.

Nitrocellulose can also be plasticized with polycaprolactone (see Chapter 22). The data given in Table IX are similar to that in Table VIII. One sees, that, as in PVC, the polycaprolactone seems to be more efficient than the EVA copolymer. An elongation of 40% can be achieved with 40% caprolactone instead of the 50% needed with the EVA copolymer.

## III.  SUMMARY

Rigid plastics are often plasticized with low molecular weight molecules to provide flexible products, usually films or sheeting. Diffusion and loss of plasticizer is a problem that can be solved in principle by use of a soft, high molecular weight polymer that is molecularly compatible with the rigid plastic.

The behavior of a compatible pair of rigid–flexible plastics in providing a flexible material is quite general, and can reasonably be predicted from one specific example to another. The behavior is best appreciated by a study of the change in $T_g$ as observed by dynamic (small-strain) modulus measurements.

Poly(vinyl chloride), plasticized with low, medium, and high molecular weight plasticizers, is an excellent example of such behavior. Low molecular weight plasticizers are more efficient but suffer from impermanence. High molecular weight plasticizers have superior permanence and better physical and electrical properties, but are less efficient. They usually do not provide the improvement in processibility provided by low molecular weight plasticizers.

## REFERENCES

1.  J. R. Darby and J. K. Sears, *Encycl. Polym. Sci. Technol.* **10**, 228–306.
2.  N. McCrum, B. Read, and G. Williams, "Anelastic and Dielectric Effects in Polymeric Solids," pp. 14, 45. Wiley, New York, 1967.
3.  F. P. Reding, E. R. Walter, and F. J. Welch, *J. Polym. Sci.* **56**, 225–231 (1962).
4.  C. F. Hammer, *Macromolecules* **4**, 69 (1971).
5.  J. J. Hickman and R. M. Ikeda, *J. Polym. Sci. Polym. Phys. Ed.* **2**, 1713–1721 (1973).
6.  J. V. Koleske and R. D. Lundberg, *J. Polym. Sci. Part A-2* **7**, 795–807 (1969).
7.  New Product Bulletin F-42501, New Poly-caprolactone Thermoplastic Polymers PCL-300 and PCL-700. Union Carbide Corp.
8.  C. F. Hammer and T. F. Martens, French Patent, 2,149,933 (to du Pont).
9.  C. F. Hammer, U.S. Patent, 3,780,140 (to du Pont).
10. C. F. Hammer, British Patent, 1,345,385 (to du Pont).
11. R. E. Godlewski and L. G. Krauskopf, Technical Service Rep., Ultrapermanent Flexible Vinyl with PCL-780. Union Carbide Corp. Tarrytown Tech. Center, Tarrytown, New York.
12. J. P. Tordella, J. T. Hyde, B. S. Gorton, and C. F. Hammer, *Soc. Plast. Eng. Annu. Tech. Conf., 33rd, May 5–8* p. 135 (1975).
13. F. Roesler and J. McBroom, *Polymer Alloys for Flexible Cable Jacketing, Proc. Int. Wire Cable Symp., 24th, Cherry Hill, New Jersey, Nov. 18–20* U.S. Army Electronics Command (1975).
14. M. Kryszewski, A. Galeski, T. Pakula, and J. Grebowicz, *J. Colloid Interface Sci.* **44**, 31 (1973).
15. J. G. Unger, U.S. Patent, 3,321,420 (to Hercules, Inc.).
16. Tech. Data CSL-190A, Nitrocellulose-Poly(ethylene-vinyl acetate), Hercules, Inc., Wilmington, Delaware.

# Chapter 18

# Block Copolymers in Blends with Other Polymers

*Gerard Kraus*

*Research and Development Department*
*Phillips Petroleum Company*
*Bartlesville, Oklahoma*

## I. INTRODUCTION

There are numerous applications in which block polymers are blended with homopolymers or random copolymers to produce useful products. In virtually all of these the block polymer consists of rubbery and glassy (or crystalline) segments. There are four basic combinations: the block polymer may be used either as the minor or major component with a rubber or a rigid (glassy or semicrystalline) polymer, respectively. All four possibilities have found applications. Some of the characteristics and uses of such blends are described in

243

this chapter. The specific application in which a block polymer is used as an interfacial agent to stabilize or "compatibilize" polymer blends is described in Chapter 12.

## II. MORPHOLOGY AND LINEAR VISCOELASTIC BEHAVIOR

Measurements of linear viscoelastic properties, together with electron and optical microscopy, have proved invaluable in characterization of poly-blends, block, and graft polymers. They are almost indispensible in the study of the morphology of the even more complex blends of homopolymers or random copolymers with block copolymers. Two cases are considered here: (1) the added polymer is chemically and structurally identical to one of the monomer sequences of the block copolymer; (2) the chemical composition of the added polymer is unlike any of the monomer sequences of the block polymer.

### A.  Case 1: One Block Chemically Identical to Added Polymer

The examples most thoroughly investigated in the literature are the block polymers of styrene (S) and butadiene (B) or isoprene (I) blended with polystyrene or the appropriate polydiene. If the monomer sequences of the block polymer are a few thousand molecular weight units in length or longer, they segregate into a two-phase domain structure. The added homopolymer, whether it is polystyrene or polydiene, is, of course, thermodynamically compatible with blocks of like composition. Nevertheless, it does not necessarily enter the domain structure of the block polymer, for the latter is governed to a large extent by consideration of space filling to maintain normal density without excessive distortion of the molecular chains into energetically unfavorable configurations. Thus, if high molecular weight polystyrene is blended with a styrene–butadiene block copolymer having styrene blocks much shorter than those of the homopolymer, the latter will not enter the polystyrene domains of the block polymer, but will form its own domains. An example is shown in Fig. 1, where the 35-Hz loss modulus of such a blend is seen to exhibit two polystyrene transitions, one for the homopolymer and one for the block polymer domains. The difference in the position of the two loss maxima arises from the significant contribution the diffuse domain interphase makes to the polystyrene loss maximum of the block polymer when the styrene block length decreases below about 20,000 molecular weight units [1].

The effects of mixing polystyrene and polyisoprene with SI diblock

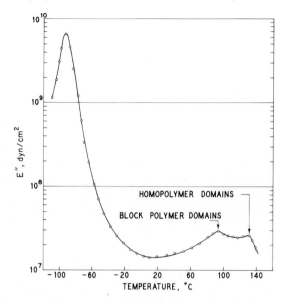

Fig. 1 Loss modulus for a blend of 75 parts of BS tetrachain block polymer (30% styrene) with 25 parts of styrene homopolymer of high molecular weight.

polymers on the morphology of the resulting blends have been studied in detail by Kawai *et al.* [2]. Their results, obtained on films prepared by solution casting from a common solvent, are illustrated schematically in Fig. 2 for the case of homopolymer molecular weights of the same order of magnitude or smaller than the corresponding block molecular weights. Note that in Fig. 2a the block polymer morphology is spheres of B in A, and adding homopolymer A simply results in its absorption by the matrix. Adding increasing amounts of homopolymer B to the block polymer, however, takes one progressively through the entire gamut of basic morphologies. This behavior with polymers of rodlike and lamellar morphologies is illustrated in Fig. 2b and c. The ternary blends inside the boundaries of the triangles have stable morphologies only outside the crosshatched areas. In this region each kind of block polymer domain accommodates homopolymer of like composition, and the immiscible homopolymers are thereby restrained from phase separation on a macroscopic scale; the block copolymer acts much like an emulsifier. Inside the shaded areas of the diagrams the amount of block polymer is no longer sufficient to prevent separation of A and B into macroscopic domains.

Experiments of Bradford [3] on blending SB and SBS block polymers with polystyrene support the general behavior depicted in Fig. 2 for the diblock polymer blends. For a 53% styrene SBS polymer, however, even addition of

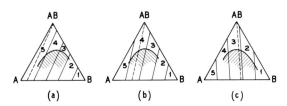

Fig. 2   Schematic phase diagrams illustrating domain formation from solution in a common solvent of block polymer AB and homopolymers A and B: (1) spheres of A in a matrix of B; (2) rods of A in a matrix of B; (3) alternating lamellae of A and B; (4) rods of B in a matrix of A; (5) spheres of B in a matrix of A. The solid lines are boundaries between the various characteristic domain structures. The dashed line represents the composition of the block polymer. (Reproduced from Inoue *et al.* [2] by permission of the publisher.)

99% low molecular weight polystyrene failed to produce spherical polybutadiene domains, but gave rise to ribbonlike structures, probably the remnants of single polystyrene lamellae. Bradford's conclusion that often the morphology of block copolymer films (and their blends) can be predicted with confidence, while in other cases totally unanticipated structures are observed appears to be still valid.

When homopolymer molecular weights greatly exceed corresponding block molecular weights, the resulting morphology of the blend depends to an important extent on which is the continuous matrix phase of the block polymer. Thus, if A is the matrix, it will accommodate homopolymer A even if the molecular weights are seriously mismatched. However, the same block polymer cannot accommodate high molecular weight homopolymer B in its discrete B domains, as in Fig. 1.

The effects of blending polybutadiene with a butadiene–continuous styrene–butadiene block copolymer on dynamic viscoelastic behavior was studied by Rollmann and Kraus [4]. The block polymer was of the $(SB)_x$ type, containing 20% styrene, $3 < x < 4$, $\overline{M}_w = 190{,}000$. The added polybutadienes were identical in microstructure to the B blocks and ranged in $\overline{M}_w$ from 10,000 to 634,000. The highest blend ratio investigated was one part by weight of homopolymer to two parts of block polymer. All blends had spherical polystyrene domains in a polybutadiene continuum. As expected, the polybutadiene glass transition temperature ($T_g$) was the same in all blends. Interesting differences were observed in the region of viscoelastic response between the principal domain transitions, where the topology of the polybutadiene entanglement network makes an important contribution to viscoelastic behavior. The added polybutadiene molecules are free to move through the entanglement network and around the polystyrene domains in much the same manner free molecules move through chemically cross-linked networks [5, 6]. The motion, which has been treated theoretically by de Gennes [7], is a

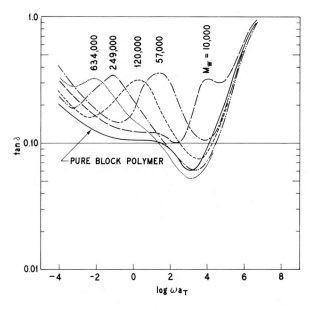

Fig. 3 Result of blending polybutadienes of different molecular weights into styrene–butadiene block polymer; blend ratio 1/2; data reduced to 25°C. The frequency ($\omega$) of the secondary loss maximum varies as $M^{-3}$. (Data from Kraus and Rollmann [4].)

snaking ("reptation") of the molecule, as if constrained in a tube, with the chain ends choosing different paths between successive forward and backward movements. This motion leads to the appearance of a secondary loss maximum, the position of which on a frequency scale is theoretically proportional to $M^{-3}$. In isochronal (constant-frequency) data the secondary maxima in $G''$ or $\tan\delta$ are generally obscured by the polystyrene glass transition except when $M$ is small. However, by extending dynamic data with 25°C stress relaxation data converted to dynamic moduli, the maximum was located in all cases and its position found to be in agreement with the prediction of de Gennes for the motion of randomly contorted macromolecules around fixed obstacles. The $\tan\delta$ versus reduced frequency plots are reproduced in Fig. 3.

The addition of minor amounts of styrene–butadiene block polymers to polystyrene is generally less eventful; the existence of discrete polybutadiene domains shows up as the expected low-temperature loss maximum [8].

McIntyre and Compos-Lopez [9] studied blends of 15,000 molecular weight polystyrene with a 10–80–10 SIS block polymer of 137,000 molecular weight by small-angle x-ray scattering, but an unequivocal morphological characterization was not realized. Effects of sample preparation and thermal history on

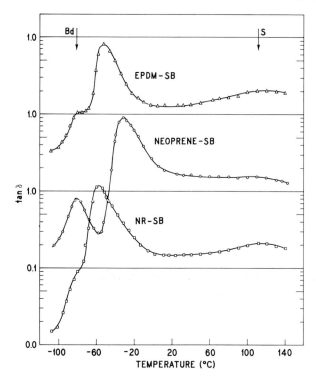

Fig. 4   Loss tangents (35 Hz) of cross-linked polyblends of 30% SB block polymer (25% styrene) with 70% of another rubber. Curves are displaced by one decade each. (Reproduced from Kraus and Railsback [14] by permission of the publisher.)

properties of a blend of an SBS triblock polymer with low molecular weight polystyrene have been described by Harpell and Wilkes [10].

High molecular weight polystyrene and poly(α-methylstyrene), though very close in their solubility parameters, do not form compatible poly-blends, but styrene–α-methylstyrene block polymers (SA) are single phase [11]. Robeson *et al.* [12] found some blends of SA diblock polymers compatible with polystyrene, and others not. Blends of SA with poly(α-methylstyrene) were compatible. In a more recent study Hansen and Shen [13] found SAS and ASA triblock polymers to form generally single-phase-blends with either of the homopolymers, except in one instance in which an ASA block polymer of about 15,000–70,000–15,000 block lengths gave evidence of incompatibility with 160,000 molecular weight α-methylstyrene homopolymer. Evidently, the closely similar solubility characteristics make this a very sensitive system in which compatibility is exceedingly difficult to predict.

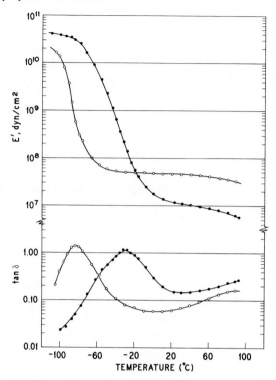

Fig. 5 Storage modulus and loss tangent (35 Hz) of 20/80 styrene–butadiene radial block polymer with polybutadiene-compatible low molecular weight resin; (○) pure block polymer; (●) blend. Blend ratio 4/3.

## B. Case 2: No Block Chemically Identical to Added Polymer

By far the most common situation resulting from blending a block polymer with a polymer differing in composition from either block is the formation of a three-phase system. Some examples are shown in Fig. 4 [14]. An SB diblock polymer containing 25% styrene was blended in 30/70 ratio with natural rubber, polychloroprene, and EPDM (ethylene–propylene–diene terpolymer), respectively. In all cases the polybutadiene and polystyrene glass transitions are detected. Of course, resolution suffers when the $T_g$ lie close together, as in the example of natural rubber and the polybutadiene phase of the block polymer.

Robeson *et al.* [12] investigated the rare case in which a single-phase compatible block polymer is blended with a homopolymer compatible with one or possibly both of the blocks. The blends prepared were styrene–α-methylstyrene block polymer with poly(2,6-dimethyl-1,4-phenylene ether).

Dynamic mechanical data showed these compositions to be one-phase systems.

Low molecular weight hydrocarbon and other resins are often blended with rubbery SBS or SIS block copolymers in the formulation of adhesives and sealants [15, 16]. These resins are frequently compatible with one or the other of the blocks of the copolymer by virtue of their low molecular weight. Resins compatible with the diene center block are usually effective tackifiers (see below). They soften the polydiene matrix by their diluent effect, but raise $T_g$ of the matrix as a result of their own high glass transition temperature (Fig. 5). This behavior has been suggested to be basic to the tackification phenomenon [17]. Polystyrene-soluble resins do not tackify but may raise $T_g$ of the hard phase, thus imparting better high-temperature properties to the blend.


### C.  Some General Remarks on Morphology and Physical Properties

Whenever one phase in a blend of a block polymer with a homopolymer or random copolymer becomes the continuum, its properties dominate the overall properties of the composite. For example, polystyrene containing a minor amount of an SBS triblock polymer will still be a resin, although its modulus will be lowered and its flexibility increased. However, whether failure properties, such as impact resistance, will be improved significantly depends on the size and shape of the rubbery inclusions, the molecular weight of the blocks, whether the rubbery domains are cross-linked further, and other details. Conversely, if an SBS block polymer is added in minor amount to a rubber, the blend will be a rubber, stiffened by the filler effect of polystyrene domains. Again, the effects on failure properties depend on more subtle structural features.

When a blend exhibits a morphology in which both phases are continuous, for examples, long cylinders or rods, alternating lamellae, or even more complex structures, its mechanical behavior will be more complicated. Consider first the example of an SB block polymer blended with either homopolymer, in which the resultant morphology is cylindrical (polystyrene rods in a rubbery matrix) or lamellar. Such a blend will be extremely weak unless the rubbery phase is cross-linked by addition of a chemical cross-linking agent. A similar blend utilizing an SBS triblock polymer, however, will retain a large amount of the high strength of the block polymer itself. This strength is derived from the fact that polystyrene domains are connected by the polybutadiene blocks into an infinite network structure, reinforced by the polystyrene domains. Extensive deformation of the blend (or pure block polymer) will cause the polystyrene rods or lamellae to be broken up. This is accompanied by necking and cold drawing of the specimen, followed by a region of strain of high rubberlike extensibility up to failure.

Another obvious consequence of rodlike and lamellar morphologies is mechanical anisotropy. Frequently, mold flow is sufficient to produce significant orientation, which shows up prominently in small-strain mechanical behavior. The breaking up of the domains at high strains and the accompanying reorientation, of course, minimizes original orientation effects.

Resin-continuous composites with rodlike rubbery inclusions display much less anisotropy. For quantitative treatments of the elastic response of both particulate and nonparticulate compositions the reader is referred to Volume 1, Chapter 8.

## III.   APPLICATIONS AND PROPERTIES

In the following a number of applications of polyblends containing a block copolymer as one of the components are described in some detail. No claim to comprehensiveness is made. Instead, the intent is to illustrate both established and unconventional applications, which, it is hoped, will suggest to the reader still further ways of employing block polymers in the formulation of useful composites.

### A.   Modification of Glassy and Semicrystalline Plastics with Block Copolymers

There are two established routes by which impact resistance may be built into a brittle plastic through inclusion of rubberlike particles. The simplest is mechanical mixing of rubber and resin. This process produces products of generally inferior quality, and virtually all large-scale commercial producers of rubber-modified resins such as impact polystyrene and ABS utilize graft polymerization (see Chapters 12–14). The use of block copolymers provides a third alternative, which offers many of the advantages of the graft polymer resins.

Riess *et al.* [18] studied the ternary system polystyrene–polyisoprene–SI diblock polymer. They were able to produce impact-resistant compositions both with binary and ternary blends. Best results were obtained with a block polymer of high molecular weight containing about 40% styrene. To achieve significant improvements over the homopolymer blends rather large amounts of block polymer were required. Some of Riess' data are shown in Table I. Hardness was reported only as exceeding 80 degrees Shore C and no other

*Gerard Kraus*

Table I

Impact Resistance of Binary and Ternary
Blends[a] (PS–PI–SI-Block)

| Composition (by weight) | | | |
|---|---|---|---|
| PS[b] | PI[c] | SI block[d] | Impact (Charpy) |
| 100 0 | 0 | 0 | 0.4 |
| 82.5 | 17.5 | 0 | 1.9 |
| 80 | 10 | 10 | 2.5 |
| 72.5 | 0 | 27.5 | >5.0 |
| 75 | 25 | 0 | 1.6 |
| 67.5 | 16.25 | 16.25 | 2.9 |
| 55 | 0 | 45 | >5.0 |

[a] Data from Kohler *et al.* [18].
[b] $\overline{M}_w = 165,000$.
[c] $\overline{M}_w = 160,000$.
[d] $\overline{M}_n = 475,000$, 38.8% styrene.

mechanical properties were given. The authors made an attempt to correlate impact resistance with morphology judged from phase contrast optical micrographs. They defined regions in composition space of compatibility and incompatibility. In the light of present knowledge it seems certain that none of the blends would have been judged compatible by electron microscopy or by dynamic viscoelastic measurements, and the "compatible" compositions are undoubtedly two-phase systems differing from "incompatible" ones only in their domain sizes and shapes.

Childers *et al.* [8, 19] blended different styrene–butadiene block polymers with polystyrene, both with and without dicumyl peroxide to cross-link the rubbery domains. Block polymer structures investigated were SB, SBS, (SB)ₓ, and S:BS, the last notation referring to diblock polymer containing a random copolymer sequence and a pure polystyrene block. Some results with block polymers of 25% styrene blended with polystyrene in a 1/3 ratio are shown in Table II. Note that every one of the block polymers produced a much greater improvement in impact strength than did polybutadiene or the random copolymer. Cross-linking the rubber phase was effective in all cases, particularly in raising the tensile strength of the block polymer blends.

The length of the styrene blocks in the examples of Table II was 18,000 molecular weight units or longer. In experiments with block polymers of 8000–9000 styrene block length a significant loss in impact resistance was observed. There appears to be no doubt that the styrene blocks attached to the

Table II

Blends of Polystyrene with Butadiene–Styrene Block Polymers[a]

| Added rubber[b] | Dicumyl peroxide (%) | Melt flow (5 kg, 200°C) | Notched Izod (ft-lb/in.) | Flexural modulus (psi) | Tensile strength (psi) | Breaking elongation (%) |
|---|---|---|---|---|---|---|
| Polybutadiene | 0 | 4.5 | 0.4 | 216,000 | 2350 | 3 |
| | 0.1 | 0.6 | 0.8 | 213,000 | 2310 | 6 |
| Random copolymer | 0 | 7.2 | 0.6 | 217,000 | 2120 | 2 |
| | 0.1 | 0.8 | 1.2 | 224,000 | 2500 | 21 |
| S:BS block (20% block S) | 0 | 2.8 | 3.8 | 204,000 | 1780 | 4 |
| | 0.1 | 1.1 | 4.3 | 232,000 | 3180 | 47 |
| BS block | 0 | — | — | — | — | — |
| | 0.1 | 0.8 | 4.2 | 243,000 | 3620 | 54 |
| SBS block | 0 | 0.8 | 3.6 | 246,000 | 2960 | 15 |
| | 0.1 | 1.1 | 4.0 | 250,000 | 3610 | 32 |
| $(SB)_x$ | 0 | 2.3 | 2.9 | 249,000 | 2350 | 12 |
| | 0.1 | 1.7 | 4.6 | 274,000 | 3460 | 7 |

[a] Data from Childers et al. [8].
[b] All copolymers 25% total styrene. Blend ratio is 3/1 PS/rubber.

rubbery domains act as anchors to the polystyrene continuum, performing the role of the grafts in conventional impact polystyrene or ABS.

Blends of polystyrene with SBS triblock polymers were studied further by Durst et al. [20]. By careful optimization of the system excellent compositions were obtained without the use of peroxide.

Table III

Blends of Styrene–Acrylonitrile Copolymer with
Block Polymers Containing
Poly(ε-caprolactone) Sequences[a]

| | S–B–CL | (S:B)–CL[b] |
|---|---|---|
| Block polymer in blend (%) | 22.5 | 25 |
| Dicumyl peroxide (%) | 0.1 | 0.1 |
| Melt flow (5 kg, 200°C) | 0.3 | 0.4 |
| Notched Izod (ft-lb/in.) | 11.1 | 15.0 |
| Flexural modulus (psi) | 298,000 | 249,000 |
| Tensile strength (psi) | 5610 | 4600 |
| Breaking elongation (%) | 37 | 79 |

[a] From Childers and Clark [21].
[b] (S:B) is a random copolymer block.

Poly (ε-caprolactone) shows a surprising degree of compatibility with a number of other polymers (see Chapter 22), including certain random copolymers of styrene and acrylonitrile (SAN resins). By blending SAN with block polymers of styrene, butadiene, and ε-caprolactone and cross-linking the polybutadiene domains, Childers and Clark [21] were able to prepare compositions resembling ABS resins prepared by latex graft polymerization (Table III). By lowering the amount of block polymer in the blend, melt flow, modulus, and tensile strength could be increased at the expense of some impact resistance.

The use of SBS block polymers in blends with graft process high-impact polystyrene (HIPS) recently has been described by Bull and Holden [22]. Further improvement in impact resistance is realized with accompanying loss in flexural modulus and hardness. However, by using ternary blends of HIPS, polystyrene, and an SBS block polymer the authors were able to maintain the modulus of HIPS while increasing impact resistance. In one example a 40–60–15 blend provided a 64% improvement in Izod impact over HIPS at equal flexural modulus.

Table IV

Blends of Polypropylene with an SBS Block Polymer[a]

| Polypropylene | 100 | 90 | 80 |
|---|---|---|---|
| SBS block polymer | 0 | 10 | 20 |
| Falling weight impact strength (kJ/m) | | | |
| At 23°C | 2.5 | 5.8 | No break |
| At −29°C | <0.8 | 1.1 | 6.4 |
| Flexural modulus (M pascal) | 1310 | 1110 | 890 |
| Rockwell A hardness | 98 | 87 | 71 |

[a] From Bull and Holden [22].

Some data of Bull and Holden [22] on the modification of polypropylene with SBS block polymer (30% styrene) are given in Table IV. These results are typical of what can be achieved by use of styrene–diene block polymers in blends with polyolefins. The writer's unpublished data show that neither polystyrene nor polybutadiene blocks are compatible with polypropylene or polyethylene. When SBS was added to low-density polyethylene even in small amounts (<10%), substantial improvements in environmental stress crack resistance were achieved [22].

A growing application for linear triblock or branched multiblock diene–styrene copolymers is in polyethylene film. Resistance to tearing and tensile impact failure are increased significantly by the use of 5–20% block polymer. The use of diblock polymers in this application has also been described [22a].

## B.  Compounding of Thermoplastic Block Polymer Elastomers

In practical applications SBS, SIS, and related elastomeric block polymers are frequently blended with other polymers. In almost all these instances cost is an important factor, the added polymer being cheaper than the structurally more complex block polymer. However, specific effects on physical properties are usually another important consideration.

Polystyrene is unquestionably the most widely used polymer in compounding SBS and $(SB)_x$ block polymers. Predictably, it increases stiffness and hardness. Most commercial polystyrenes also increase melt flow. An important application of these blends is in injection-moldable shoe soles. The complete formulation also includes large quantities of plasticizing oils and fillers, along with processing aids and other additives [23]. These compounds

Table V

Some Formulations Involving Blends of Butadiene–Styrene Block
Polymer Elastomers with Other Polymers[a]

| | Injection-molded shoe soling | Extruded slab soling | Hose or tubing | Ozone-resistant molding or extrusion compound |
|---|---|---|---|---|
| Block copolymer | 150[b] | 100[c] | 100[d] | 150[b] |
| Polystyrene | 80 | — | 40 | 40 |
| Polyethylene | — | 5 | — | — |
| Polyindene | — | — | 20 | — |
| EPDM rubber | — | — | — | 20 |
| Hard clay | 50 | 40 | — | 80 |
| Calcium carbonate | 50 | — | 120 | — |
| Titanium dioxide | 10 | — | 5 | — |
| Carbon black | — | — | — | 5 |
| Yellow iron oxide | — | 2 | — | — |
| Naphthenic oil | 80 | — | 80 | 50 |
| Stearic acid | 3 | — | 3 | 3 |
| Rosin | — | 20 | — | — |
| Wax | — | — | — | 6 |

[a] From Haws and Middlebrook [23].
[b] Butadiene–styrene thermoplastic elastomer (40% styrene); masterbatch with 50 phr naphthenic oil.
[c] Butadiene–styrene thermoplastic elastomer (40% styrene).
[d] Butadiene–styrene thermoplastic elastomer (30% styrene).

are injection molded directly onto the canvas uppers of tennis shoes, deck shoes, and other casual footwear. A very real advantage over conventional formulations employing SBR and high styrene resin, both random copolymers, is improved low-temperature flexibility derived from the very low $T_g$ of the polybutadiene center blocks ($-98°C$) with no sacrifice in (desirable) hardness at higher temperatures. Of course, vulcanization is unnecessary and scrap can be reprocessed. Polystyrene blends are also used in extrusion compounds for such items as tubing, hose, or cove base.

Polyethylene is sometimes added to styrene–butadiene thermoplastic elastomers in minor amounts. It increases hardness and resistance to abrasion [23] and often gives the compound a satin finish [16]. Polyindene has been reported to increase flex life, while coumarone–indene resins increase tensile strength at elevated temperatures [23]. The latter effect is believed to be the result of compatibility with the styrene domains of the block polymer [15].

Ethylene–propylene copolymers or terpolymers with a small quantity of an unsaturated comonomer (EPDM rubber) impart ozone resistance to styrene–butadiene block polymers [23], as do copolymers of ethylene and vinyl acetate [16]. Ethylene–propylene rubber and EPDM perform the same function in other unsaturated elastomers; the amounts used are usually 20–25% of the blend (see Chapter 19).

A number of actual formulations illustrating some of the above applications are given in Table V. For details and property evaluations the reader is referred to Haws and Middlebrook [23].

## C.  Block Polymer-Based Adhesives

One of the principal advantages of SBS, SIS, and related block polymers in the formulation of pressure-sensitive and hot-melt adhesives is resistance to flow at ambient temperature, commonly termed *holding power*. Below the glass transition of the polystyrene domains, true viscous flow cannot occur except by disruption of the domain structure. This involves pulling styrene blocks out of the glassy polystyrene domains, undoubtedly with subsequent attachment to other domains. Recoverable creep plays only a minor part in the holding power of adhesives; for a tape carrying a 0.002-cm-thick adhesive layer to slip 1 mm a shear strain of 50 is required, which is certainly greater than the recoverable shear strain of the adhesive. The other requirements of pressure sensitive adhesives—tack and peel strength—are readily achieved by the use of tackifying resins effective also for rubbery homopolymers or random copolymers, for instance natural rubber or SBR.

Dahlquist [24] has shown that a necessary condition for the development of satisfactory pressure sensitive tack is the establishment of full contact with the microscopically rough substrate in the time allowed for formation of the bond, that is, on a time scale of 1 sec. This is usually achieved if the 1-sec compressive creep compliance of the adhesive is about $10^{-7}\,\mathrm{cm^2/dyn}$ or greater; serious loss of tack by contact limitation results when this compliance falls below $10^{-8}\,\mathrm{cm^2/dyn}$. For this reason the best SIS or SBS block polymers for use in pressure–sensitive adhesives contain 30% styrene or less, preferably 15–20%. Polystyrene domain connectivity associated in the higher styrene contents is undesirable since it makes it exceedingly difficult if not impossible to meet the above criterion at room temperature. With polymers of 30–40% styrene content, properly formulated, the criterion can be met at elevated temperatures, and such polymers are useful in hot-melt adhesives.

The role of the tackifier is not merely to increase the compliance of the polymer during bonding, but also to increase the modulus and the dissipation of strain energy at shorter time scales if the adhesive bond is to withstand large forces at high strain rates. The tackifier does this by increasing $T_g$ of the rubbery matrix with which it is compatible (see Fig. 5). Excessive amounts of tackifying resin, however, will raise the $T_g$ (and with it the modulus) so much that the full contact requirement for bonding can no longer be fulfilled. Thus, there is an optimum level of tackifier for any given adhesive formulation. In general, this optimum lies in the vicinity of 100 phr for both SIS and SBS block polymers.

Among tackifying resins used with block polymers are polyterpenes and partly or fully hydrogenated esters of glycerol and rosin acid. They are compatible with the diene center blocks, although exceptions have been noted [25]. Polystyrene-compatible resins are poly($\alpha$-methylstyrene), styrene–$\alpha$-vinyl toluene copolymers, coumarone–indene resins, and others [15, 16]. Although these resins do not impart tack, they are sometimes used in adhesives to improve other properties, for example, high-temperature strength. Ordinary plasticizers are also used in formulating adhesives. The essential part of the formulation, however, consists only of block polymer, tackifier, and one or more antidegradants.

The simultaneous achievement of high tack, high holding power, and a practical adhesive formulation viscosity requires compromises in the design of block polymers for adhesive uses. Thus, high holding power is achieved through high molecular weight and high styrene content. Both variables effectively increase the polystyrene block length. Unfortunately, high molecular weight leads to high formulation viscosity and high styrene content to low tack. The use of radial structure block polymers $(SB)_x$ or $(SI)_x$ is of some advantage, since their melt viscosity at equal molecular weight is smaller than that of the linear block polymers, permitting the use of higher molecular weights [26].

### D.  Block Polymers in Rubber Compounding

There is a variety of applications of block polymers in the compounding of rubbers. The preferred type of block polymer is the AB diblock, for example, butadiene–styrene with a monomer ratio of 75/25. Since the final compound must be vulcanized, no significant advantages result from the use of SBS or (SB)$_x$; in either case the polybutadiene blocks are cross-linked into the network, and the network-forming ability of the tri- and multiblock polymers is no longer of much consequence to the physical properties of the blend.

At fixed molecular weight and styrene content the melt viscosity of an SB diblock polymer is much larger than that of a random copolymer, and its temperature coefficient is also greater [14]. This permits the use of relatively low molecular weight block polymers ($\overline{M}_w \approx 10^5$) as modifiers of rheological properties of unvulcanized rubber compounds. Because the SB block polymers are products of alkyllithium polymerization, their molecular weight distribution is narrow, so that their number-average molecular weight is not so small as to cause significant deterioration of vulcanizate properties. Desirable rheological effects of SB-block polymers in blends with general purpose unsaturated rubbers (natural rubber, SBR) are reduction in mill shrinkage, increased extrusion rates, reduced die swell, and smoother surfaces of extrudates [14, 27]. For example, as little as 5–10 phr of a 75/25 SB block polymer has been found to reduce the mill shrinkage of natural rubber mechanical goods compounds from 35 to 10–15% without much effect on vulcanizate properties. Excellent extrusion compounds can be formulated from blends with SBR random copolymers. In injection-molded compounds, injection times can be reduced significantly, with resulting gains in production rates.

Another interesting property of SB block polymers that can be exploited in blends with other rubbers is incompatibility with the blend rubber on a molecular scale (see Fig. 4). The result of this is the felicitous combination of greater hardness at ambient temperatures and the better low-temperature flexibility already referred to. On the other hand, the combination of improved extrudability and low-temperature flexibility is attractive in wire insulation [14].

Use of SB block polymers for blending is not recommended in applications requiring low energy dissipation, such as in automotive tire treads and carcasses. The relatively low molecular weight and the loss maximum of the polystyrene domains near 100°C are both sources of additional hysteresis.

### E.  Other Applications

An interesting and unusual application of block polymer blends has been described by Schramm and Blanchard [28] and by Paul *et al.* [29, 30]. Noting that polyethylene chlorinated by a slurry process is a block polymer in the sense that chlorine is incorporated nonrandomly along the chain, they investigated its use as a modifying resin for recycled waste plastics. Such waste streams are likely to contain high proportions of polyethylene (PE), poly(vinyl chloride) (PVC), and polystyrene (PS), but owing to the incompatibility of these polymers, the mechanical properties of the resulting blends are poor, lacking particularly in ductility. The chlorinated polyethylene (CPE) serves to promote interfacial adhesion between phases through its PVC-like and polyethylenelike segments and to decrease domain sizes. Ductility and toughness are increased at a sacrifice in modulus and tensile strength. From a practical standpoint the gain in the first two properties outweighs the disadvantages. Best results were obtained with PE–PVC blends; such waste might be generated by the wire and cable industry.

Hydrogenated SBS triblock polymers are now available, with a center block that is, in effect, a random copolymer of ethylene and 1-butene [16]. These polymers show greatly improved resistance to oxidation and ozone attack. Their environmental stability may be exploited in blends. For example, Davison and Wales [31] describe blends of such block polymers with unsaturated carboxylic acid-modified polypropylene or its salts.

It is to be expected that many other applications will be developed in this rapidly growing area of polymer technology. Moreover, although this discussion has been limited to the use of block copolymers in polyblends, similar effects can generally be achieved with graft polymers of the appropriate composition and structure.

## IV.  SUMMARY

Block copolymers can be blended with other polymers to produce materials with unique physical properties. The resulting blends are almost invariably multiphase systems, but compatibility of one of the blocks with the added homopolymer or random copolymer is frequently the key to the performance of the blend. Styrene (S)–diene (D) copolymers of the SD, SDS, or radial $(SD)_x$ type are by far the most common class of block polymers employed in blending. Applications include modification of brittle plastics, notably polystyrene, to increase impact resistance and improvement of tear strength and impact resistance of polyolefin films. Minor amounts of block polymers in rubbers are used to improve processing behavior, increase hardness, and impart low-temperature ductility. In most applications calling for block polymers as a major constituent, they are compounded with other polymers. The latter may be low molecular weight resins, such as tackifiers used in adhesive compositions, or high molecular weight polymers. Use of polystyrene in shoe sole stocks formulated from thermoplastic butadiene–styrene block polymers (plastomers) is particularly widespread.

## REFERENCES

1. G. Kraus and K. W. Rollmann, *J. Polym. Sci. Polym. Phys. Ed.* **14**, 1133 (1976).
2. T. Inoue, T. Soen, T. Hashimoto, and H. Kawai, *in* "Block Polymers" (S. L. Aggarwal, ed.), p. 73. Plenum, New York, 1970.
3. E. B. Bradford, *in* "Colloidal and Morphological Behavior of Block and Graft Polymers" (G. E. Molau, ed.), pp. 29–31. Plenum, New York, 1971.
4. G. Kraus and K. W. Rollmann, *J. Polym. Sci. Polym. Phys. Ed.* **15**, 385 (1977).
5. O. Kramer, R. Greco, R. A. Neira, and J. D. Ferry, *J. Polym. Sci. Polym. Phys. Ed.* **12**, 2361 (1974).
6. O. Kramer, R. Greco, J. D. Ferry, and E. T. McDonel, *J. Polym. Sci. Polym. Phys. Ed.* **13**, 1075 (1975).
7. P. G. de Gennes, *J. Chem. Phys.* **55**, 572 (1971).
8. C. W. Childers, G. Kraus, J. T. Gruver, and E. Clark, *in* "Colloidal and Morphological Behavior in Block and Graft Polymers" (G. E. Molau, ed.), pp. 193–207. Plenum, New York, 1971.
9. D. McIntyre and E. Campos-Lopez, *in* "Block Polymers" (S. L. Aggarwal, ed.), p. 29. Plenum, New York, 1970.
10. G. A. Harpell and C. E. Wilkes, *in* "Block Polymers" (S. L. Aggarwal, ed.), pp. 31–41. Plenum, New York, 1970.
11. M. Baer, *J. Polym. Sci. Part A* **2**, 417 (1964).
12. L. M. Robeson, M. Matzner, L. J. Fetters, and J. E. McGrath, *in* "Recent Advances in Polymer Blends, Grafts and Blocks" (L. H. Sperling, ed.), pp. 281–300. Plenum, New York, 1973.
13. D. R. Hansen and M. Shen, *Macromolecules* **8**, 903 (1975).
14. G. Kraus and H. E. Railsback, *in* "Recent Advances in Polymer Blends, Grafts and Blocks" (L. H. Sperling, ed.), pp. 241–267. Plenum, New York, 1973.
15. G. Holden, *in* "Block and Graft Polymerization" (R. I. Ceresa, ed.), Vol. I, Chapter 6. Wiley, New York, 1973.
16. G. Holden, *in* "Recent Advances in Polymer Blends, Grafts and Blocks" (L. H. Sperling, ed.), pp. 269–279. Plenum, New York, 1973.
17. M. Sherriff, R. W. Knibbs, and P. G. Langley, *J. Appl. Polym. Sci.* **17**, 3495 (1973).
18. J. Kohler, G. Riess, and A. Banderet, *Eur. Polym. J.* **4**, 173, 187 (1968).
19. C. W. Childers, U.S. Patent, 3,429,951 (to Phillips Petroleum Company) (February 25, 1969).
20. R. R. Durst, R. M. Griffith, A. J. Urbanic, and W. J. Van Essen, *Amer. Chem. Soc. Organ. Coatings Plast. Div. Prepr.* **34**(2), 320 (1974).
21. C. W. Childers and E. Clark, U.S. Patent, 3,649,716 (to Phillips Petroleum Company) (March 14, 1972).
22. A. L. Bull and G. Holden, paper presented at Amer. Chem. Soc. Rubber Division meeting, Minneapolis, Minnesota (April 1976).
22a. H. Asai and A. Wada, Japanese Patent, 73 76,939 (to Nippon Zeon Ltd.) (Oct. 16, 1973).
23. J. R. Haws and T. C. Middlebrook, *Rubber World* **167**(4), 27 (1973).
24. C. A. Dahlquist, *in* "Treatise on Adhesion and Adhesives" (R. L. Patrick, ed.), p. 219. Dekker, New York, 1969.

25. G. Kraus, F. B. Jones, O. L. Marrs, and K. W. Rollmann, *J. Adhesion* **8**, 235 (1977).
26. O. L. Marrs, R. P. Zelinski, and R. C. Doss, *J. Elastom. Plast.* **6**, 246 (1974).
27. H. E. Railsback and G. Porta, *Mater. Plast. Elastom.* **35**(1), 63 (1969).
28. J. N. Schramm and R. R. Blanchard, *Tech. Papers, Regional Tech. Conf., Palisades Sect. Soc. Plast. Eng.* (October 1970).
29. D. R. Paul, C. E. Locke, and C. E. Vinson, *Polym. Eng. Sci.* **13**, 202 (1973).
30. C. E. Locke and D. R. Paul, *Polym. Eng. Sci.* **13**, 308 (1973).
31. S. Davison and M. Wales, German Patent, 2,312,752 (to Shell International) (September 27, 1973).

# Chapter 19

# Elastomer Blends in Tires

## E. T. McDonel, K. C. Baranwal, and J. C. Andries

*The B. F. Goodrich Company*
*Brecksville, Ohio*

# I.  INTRODUCTION

## A.  Background on the Tire Industry

The motor vehicle transportation industry, as we know it today, would not be possible without the pneumatic tire. In 1975 the United States tire industry used 64% of the total new rubber consumed in the United States to make 208.5 million tires, 95% of which were passenger and truck tires [1]. Recent consumption of the rubbers important to the tire industry in the United States is listed in Table I. The importance of elastomer blends becomes manifest when we realize that nearly every major rubber component in a tire constitutes a blend of two or more of these rubbers.

Table I

Consumption of New Rubber in the United States (1000 metric tons)[a]

|  | 1975 | 1976 (est.) |
|---|---|---|
| Styrene butadiene rubber (SBR) | 1158 | 1325 |
| Natural polyisoprene rubber (NR) | 620 | 740 |
| Synthetic polyisoprene rubber (IR) | 62 | 100 |
| Polybutadiene rubber (BR) | 321 | 335 |
| Butyl rubber (IIR) | 110 | 140 |
| Ethylene propylene rubber (EPDM or EPM) | 92 | 100 |

[a] Data from *Rubber Age* [2] and Rubber Manufacturers' Association [3].

## B.  Scope of This Review

The pneumatic tire has a wide range of diverse applications; it is used in aircraft, bicycles, farm tractors, etc. This chapter will be restricted to an examination of critical performance criteria of passenger and truck tires.

The rubber blends reviewed are those types normally obtained in tire manufacturing, that is, by physical blending in standard rubber equipment rather than by such preparative techniques as block or graft polymerization.

## C.  Operating Features of Pneumatic Tires

To appreciate fully the performance required of rubber blends in tires, we should review what the tire does. Briefly stated, the tire supports the vehicle, absorbs road shocks, transmits tractive power, steers, and brakes the vehicle. It is required to be rigid and flexible. The requirements of a structure necessary to survive these conflicting demands have been reviewed [4]. It is a thin-shelled structure of double curvature made of high-modulus filaments (textile) embedded in a low-modulus matrix (rubber). In addition to being rigid enough to develop substantial forces in any direction, the tire must be flexible enough to deform repeatedly from its doubly curved surface to a plane surface when it contacts the road without being stressed past its yield point or sustaining fatigue failure.

The conventional tire has three distinct structural features: (1) rubber compounds to provide road contact, to contain air, and to insulate the textile cords; (2) textile cords for rigidity and strength; (3) steel beads to connect the tire to the wheel of the car.

A schematic diagram of a typical tubeless, belted, passenger tire section is shown in Fig. 1. Rubber blends are utilized to assist in achieving the best balance of properties in each portion of the tire. Balance is emphasized since compromises in properties are usually needed. The tread must provide wear resistance, good friction properties, fatigue resistance, and resilience to minimize cuts and tears. Low hysteresis loss is useful, especially in truck tires, since lower operating temperatures prolong tire life. The belt coat compound, the rubber surrounding the high-modulus cords in the restrictor band, must have good adhesion to the textile, and must be high modulus to provide textile load transfer as well as insulation. The carcass coat compound, the rubber surrounding the textile fabric in the carcass stock, must also insulate the cords to maintain cord bundle integrity during bending or compression. The compound must have low hysteresis losses, excellent flex fatigue resistance, and low modulus to accommodate slight shifts in cord path. The inner liner compound again must have good flex fatigue resistance and must minimize diffusion of the inflating air. The blends used in liner compounds are usually formulated for minimum cost as well as low diffusion. They will not be discussed in this chapter.

The sidewall compound must be especially resistant to flex fatigue and cut growth since it is exposed to maximum flexing. It must also resist cracking initiated by sunlight and atmospheric exposure (e.g., ozone). The decorative

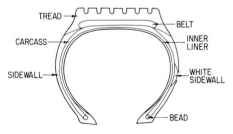

Fig. 1  Schematic diagram of a typical belted passenger tire.

white sidewall is especially interesting since ozone cracking resistance must be obtained by special elastomer blends rather than by the more traditional technique of chemical antiozonants, which tend to discolor the white compound.

## D. An Overview of Elastomer Blends in Tires

Elastomer blends are used in tire rubber compositions for three reasons: (a) the compound cost may be lowered; (b) during production, the tire may be more easily fabricated in the complex shaping, forming, and building operations used to make a tire; (c) the final product performance can be modified beneficially.

Natural rubber and styrene–butadiene copolymer have long been blended for these reasons. The last 15 years has seen a very rapid growth in the use of blends with stereopolybutadiene to improve wear and flex resistance in treads and to improve cut and flex resistance in carcasses and sidewalls. Table II is a summary of typical rubber combinations currently being used in tires.

A unique factor pervading rubber compounding, particularly tires, is carbon black. It is the essential ingredient contributing the property of "reinforcement," which is reflected in increased hardness, modulus, tear resistance, abrasion resistance, and hysteresis loss. Carbon black–rubber reinforcement have been exhaustively studied [5] and it is not within the scope of this chapter to do this. However, many of the features of rubber compounds are determined by the way in which carbon black is distributed between the elastomer phases in blends. Carbon black poses a thorny experimental problem. Its presence mutes or obscures many of our key experimental observations.

Table II

Typical Elastomers in Passenger and Truck Tires

|  | Passenger | Truck |
|---|---|---|
| Tread | SBR–BR | NR[a]–BR or SBR–BR |
| Belt | NR | NR |
| Carcass | NR–SBR–BR | NR–BR |
| Black sidewall | NR–SBR or NR–BR | NR–BR |
| White sidewall | NR–SBR–EPDM–IIR[b] | — |
| Liner | NR–SBR or NR–SBR–IIR | NR–IIR |

[a] Includes synthetic polyisoprene IR.
[b] Includes halobutyl.

A comprehensive review of elastomer blends [6] has appeared recently that provides the basis for a review of the literature in general.

## II. CHARACTERIZATION OF ELASTOMER BLENDS IN TIRES

A large variety of polymer characterization techniques have been used to identify the individual polymers, and to investigate the morphologies seen with elastomer blends. The most important of these methods are discussed, including phase-contrast and electron microscopy, thermal analysis, spectroscopy, and pyrolysis gas chromatography.

### A. Microscopy

Optical and electron microscopy provide methods to examine the morphology of elastomer blends, that is, the study of the zone sizes and distribution of each polymer component in the blend and of the interfaces between them. Simply put, microscopy allows us to see what component parts a blend contains. An excellent review article covering the entire area of rubber microscopy has been written by Kruse [7]. In the area of rubber blends Walters and Keyte [8] used phase-contrast optical microscopy to differentiate the components in blends involving natural polyisoprene rubber (NR), butadiene rubber (BR), and styrene–butadiene rubber (SBR). They found separate zones of one polymer dispersed in another with the size of the zones being influenced by method of blending, mixing time, and polymer viscosities.

Electron microscopy is more suitable than optical microscopy for studying elastomer blend morphologies due to the development of very small, dispersed-phase, zone sizes, as low as about 0.5 $\mu$m. Hess and co-workers [9] have obtained electron micrographs showing good phase contrast of NR blends with SBR, BR, and chlorobutyl rubber (Cl-IIR) by swelling the cured blend with methyl methacrylate, butyl methacrylate, and styrene. Subsequent polymerization of these monomers produces a hard material that can be successfully microtomed for electron microscopy. Differential swelling between the components of the blends gives rise to the phase contrast. Marsh *et al.* [10, 11] also obtained good phase differentiation with electron microscopy by using another swelling technique, in which specimens swollen in naphtha are brushed upon a substrate and observed in the stretched state after the solvent has evaporated.

Both of these swelling methods distort the zone sizes of the dispersed phase by virtue of the gross differential swelling that occurs, and, therefore, accurate spatial measurements may be difficult to make. Smith and Andries [12] avoided this problem by adopting an ebonite technique to examine rubber blends, in which the unvulcanized blend is hardened in a sulfur–sulfenamide–zinc stearate mixture, then microtomed. A preferential curing takes place in which one phase becomes more electron dense than the other, thereby developing good phase differentiation. Zone size distortion can also be avoided by staining the blends with osmium tetroxide, which selectively stains the polymer in a blend with the greater unsaturation. Andrews [13] has used this electron microscopic technique with NR–EPDM (ethylene–propylene–diene terpolymer) blends, and Avgeropoulos et al. [14] have used it with EPDM–BR blends. Finally, Lewis and co-workers [15] have combined autoradiography with electron microscopy to identify accurately the phases in SBR–NR blends.

## B.  Infrared Spectroscopy

Infrared spectroscopy has been extensively used as a qualitative and quantitative identification technique for rubber, including elastomer blends. Hampton [16] has published a detailed review article on the application of infrared spectroscopy to elastomer systems. In general, infrared spectroscopy is useful in gross identification but does not characterize significant morphology.

## C.  Thermal Analysis

Thermal analysis techniques such as differential thermal analysis (DTA), derivative thermogravimetry (DTG), and differential scanning calorimetry (DSC) are particularly useful tools for studying elastomer blends. Morris [17] and de Decker [18] have applied DTA to elastomer blends and have suggested that for heterogeneous blends (presence of a dispersed phase), DTA curves show the unchanged summation of the individual responses of the blend polymers. Homogeneous blends (no dispersed phase), however, may give either a single glass transition temperature ($T_g$) of one of the blend components or a single, intermediate $T_g$. Brazier and Nickel [19] determined the heat of vulcanization ($\Delta H_v$) of NR–BR blends with DSC; they studied the effects of polymer ratio, sulfur level, carbon black content, and stearic acid level on $\Delta H_v$. Sircar and Lamond [20] used DSC and DTG along with $T_g$ measurements to make polymer identification of tread and black sidewall compounds containing blends of either NR, IR (synthetic polyisoprene rubber), SBR, or BR. Sircar and Lamond also used these same techniques to identify all polymers in white sidewall compounds, [21, 22], and to identify polymer composition of butyl and halogenated butyl blends used in innerliner compounds [23]. Using a similar total thermal analysis method, Brazier and Nickel [19] analyzed fully compounded rubber stocks for all the main constituents.

## D. Gas Chromatography

Curie-point pyrolysis gas chromatography has recently been used by Krishen and Tucker [24] to make quantitative determinations of the elastomer constituents in compounded stocks of SBR, BR, EPDM, and NR. Sugiki [25] has used pyrolysis gas chromatography with a radiofrequency induction apparatus to examine three component blends of NR, SBR, and BR.

## E. Solubility Differentiation

Solubility differences between polymers in a common solvent have been used to measure blend compatibility [26]. Although these solubility differences may give some information concerning compatibility, dilute solutions do not provide the environment that exists when two solid polymers are blended together.

## F. Nuclear Magnetic Resonance Spectroscopy

The use of nuclear magnetic resonance (NMR) in the study of vulcanized, filled elastomer blends has been limited owing to line broadening of the spectra, but Fujimoto [27] has used NMR in conjunction with other techniques to study the heterogeneity of filled elastomer systems.

## III. PRACTICAL MIXING TECHNIQUES

### A. Latex and Solution Blending

Methods that are used to mix elastomer blends include latex blending, solution blending, and mechanical blending. The first two are not practiced in the tire industry, but are of interest as alternatives to mechanical mixing. The processing, properties, and economics of latex blending have been generally reviewed [28]. In his investigations of latex blending [29, 30] Blackley has examined the effects of blend composition on viscosity and the mechanical stability of NR–SBR latex blends. At this time, latex blending does not offer any property advantage for tire blends over the other mixing methods, but it has been reported [31] that a BR latex blend with SBR or NR gives an improved dispersion of ISAF carbon black.

In solution blending, the high mobility in solution prior to evaporation of the solvent allows blends to separate and causes a macroheterogeneous morphology consisting of rather large zones of dispersed polymer. Walters and Keyte [8] mixed SBR–NR and BR–NR blends in a 5% solution of benzene, evaporated the solvent, and examined the blend morphology with phase-contrast microscopy. The resultant polymer blend consisted of large isolated flat areas of SBR or BR, often 400 µm thick surrounded by NR. As discussed later, these dispersed zones are much larger than those obtained with mechanical mixing. Changing solvent or concentration had little effect on the morphology. Gardiner [32] studied solution mixing of blends containing various amounts of NR, SBR, BR, EPDM, and IIR (butyl rubber) in toluene by phase contrast microscopy. He found dispersed zones of 1–100 µm and suggested that the smaller the difference in critical surface tension, solubility parameter, and solvent interaction parameter between the polymers, the smaller the zone size will be. Shundo et al. [33] have prepared solution blends of NR and SBR by coprecipitation with addition of alcohol and found very uniform blends.

## B.  Morphology of Mechanical Blending

Mechanical blending is the most widely used method for mixing elastomer blends in the tire industry. Only heavy equipment such as a Banbury internal mixer, the open-roll mixer, and specially designed screw extruders are capable of developing the high shearing forces necessary to blend high molecular weight elastomers and especially to incorporate and disperse carbon black.

Walters and Keyte [8] were the first to make an extensive study of elastomer blend morphology by studying the factors that affected the zone size of the dispersed phase in polymer blends. They examined two component blends of SBR, NR, BR, IIR, and neoprene (CR) with phase-contrast microscopy. They found a heterogeneous morphology consisting of discrete zones from about 0.5–100-µm diameter of the dispersed polymer with a continuous matrix of the second polymer. Occasionally they found cocontinuous phases with no distinct disperse or continuous phases. These polymer zones were reduced in size with increased milling time or temperature, until a minimum size zone was reached. The smallest zone sizes of the dispersed polymer were obtained when the blended polymers had similar viscosities or molecular weights. With dissimilar viscosities, the higher molecular weight polymer was dispersed in a continuous matrix of the lower viscosity polymer. Similarity of molecular structure also reduced zone size, but the effect was not as great as that of viscosity. Under all conditions tested, homogeneous blends or molecular solutions were never obtained; that is, either a distinct, dispersed phase or cocontinuous phases occurred.

Callan [34] has reported similar morphology, with IIR/EPDM blends, consisting of dispersed zones of the minor polymer within a continuous matrix of the major polymer. Phase inversion occurred at the 50/50 ratio, the mixture then consisting of cocontinuous polymer zones. Hess [9] used electron microscopy to investigate NR blends with SBR, BR, and Cl-IIR rubber. He observed similar dispersed phase morphology and confirmed the importance of the viscosity difference of the polymers on the zone size of the dispersed phase.

Marsh *et al.* [10], using an electron microscopic technique discussed

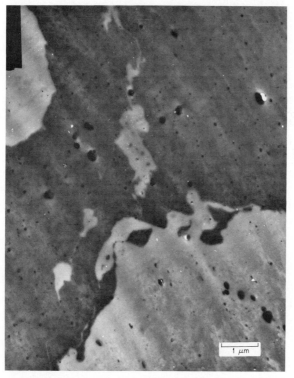

Fig. 2   Zone size of disperse elastomer phase: 75/25 SBR–BR, 1-min mill mixing at 25°C; BR is light phase. (From Smith and Andries [12], Fig. 9.)

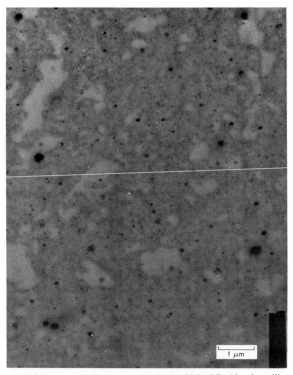

Fig. 3    Zone size of disperse elastomer phase: 75/25 SBR–BR, 10-min mill mixing at 25°C; BR is light phase. (From Smith and Andries [12], Fig. 10.)

earlier, found distinct phase differentiation for all elastomer blends studied, except SBR–BR. They concluded that SBR–BR blends were homogeneous, that is, they formed a molecular solution. Later Marsh [11] confirmed these results and reported that DTA of cross-linked SBR–BR blends gave one $T_g$, indicating that they were indeed homogeneous. Callan [38], using a different electron microscopic technique from Marsh, found that SBR blends with three types of BR showed the existence of discrete polymer zones under all mixing conditions. He supported this work with DTA data on uncross-linked samples showing the presence of $T_g$'s typical of each individual polymer. Smith and Andries [12] further examined SBR–BR blends with a nonswelling electron microscopic technique that gives good phase contrast between the two polymers. They found that under all conditions studied, SBR–BR blends were heterogeneous. They concluded that earlier work showing SBR–BR blends to be homogeneous were probably due to the inability of the technique to differentiate between the two polymers. In agreement with earlier work, the zone size of the dispersed phase decreases with increasing milling time (as seen by comparing Fig. 2 with Fig. 3); increasing similarity of viscosity (as seen by comparing Fig. 3 with Fig. 4); increasing temperature (as seen by comparing Fig. 4 with Fig. 5); and increasing similarity of molecular structure. From this work and that of many others, all blends of elastomers used in tires should be characterized by a certain degree of heterogeneity. Thermodynamics also predicts that two polymers do not have to be very dissimilar before they are insoluble in each other and lead to the existence of discrete zones upon mixing (see Volume 1, Chapters 2 and 3).

Recently, Avgeropoulos *et al.* [14] have examined the relationship between the rheology of EPDM–high vinyl content BR blends and their morphology. They found that the morphology of the blend is dependent upon the composition of the blend and upon the ratio of the viscosities of the two polymers (see Fig. 6). For a given EPDM–BR blend, the continuous-phase

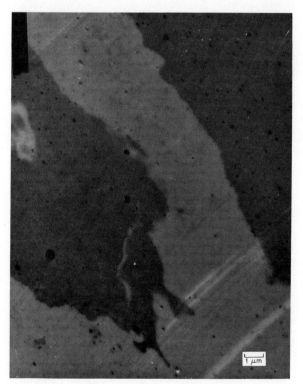

Fig. 4  Zone size of disperse elastomer phase: 75/25 SBR/high molecular weight BR, 10-min mill mixing at 25°C; BR is light phase. (From Smith and Andries [12], Fig. 11.)

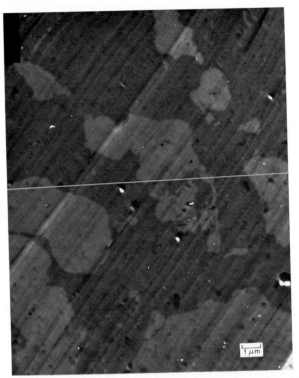

Fig. 5    Zone size of disperse elastomer phase: 75/25 SBR/high molecular weight BR, 10-min mill mixing at 65°C; BR is light phase. (From Smith and Andries [12], Fig. 12.)

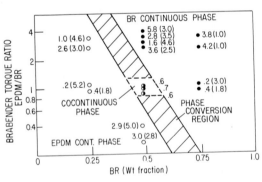

Fig. 6    Morphology changes as a function of viscosity (Brabender torque ratio) and composition of EPDM–BR blends. The numbers show the mean domain diameter in microns and the length to diameter ratio (in parentheses). Shaded symbols indicate EPDM as the disperse phase; open symbols indicate BR. (From Avgeropoulos *et al.* [14], Fig. 8.)

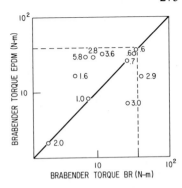

Fig. 7   Domain size (mean diameter in microns) as a function of viscosity (Brabender torque) and torque ratio for 50/50 EPDM–BR systems. The dashed lines indicate the entanglement plateau region. (From Avgeropoulos *et al.* [14], Fig. 10.)

polymer will be the one that has the lower viscosity as long as the composition is within a certain range. Avgeropoulos confirmed the fact that the zone size tended to decrease as the viscosity ratio approached unity for the two polymers (see Fig. 7), but he also observed the zone size to decrease with increasing viscosity of the components (see Volume 1, Chapter 7). That is, for blends the constituents of which have similar viscosities, the blend with the higher viscosity will have the smaller zone size. However, the change in zone size with the magnitude of the viscosity occurs at a lower rate than the change with the viscosity ratio.

## C.   Carbon Black Distribution among Elastomer Phases

As previously mentioned, carbon black is added to all practical tire compounds as a reinforcing filler. It is apparent that variations in carbon black distribution among the individual phases and possible carbon black transfer from one phase to another could have profound effects on the physical properties of blends. Much research has been done in this area. Because of difficulties in experimental techniques, many inconsistencies have been reported.

Walters and Keyte [8], using phase-contrast microscopy, studied carbon black distribution and found black to initially prefer the NR phase in NR–high molecular weight BR blends before eventually going into both polymers. It appeared that the black mixed initially with the least viscous polymer until its viscosity approached that of the more viscous polymer; then the black was taken up by both polymers. Hess [9] confirmed these findings with electron microscopy, observing that the carbon black located preferentially in the lower-viscosity polymer of a NR–BR blend, and suggested the possibility of carbon black transfer from one phase to another. Hess [36] later reported that carbon black did migrate into the BR phase from NR in solution masterbatch blends of BR–NR. Marsh [11] observed that carbon black preferred polychloroprene (or neoprene (CR) in NR–CR blends, but concluded that actual carbon black transfer from one polymer to another does not occur. Marsh [37] also studied the carbon black distribution in a triblend of NR–EPDM–neoprene and concluded that carbon black does not transfer from one elastomer phase to another in this blend either.

The conflicting information concerning carbon black distribution and transfer was essentially resolved by Callan et al. [38]. They found that carbon black affinity in 50/50 blends decreased in the order of BR, SBR, NR, EPDM, and IIR. The carbon black affinity was dependent upon unsaturation, viscosity, polarity, and mix procedure. Carbon black transfer occurred only when there was a low degree of interaction between the carbon black and the elastomer. This situation exists when the masterbatch has minimum heat or mechanical history (such as solution masterbatch blending) or involves a low molecular weight or low-unsaturation elastomer. This situation would not exist when a rubber blend has been mixed by standard high-shear techniques, such as a Banbury mixer or mill. In essential agreement with this work, Sircar [35] used DTA crystallization endotherm of BR, elastic modulus measurements, and electrical conductivity to show that carbon black does not transfer from SBR, BR, and NR (highly unsaturated polymers) when made by conventional Banbury mixing procedures. However, carbon black transfer was found to take place from a low unsaturation polymer (Cl-IIR) to a high unsaturation polymer (BR) in these blends. It is evident that carbon black transfer is the exception rather than the rule. The more important factor, as will become evident in discussions of vulcanized physical properties, is the distribution of carbon black between, and the dispersion of carbon black within, the individual elastomer phases.

## IV.   PROPERTIES OF UNVULCANIZED ELASTOMER BLENDS

### A.   Flow Properties

Elastomer blends are not homogeneous. Even for the "most" compatible system, for example, SBR–BR, polymers are incompatible at the molecular level in a well-mixed blend [12, 38]. The development and size of this microheterogeneity affect the flow as well as vulcanizate properties. Such factors as viscosity and $T_g$ of elastomers [14], blending conditions [9, 14, 39], for example, shear rate and mixing temperature, and polymer ratios affect zone sizes. Therefore, these parameters also affect flow properties.

## *1. Nonblack Elastomer Blends*

For NR–BR blends, Hess and co-workers [9] observed that mixing torque for a given polymer blend ratio was not an additive value of 100% NR and BR mixing torques. Rather, blend torque values were higher than for either elastomer by itself. Their solubility measurements on blends mixed at high shear indicated that up to 40% of the polymers were not dissolved. They concluded that during mixing high molecular weight graft copolymers are formed to give insoluble material and, therefore, greater resistance to flow.

A similar phenomena was observed by Folt and Smith [39] for NR–BR blends (see Fig. 8). However, their blends were completely soluble. No explanation of torque deviations from linearity was given. In the same study Folt reported Brabender mixing torque–time plots for NR–TPP (*trans-*polypentenamer) [39] blends for different ratios that gave an initial minimum after 30–60 sec of mixing. After that the torque increases rapidly to a maximum value with about 2 min of mixing. Further mixing lowers the torque. This phenomena was attributed to abrupt changes in morphology. With electron microscopy macrozone sizes are observed up to 2 min of mixing. Thereafter, microzone sizes are observed.

Similar effects of polymer ratios on NR–BR blend rheology have been shown by Mooney viscosity and the force required to extrude through a capillary at different shear rates up to $1.5 \text{ sec}^{-1}$ [39]. The minimum occurred between 30 and 40 phr BR. Over 60 phr BR, the flow property followed the linear additivity line (see Fig. 9). At high shear rates, the flow plots are similar to the torque–composition curves. At very high shear rates, viscosities become independent of composition [14, 39, 40], indicating an "entanglement plateau" region.

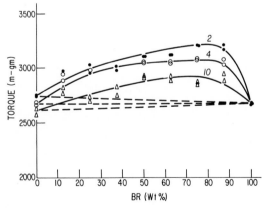

Fig. 8 Plastograph torque as a function of NR–BR blend composition at various mixing times in minutes. (From Folt and Smith [39], Fig. 13.)

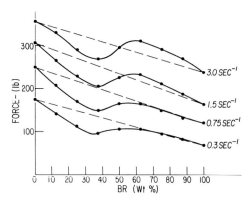

Fig. 9   Capillary extrusion force as a function of NR–BR blend composition at various shear rates. (From Folt and Smith [39], Fig. 14.)

In a two-polymer blend system, the higher content polymer becomes the continuous phase while the minor component becomes the disperse phase. For a 50/50 elastomer blend, the low-viscosity component becomes the continuous phase and surrounds the high-viscosity polymer zones. For a blend system with two different $T_g$ polymers, the flow properties depend on mixing and extrusion conditions. The polymer with the largest $T-T_g$ will be the fluid phase. This has been clearly shown for an EPDM and high-vinyl BR (98% vinyl) polymer [14]. The $T_g$ of EPDM and high vinyl BR were $-60$ and $-5°C$, respectively. At low temperature (20°C), EPDM showed lower mixing torque and was the continuous phase (see Fig. 10). At 112°C, BR was the continuous phase.

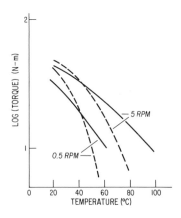

Fig. 10   Typical Brabender mixing behavior as a function of temperature of EPDM and BR polymers: (———) EPDM; (– – –) BR. (From Avgeropoulos et al. [14], Fig. 3.)

## 2. Black-Filled Elastomer Blends

Flow properties of black-filled elastomer blend compounds are much more complicated than those of gum blends. In addition to those factors affecting the flow behavior of the gum blend, the relative carbon black distribution in the polymer phases, viscosity changes due to carbon black reinforcement and mixing methods affect rheological behavior of filled blends.

Styrene–butadiene rubber and BR have the greatest affinity for carbon black; NR is next, and EPDM and butyl rubbers have the lowest carbon black reactivity [38]. Sircar and co-workers have tried to correlate filler–blend die swell to individual polymer die swell and viscosity of gum polymers. Individual die swell behavior is shown in Fig. 11.

Sircar made blends in four different ways: (1) all carbon black, 50 phr, was added to Polymer A, then this mix was diluted with B; (2) reverse of (1); (3) all carbon black was added to a preblend of the polymers; (4) polymers and carbon black were mixed together. The die swell of these four blends varied in an irregular way, and no trend was found. Their explanation was that the die swell of BR compounds is lower than those of SBR and NR compounds with less than 50-phr carbon black. Therefore, blend compounds in which the unfilled BR is the outer layer will have less die swell than when NR or SBR are the outer layers. This was found to be true for SBR–BR blends (see Table III) but not with NR–SBR or NR–BR blends. The Mooney viscosities followed the die swell pattern. The SBR–BR blend extrudates had rough surface appearances, whereas the extrudates with NR as the outer layer had smooth surface appearances. This was attributed to lower NR viscosity resulting from excessive breakdown during mixing.

There is a definite need for additional work on the rheology of tire

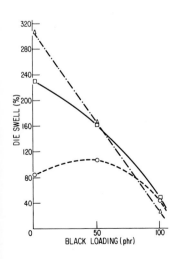

Fig. 11 Effect of carbon black loading on die swell: (— · —) NR; (——) SBR; (– – –) BR. (From Sircar *et al.* [41], Fig. 1.)

Table III

Physical Properties of BR, SBR, and SBR–BR Blends[a]

| Sample no. | 50/50 blends | Black, phr, method of incorporation | Die swell (%) | Mooney viscosity (ML-4, 212°F) | Heat generation (HBU) (°F) | Rebound (%) | Tensile (psi) | Elongation (%) | 300% modulus (psi) | Shore A hardness |
|---|---|---|---|---|---|---|---|---|---|---|
| 632 | BR gum | No black | 83 | 27 | — | — | 150 | 220 | — | 27 |
| 628B | SBR gum | No black | 230 | 32 | — | — | 280 | 520 | 90 | 35 |
| 447 | BR | 50 Phr straight | 108 | 43 | 249 | 60.2 | 1650 | 610 | 480 | 47 |
| 449 | SBR | 50 Phr straight | 161 | 40 | 248 | 52.9 | 3290 | 750 | 700 | 48 |
| 587 | SBR–BR | 50 Phr all black in BR | 152 | 35 | 213 | 62.6 | 2430 | 580 | 750 | 49 |
| 588 | SBR–BR | All black in SBR | 114 | 33 | 218 | 60.7 | 2280 | 665 | 700 | 48 |
| 589 | SBR–BR | Black in blend | 147 | 39 | 238 | 57.2 | 2720 | 730 | 660 | 47 |
| 590 | SBR–BR | Black premixed in elastomers | 142 | 38 | 232 | 59.8 | 2330 | 650 | 710 | 46 |

[a] From Sircar et al. [41].

elastomer blends. This work should concentrate on how to get the best balance of flow and vulcanizate properties with minimum processing energy input.

## B. Tack Properties

Tack, the ability of a material to adhere to itself but not to other materials, is one of the most important rubber properties for tire fabrication and building. Too tacky a stock traps air between tire components, resulting in poor durability. Too little tack makes it impossible to hold the component parts together until the tire is vulcanized. Adequate tack is a must for carcass and belt compounds to get proper ply turn-up and good uncured adhesion between plies, between plies and belts, and between plies and liner compounds.

Natural rubber has an excellent tack strength [26, 42], whereas synthetic rubbers show poor tack [42, 43]. Radial tire belt compounds, in general, contain primarily NR, and therefore have satisfactory building tack. Carcass stocks, on the other hand, are blends of SBR, BR, and NR.

Tackiness of synthetic rubbers is enhanced either by adding tackifying resins or NR and often both are used. One is limited on the amount of a tackifier used; too much resin becomes detrimental to vulcanizate properties. Therefore, a combination of NR and a tackifier is usually used in practice.

Blending two or more low-tack polymers, for example, SBR and BR, or SBR and EPDM, does not impart any tack improvement. Blending NR with these synthetic polymers increases tack almost linearly as a function of NR content [44]. A minimum of 30% is required to have an adequate building tack for standard equipment. Passenger tire carcasses generally have 30–60% NR.

Other factors [42] such as oil and carbon black content, antioxidants, surface aging and test conditions affect tack. However, these are beyond the scope of this text.

## C. Vulcanization Properties: Cure Compatibility

As mentioned earlier, the most commonly used elastomers in tire compounds are SBR, BR, NR, IIR, and EPDM. The cure rates of BR, SBR, and NR are about equal and, therefore, blends of SBR–BR, SBR–NR, and NR–BR are relatively cure compatible. The physical properties of these polymers follow almost linear additivity of the properties of the parent elastomers. Poor cure compatibility results in blends of these high un-saturation rubbers with EPDM, a low-unsaturation elastomer, or rubber blends such as NR and Cl-IIR, can cure by two different cross-linking mechanisms.

## V. PROPERTIES OF VULCANIZED ELASTOMER BLENDS

### A. Strength Properties

Unfilled elastomer blends have very low strength and therefore need carbon black reinforcement. Properties of such filled vulcanizates are complicated owing to different affinities of elastomers toward carbon black. Vulcanizate properties are not simple additive functions of parent polymers.

Corish [45] showed that, for NR–SBR blends, compounds containing all the ingredients in the NR phase with raw SBR added later had much higher tensile strength than compounds in which raw NR was added to a precompounded SBR. He concluded that SBR may have a greater affinity for black than NR. Hess *et al.* [9] found that tear and tensile strength of a 50/50 NR–BR blend containing 40 phr ISAF increased as they increased carbon black in the BR phase. As seen in Fig. 12, these properties were optimized at about 60% carbon black in the BR phase. Sircar and co-workers [41] observed a similar effect. As shown in Table III they found that in SBR–BR blends, the compound with black in the BR phase gave slightly higher tensile strength.

Besides the relative distribution of carbon black in different polymer phases, a good black dispersion in each polymer phase, good dispersion of polymer phases, relative distribution of cross-linking agents, molecular

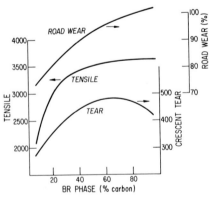

Fig. 12   Effect of carbon black distribution on vulcanizate properties, 40 phr ISAF, 50/50 NR–BR. (From Hess *et al.* [9], Fig. 14.)

weights of polymers, and cure compatibility are also very important for strength development. If two polymers do not have the same cure rates, the faster-curing elastomer will use up a major portion of curing agents, leaving the other polymer essentially "starved," which, in turn, will give poor properties. All high diene rubbers used in tires are relatively cure compatible. The problem arises when these elastomers are used in blends with low unsaturation rubbers, such as EPDM, which have much slower cure rates. Retaining accelerators in EPDM either by grafting [46], or by using high alkyl-substituted accelerators [47] gives improved diene rubber–EPDM strength.

### B.  Abrasion and Skid Resistance

Natural rubber, SBR and BR are the most commonly used polymers in tire treads. BR has higher wear resistance [48] than NR and SBR. The presence of BR improves the treadlife of NR [49] and SBR [50] compounds. This is attributed to the low glass transition temperature $(T_g)$, of BR, $-105°C$, whereas $T_g$'s of NR and SBRs are around $-57°C$. Blending these rubbers with BR lowers blend $T_g$'s and improves wear resistance. Unfortunately, poorer wet traction performance also accompanies the lower-tread $T_g$ [50]. The best trade-off between abrasion and skid resistance is obtained for SBR–CB blends in the range of a 60/40 ratio. Solution SBRs with linear molecular structure have better wear resistance than equivalent emulsion SBRs having branched molecules. Blending with BR does not improve the tread life of solution SBRs [51].

In addition to BR blending, the distribution of carbon black in blend polymers also affects the treadwear. As shown in Fig. 12, Hess *et al.* [9] reported that for a 50/50 NR–BR blend with 40 phr ISAF black, the treadlife improved with an increase of carbon black content in BR phase. As more black is located in the BR phase, it becomes more reinforced and gives better wear resistance. Reinforcement of NR is primarily by stress crystallization.

For SBRs, abrasion resistance is reduced as the styrene content increases; however, the wet coefficient of friction is increased [52]. When these polymers are blended with BR, $T_g$'s are lowered and the wet coefficient of friction is reduced in proportion to BR content. Veith [53] found a similar effect of BR on wet cornering coefficient of SBR in tires on smooth pebbles at 20, 30, and 40 mph. However, at 50 mph, BR showed no effect.

Increasing filler loadings improves wet coefficient of friction [52] of SBR and SBR–BR compounds. This effect of loading is more of a hardness effect. Veith [53] observed a similar effect of hardness. His hypothesis is that harder composites probably squeeze the water from between the rubber and the test surface much more easily. As a result, better adhesion is developed between two surfaces giving coefficients of friction. However, at very high loadings, heat buildup, dry friction, and abrasion resistance are impaired.

Grosch [54] reported that, while oil-extended NR was equivalent to oil-extended SBR on wet surfaces, it was much superior for skid resistance on icy surfaces. Tires made from 80/20 NR–BR offered a very good balance between wear and traction on wet and icy surfaces. Gnörich and Grosch [55] more recently reported that 60/40 SBR–BR blends with 50 phr HAF black had higher coefficients of friction over the temperature range of $-25$ to $-5°C$ than either polymer alone. Friction was maximum at $-15°C$ and was further increased by an increase in oil content. Black distribution between elastomer phases could be a factor in these results, but this line of experimentation has not been studied.

Other factors such as pavement type, tire type, test speed, and test temperature also affect wear and skid resistance. These are not included in this discussion, and details can be found elsewhere [56, 57].

## C. Dynamic Properties

Heat generation and dissipation in tires has been the subject of a recent review [58]. Lower tire operating temperatures are preferred. The tire temperature has a marked effect on its running characteristics, its abrasion resistance, and, in cases of extreme speeds or deflections, its life. When drive energy is transmitted from the engine to the tire, most is recovered in retraction. Some is lost as heat. The ratio of the energy lost to the energy recovered is thermal hysteresis. In the rubbery part of a tire the resistance to deformation is measured as the complex tensile modulus $E^*$ (or $G^*$—shear modulus), which is made up of an elastic response $E'$ (or $G'$) and a lagging viscous response $E''$ (or $G''$). The hysteresis is the ratio $E''/E'$ or $\tan \delta_E$ (tangent of the angle of lag).

Studebaker and Beatty [59] have pointed out that rubber compounds can be formulated for a wide range of stiffness and hysteresis. The major factors are: (1) the nature of the rubber, (2) the type and amount of cross-linking, (3) the type and amount of reinforcing filler, and (4) the plasticizer. The cross-link density and the type and amount of carbon black most profoundly influence the dynamic properties of typical tire compounds.

The consideration of the dynamic properties of rubber blends is complicated by the relative affinity of carbon black for the elastomer phases and the relative affinity of cross-linking agents for the two phases. The relative cross-linking between heterogeneous phases with differing

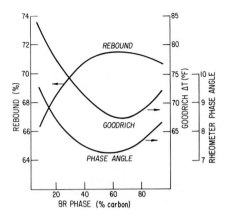

Fig. 13   Effect of carbon black distribution on hysteresis, 40-phr ISAF, 50/50 NR–BR. (From Hess *et al.* [9]. Fig. 13.)

cross-linking reactivities has been discussed by Rehner [60] and Zapp [61]. However, most tire blends involve rubbers of relatively equal cross-linking reactivity. The key influence in these elastomer blends is the way in which the carbon black is distributed between the distinct rubbery phases.

The effect of carbon black distribution and dynamic properties has been studied by Hess *et al.* [9]. The carbon black content in the BR phase of NR–BR blends was systematically varied by mixing techniques. Hysteresis loss, as shown in Fig. 13, was optimized at about 60% black in the BR phase. With the standard mixing technique of adding carbon black to a preblend of the elastomers, the carbon black tended to locate about 75% in the BR phase.

Similar work has been reported by Sircar *et al.* [41]. In NR–BR, NR–SBR, and SBR–BR blends they report a sharply lower heat generation (see Table III) when a black-loaded mix was diluted with an unloaded elastomer. Thus, the continuous phase outer layer of the soft second polymer is exerting a considerable effect on the overall properties of the vulcanizates. In further work with dynamic shear testing [62] Sircar and Lamond found that the properties of blends were influenced by the distribution of black in the phases and the degree of dispersion of black in phases, but, that once this had been stabilized the blend behaved as a single elastomer.

## D.   Fatigue Resistance

In this discussion rubber fatigue is defined as a progressive failure (crack or cut initiation and growth) resulting during repeated stressing below the ultimate stress of the sample. Rubber fatigue in tires is of concern in bias-tire treads as tread groove cracking, in radial-tire sidewalls as flex cracking, and in carcass stocks as tear and adhesion loss around the cords.

The influence of changes in rubber compound on tread groove cracking has been reviewed [63, 64]. Lake [65] has developed a fracture mechanics approach for predicting rubber fatigue life based on tearing energy and energy dissipation at the tip of a growing crack. He has applied this approach to truck tire groove cracking [66], where he showed that NR–BR blends were superior to NR alone.

Beatty [67] has examined the effect of cyclic fatigue on rubber blends. Using a 75% total stroke on a sample, he examined blends over a range of minimum strokes, including complete compression ($-75\%$ to 0), compression into tension (e.g., $-37\frac{1}{2}$ to $+37\frac{1}{2}\%$) and complete tension (e.g., $+50$ to $+125\%$). Results for NR–SBR blends are shown in Fig. 14. Natural rubber has superior fatigue life in tension, and this can be improved by blending with SBR. Styrene–butadiene rubber has superior fatigue in compression and this can be improved by blending with NR. A dramatic shift in tension fatigue between 25/75 NR–SBR and 50/50 NR–SBR is indicative of the expected phase inversion from SBR continuous to NR continuous.

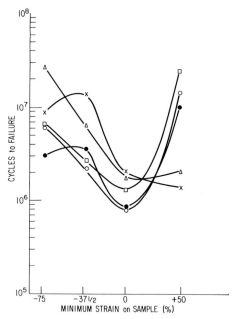

Fig. 14  Effect of NR–SBR blend on fatigue life during 75% cyclic strain in tension and compression (□) 50/50 NR–SBR; (○) 75/25 NR–SBR; (●) 100 NR; (△) 25/75 NR/SBR; (×) 100 SBR. (From Beatty [67], Fig. 7.)

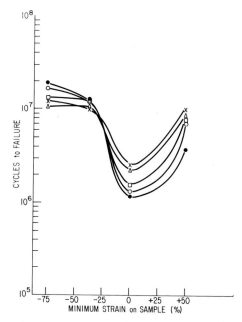

Fig. 15   Effect of OESBR/OECB blend ratio on fatigue life during 75% cyclic strain in tension and compression (OE, oil-extended): (×) OEBR; (△) 25/75 OESBR–OEBR; (□) 50/50 OESBR–OEBR; (○) 75/25 OESBR–OEBR; (●) OESBR. (From Beatty [67], Fig. 17.)

Equivalent data for blends of oil-extended BR–SBR are given in Fig. 15. Here the fatigue resistance of the oil-extended SBR is continuously improved with increasing amounts of oil-extended BR.

Sircar *et al.* [41] have reported on the strong influence of carbon black distribution in the elastomer phases on Ross flex fatigue life. They show a marked flex improvement in NR–BR and SBR–BR blends when BR forms an outer low-carbon black phase. Similar results with ring-flex samples have been obtained by Lee [68]. As shown in Fig. 16, crack growth was slowed by forcing more carbon black into the SBR phases of an SBR–BR blend by mixing varying amounts of carbon black into the individual rubbers and cross-blending. Interestingly, when free carbon black is added to an SBR–BR blend the fatigue life fell very close to that of Curve 1 representing 60% carbon black in the BR phase.

### E.   Aging and Ozone Resistance

Rubber compounds have a propensity to deteriorate in properties with exposure to heat, oxygen, and ozone. Considerable attention has been given

*E. T. McDonel, K. C. Baranwal, and J. C. Andries*

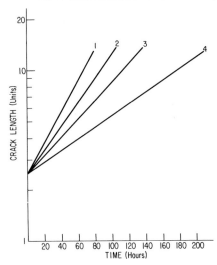

Fig. 16   Effect of carbon black distribution on fatigue life crack growth in SBR–BR blends (60 SBR–40 BR–85 carbon black). Percent carbon black in BR phase: 1, 60%; 2, 40%; 3, 25%; 4, 10%. (From Lee, private communication.)

to the mechanism and prevention of heat and oxygen aging of rubber compounds in general [69]. Little work has been devoted expressly to elastomer blends. In general, aged tensile properties of blends will fall between those of the parent elastomers with the blend displaying properties closest to the parent elastomer occupying the continuous phase.

Attack on rubber by atmospheric ozone is manifested by surface crack formation and growth. For ozone cracking to occur a critical strain (or stored energy density) must be imposed on the sample [70, 71]. The use of low-unsaturation rubbers, EPDM and Cl–IIR, to improve the ozone crack resistance of tire sidewalls, both black and white, has been the subject of a number of studies [72–74]. Here, EPDM is more effective as an antiozonant than Cl–IIR. All commercially available EPDMs at 35–40% levels in NR effectively stop all ozone cracking. This is the concentration at which EPDM approaches becoming a cocontinuous phase with the NR phase. Because of cure incompatibility, lower EPDM concentrations in conjunction with Cl-IIR are often used in practical recipes. While less effective for ozone protection, Cl-IIR confers improved physical properties owing to its better cure compatibility.

Andrews [70] has proposed a mechanism for the protection of unsaturated rubbers by low-unsaturation blends. A microscopic crack is hindered in its development to become a macroscopic one because of the physical barrier of the inert elastomer. This barrier may be circumvented by "crack jumping" of the inert particle in certain circumstances. Ambelang *et al.* [75] found results consistent with this mechanism. They explained the fact that EPDM (terpolymer) in blends gave improved protection compared with ethylene-propylene copolymer (EPM) on the basis that the EPDM can be more finely dispersed in the matrix. In a 30/70 blend of EPDM with SBR the maximum distance between EPDM particles was 42 $\mu$m, while for EPM this distance was 64 $\mu$m.

The importance of dispersibility and flow of the dispersed EPDM phase in the continuous, unsaturated polymer matrix for ozone protection is extremely critical and cannot be overrated. Differences in behavior of EPM and EPDMs made with various termonomers can be magnified by comparing them at the threshold protection concentration of 20 pph total rubber. The effect of EPM–EPDM type and mixing thoroughness on ozone cracking of a typical white sidewall compound is shown in Fig. 17 [76]. Thorough blending of the elstomers can improve performance of any given blend. However, EPDM containing dicyclopentadiene (DCPD) as a termonomer is inherently better than the others. No consistent correlation between ozone protection and molecular weight or molecular weight distribution has been found in studies of this type.

When these polymers are evaluated in a standard extrusion plastometer test (ASTM Standard D 1238-73), they conform to the equation:

$$Q = Kp^m$$

where $Q$ is the quantity of polymer extruded per unit time, $p$ is the imposed pressure, and $K$ and $m$ are characteristic of the polymer. An increasing exponent $m$ value has been interpreted as indicating an increasing degree of branching. The results shown in Fig. 17 are consistent with increasing degree of branching (below a gel point) conferring improved ozone protection since the $m$ for EPM is 1.5, for 1,4-hexadiene EPDM it is 1.8 and for

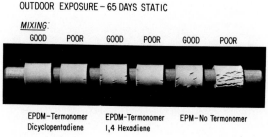

OUTDOOR EXPOSURE – 65 DAYS STATIC

*MIXING:*
GOOD   POOR   GOOD   POOR   GOOD   POOR

EPDM–Termonomer   EPDM–Termonomer   EPM–No Termonomer
Dicyclopentadiene   1,4 Hexadiene

Fig. 17   Effect of EPM–EPDM type and the mixing intensity on static ozone protection of a typical white sidewall compound.

dicyclopentadiene EPDM it is 2.5. This is also consistent with poly-merization kinetics, which predict that the EPM should be linear whereas dicyclopentadiene should branch slightly because its second unsaturation is 10% as reactive as the first.

A study [76] listed in Table IV reinforces the correlation between EPDM branching and ozone protection. The EPDM was lightly branched to varying degrees by heating with dicumyl peroxide and evaluated at 20 pph in a typical white sidewall compound. The ozone protection improved directly with the $m$ value. While not shown in this work, EPMs with excessive heat treatment give gel, poor dispersion, and consequently poor ozone protection. Although this correlation exists, no reported study has as yet dealt with the disperse phase morphology, giving maximum ozone protection at a given blend level.

Table IV

Relationship of Extrusion Plastometer $m$ to Ozone Crack Resistance
of a 20-phr EPDM White Sidewall

|  | $m$ Exponent from extrusion plastometer | Ozone chamber, 50 pphm $O_3$, 20% elongation (hr to first visible crack) |
|---|---|---|
| EPM | 1.5 | 2 |
| EPM treated with 0.5% | 1.5 | 2 |
| dicumyl peroxide to varying | 1.8 | 24 |
| degrees of branching | 2.1 | 120 |
|  | 2.4 | >168 |

ACKNOWLEDGMENTS

The authors wish to thank the B. F. Goodrich Company for permission to publish this chapter and Mrs. Gail Tirpak for extensive assistance in the preparation of the manuscript.

REFERENCES

1. *Modern Tire Dealer* p. 62 (January 1976).
2. *Rubber Age* **108**, 26 (1976).
3. Rubber Manufacturers Association, Statistical Report (1975).
4. V. E. Gough, *in* "Mechanics of Pneumatic Tires" (S. K. Clark, ed.), p. 356. Nat. Bur. Std. Monograph 122, Nat. Bur. Std., Washington, D.C., 1971.
5. E. M. Dannenberg, *Rubber Chem. Technol.* **48**, 410 (1975).
6. P. J. Corish and B. D. W. Powell, *Rubber Chem. Technol.* **47**, 481 (1974).
7. J. Kruse, *Rubber Chem. Technol.* **46**, 653 (1973).

8.  M. H. Walters and D. N. Keyte, *Trans. Inst. Rubber Ind.* **38**, 40 (1962); *Rubber Chem. Technol.* **38**, 62 (1965).
9.  W. M. Hess, C. E. Scott, and J. E. Callan, *Rubber Chem. Technol.* **40**, 371 (1967).
10. P. A. Marsh, A. Voet, and L. D. Price, *Rubber Chem. Technol.* **40**, 359 (1967).
11. P. A. Marsh, A. Voet, and L. D. Price, *Rubber Chem. Technol.* **41**, 344 (1968).
12. R. W. Smith and J. C. Andries, *Rubber Chem. Technol.* **47**, 64 (1974).
13. E. H. Andrews, *J. Appl. Polym. Sci.* **10**, 47 (1966).
14. G. N. Avgeropoulos, F. C. Weissert, P. H. Biddison, and G. G. A. Böhm, *Rubber Chem. Technol.* **49**, 93 (1976).
15. E. Lewis, H. V. Mercer, M. L. Deviney, L. Hughes, and J. E. Jewell, paper presented to Div. of Rubber Chem., Amer. Chem. Soc., Chicago, Illinois (October 1970); Abstract, *Rubber Chem. Technol.* **44**, 855 (1971).
16. R. R. Hampton, *Rubber Chem. Technol.* **45**, 546 (1972).
17. M. C. Morris, *Rubber Chem. Technol.* **40**, 341 (1967).
18. H. K. deDecker and D. J. Sabatine, *Rubber Age* **99**, 73 (1967).
19. D. W. Brasier and G. H. Nickel, *Rubber Chem. Technol.* **48**, 26 (1975).
20. A. K. Sircar and T. G. Lamond, *Rubber Chem. Technol.* **48**, 301 (1975).
21. A. K. Sircar and T. G. Lamond, *Rubber Chem. Technol.* **48**, 631 (1975).
22. A. K. Sircar and T. G. Lamond, *Rubber Chem. Technol.* **48**, 640 (1975).
23. A. K. Sircar and T. G. Lamond, *Rubber Chem. Technol.* **48**, 653 (1975).
24. A. Krishen and R. G. Tucker, *Anal. Chem.* **46**, 29 (1974).
25. S. Sugiki *et al.*, *J. Soc. Rubber Ind. Jpn.* **45**, 299 (1972).
26. S. S. Voyutsky, "Auto Adhesion and Adhesion of High Polymers," p. 141. Wiley (Interscience), New York, 1963.
27. K. Fujimoto, *J. Soc. Rubber Ind. Jpn.* **43**, 54 (1970).
28. S. N. Angone, *Rubber J.* **149**(5), 37 (1967).
29. D. C. Blackley and R. C. Charnock, *J. Inst. Rubber Ind.* **7**, 60 (1973).
30. D. C. Blackley and R. C. Charnock, *J. Inst. Rubber Ind.* **7**, 113 (1973).
31. Japan Synthetic Rubber Co. Ltd., British Patent, 1,046,215 (October 19, 1966).
32. J. B. Gardiner, *Rubber Chem. Technol.* **43**, 370 (1970).
33. M. Shundo, M. Imoto, and Y. Minoura, *J. Appl. Polym. Sci.* **10**, 939 (1966).
34. J. E. Callan, B. Topcik and F. P. Ford, *Rubber World* **151**, 60 (1965).
35. A. K. Sircar and T. G. Lamond, *Rubber Chem. Technol.* **46**, 178 (1973).
36. C. E. Scott, J. E. Callan, and W. M. Hess, *Natural Rubber Conf.*, *Kuala Lumpur, Malaysia* (1968).
37. P. A. Marsh, T. J. Mullens and L. D. Price, *Rubber Chem. Technol.* **43**, 400 (1970).
38. J. E. Callan, W. M. Hess, and C. E. Scott, *Rubber Chem. Technol.* **44**, 814 (1971).
39. V. L. Folt and R. W. Smith, *Rubber Chem. Technol.* **46**, 1193 (1973).
40. W. W. Graessley, *Adv. Polym. Sci.* **16**, 38 (1975).
41. A. K. Sircar, T. G. Lamond, and P. E. Puiter, *Rubber Chem. Technol.* **47**, 48 (1974).
42. K. C. Baranwal, *Rubber Age* **102**, 52 (1970).
43. H. E. Railsback, W. S. Howard, and N. A. Stumpe, Jr., *Rubber Age* **106**, 46, (1974).
44. R. T. Morrissey, *Rubber Chem. Technol.* **44**, 1029 (1971).
45. P. J. Corish, *Rubber Chem. Technol.* **40**, 324 (1967).
46. K. C. Baranwal and P. N. Son, *Rubber Chem. Technol.* **47**, 88 (1974).
47. M. E. Woods and T. R. Mass, "Copolymers, Polyblends, and Composites" (N. A. J. Platzer, ed.), Amer. Chem. Soc. Adv. in Chem. Ser. 142. Amer. Chem. Soc., Washington, D.C., 1975.
48. L. P. Gelinas, I. W. E. Harris, and E. B. Storey, *Rubber Chem. Technol.* **35**, 354 (1962).
49. L. H. Krol, *Rubber Chem. Technol.* **39**, 452 (1966).

50. B. Kastein, paper presented at the *Tire Mater. Decisions Symp.*, *Minneapolis, Minnesota* (August 29, 1972).

51. K. C. Baranwal and E. T. McDonel, Unpublished data.

52. K. C. Baranwal, paper presented at the *Rubber Div. Amer. Chem. Soc. Meeting, Miami Beach, Florida* (April 1971).

53. A. G. Veith, *Rubber Chem. Technol.* **44**, 962 (1971).

54. K. A. Grosch, *Rubber Age* **99** (October 1967).

55. W. Gnörich, and K. A. Grosch, *Rubber Chem. Technol.* **48**, 527 (1975).

56. A. Schallamach, *Rubber Chem. Technol. Rubber Rev.* **41**, 209 (1968).

57. K. A. Grosch and A. Schallamach, *Rubber Chem. Technol. Rubber Rev.* **49**, 862 (1976).

58. P. Kainradl and G. Kaufman, *Rubber Chem. Technol.* **49**, 283 (1976).

59. M. L. Studebaker and J. R. Beatty, *Rubber Chem. Technol.* **47**, 803 (1974).

60. J. Rehner, Jr. and P. E. Wei, *Rubber Chem. Technol.* **42**, 985 (1969).

61. R. L. Zapp, *Rubber Chem. Technol.* **46**, 251 (1973).

62. A. K. Sircar and T. G. Lamond, *Rubber Chem. Technol.* **48**, 89 (1975).

63. J. R. Beatty, *Rubber Chem. Technol.* **37**, 1341 (1964).

64. M. L. Studebaker, *Rubber Chem. Technol.* **41**, 373 (1968).

65. G. L. Lake, *Rubber Chem. Technol.* **45**, 309 (1972).

66. B. E. Clapson and G. J. Lake, *Rubber Chem. Technol.* **44**, 1186 (1971).

67. J. R. Beatty, paper presented at the *Int. Rubber Symp., Rubber Div., Amer. Chem. Soc., San Francisco, California* (October 5–8, 1976).

68. Biing-Lin Lee, Private communication.

69. J. R. Shelton, *Rubber Chem. Technol.* **30**, 1251 (1957).

70. E. H. Andrews, *J. Appl. Polym. Sci.* **10**, 47 (1966).

71. Z. W. Wilchinskey and E. N. Kresge, *Rubber Chem. Technol.* **47**, 895 (1974).

72. A. H. Speranzini and S. J. Drost, *Rubber Chem. Technol.* **43**, 482 (1970).

73. Z. T. Ossefort and E. W. Bergstrom, paper presented at a *Meeting Rubber Div., Amer. Chem. Soc., Atlantic City, New Jersey* (September 10–13, 1968).

74. L. Gursky, F. J. Masiello, and A. J. Wallace, presented at a *Meeting Rubber Div., Amer. Chem. Soc., Cleveland, Ohio* (May 6–9, 1975).

75. J. C. Ambelang, F. H. Wilson, Jr., L. E. Porter, and D. L. Turk, *Rubber Chem. Technol.* **42**, 1186 (1969).

76. E. T. McDonel, unpublished results.

# Chapter 20

# Rubbery Thermoplastic Blends

## E. N. Kresge

*Exxon Chemical Company*
*Elastomers Technology Division*
*Linden, New Jersey*

## I. INTRODUCTION

Rubbery thermoplastic blends are one type of thermoplastic material that are processable as a "melt" at an elevated temperature but exhibit properties similar to a vulcanized elastomer at use temperature. This chapter deals with the structural aspects of rubbery thermoplastic blends, such as the control of phase morphology and the relationship between morphology and physical properties. Technologically useful blends are discussed. High-modulus, nonrubbery blends used as rubber-modified plastics are described in Chapter 13.

There are two basic types of rubbery thermoplastic blends: modified thermoplastic elastomers and thermoplastic elastomers by polymer blending. The modified thermoplastic elastomers are blends of a compositionally homogeneous thermoplastic elastomer (TPE) such as a thermoplastic polyurethane (TPU) with a second polymer to alter properties. For example,

293

TPU can be blended with plasticized poly(vinyl chloride) (PVC) to decrease hardness or to provide a range of hardnesses. Thermoplastic polyurethane can also be blended with styrene acrylonitrile copolymers to increase modulus. The TPEs by polymer blending are polymer combinations that produce a rubbery thermoplastic from nonthermoplastic elastomer and plastic materials. A blend of semicrystalline isotactic polypropylene, a high-modulus plastic, with an amorphous ethylene propylene copolymer, an uncross-linked elastomer, is a rubbery thermoplastic.

The distinction between thermoplastics as rubbery and nonrubbery is quite arbitrary. The three typical types of stress–strain properties for polymeric materials are illustrated in Fig. 1. Polymer composites and blends have been developed which encompass the complete spectrum of stress–strain properties from high-modulus materials for structural applications to cross-linked elastomers, which are low modulus and can be extended to several hundred percent and return to their original dimension on release of the applied stress. In general, however, materials are considered rubbery if they can be extended over 100% without failure and return to nearly their original dimensions in a short period of time. Modified thermoplastic elastomers and thermoelastic blends exhibit stress–strain characteristics that fall between the low modulus cross-linked elastomers (Curve C) and the high-modulus-yielding materials that undergo a high degree of plastic deformation (Curve B).

By the use of polymer blends physical properties can be altered to produce useful materials with a wide range of applications. Desirable properties

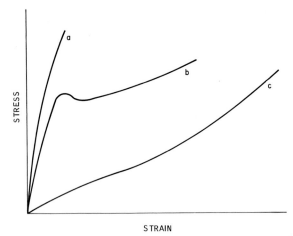

Fig. 1    Three typical types of stress–strain curves: (a) high modulus, plastic; (b) high modulus, with yielding and poor recovery on return cycle; (c) low-modulus, rubbery.

can be achieved more easily in many instances by proper blend selection than by polymerization of a new polymer. This being the case, a major thrust of the technology on rubber thermoplastic blends is to provide materials to meet specific applications as efficiently as possible. Moreover, certain rubbery thermoplastic blends possess a combination of physical and rheological properties that are unavailable in a single polymer. In addition, certain environmental properties, such as the heat and moisture resistance of polyolefins, can be obtained more easily in TPE blends than in block copolymers.

## II. MODIFIED THERMOPLASTIC ELASTOMERS

### A. Morphology Requirements

#### 1. Thermoplastic Elastomers

Thermoplastic elastomers exhibit some of the physical properties of elastomers at use temperature and are processable at elevated temperature. The elastic or rubbery nature of these materials depends on the presence of a three-dimensional polymeric network resembling the structure that results from cross-linking an elastomer by chemical means, but the plasticlike flow depends on the absence of cross-links at processing temperature. Various mechanisms are used to produce the temporary cross-linking, and there are several available polymeric structures that produce thermoplastic elastomers. Basic requirements are amorphous rubber chains "cross-linked" by a thermally reversible mechanism. Several examples of the amorphous chains and cross-linking mechanism employed are given in Table I.

Thermoplastic block copolymers of styrene–butadiene–styrene depend on a glassy styrene phase, and certain block copolyesters depend on a crystalline phase to act as thermally labile cross-links. Ion clustering of metal carboxylate groups in poly[(butadiene–(methacrylic acid)] or nickel chloride coordinated with poly[(butadiene-*co*-styrene-*co*-vinylpyridine)] provide a thermally reversible cross-linking mechanism in ionomeric materials.

#### 2. Blends

To retain its rubbery nature when blended with other polymers the thermoplastic elastomer must be retained as a continuous phase. The rubbery nature of the blends still depends on properties of the thermoplastic elastomer but the properties are altered by modifying one or both of the phases or

Table I

Thermoplastic Elastomers

| Cross-link type | Typical structure | Reference |
|---|---|---|
| Glassy block polymer | Styrene–butadiene–styrene block copolymer | [1] |
| | Styrene–dimethylsiloxane block copolymer | [2] |
| | α-Methylstyrene–dimethylsiloxane block copolymer | [3] |
| Cyrstalline block polymer | Tetramethyl-*p*-silphenylenesiloxane–dimethylsiloxane block copolymer | [4] |
| | [(Tetramethylene ether) glycol terephthalate]–(tetramethylene terephthalate) block copolymer (segmented polyester) | [5] |
| H bonding with crystalline or glassy blocks | (Butylene adipate)–(*p,p'*-diphenylmethene diisocyanate)–(1,4-butane diol) block copolymer (segmented polyurethane) | [6] |
| Ionic sites | Butadiene–methacrylic acid copolymer neutralized with metal salts | [7] |

introducing another phase. This is illustrated by considering the modification of styrene–butadiene–styrene block copolymers by addition of butadiene or styrene homopolymers. The butadiene homopolymer resides in the polybutadiene phase of the block copolymer and the modulus is reduced. Styrene homopolymer is incorporated in the polystyrene phase and the modulus is increased. Marked changes in properties are produced by introducing a separate phase in styrene–butadiene–styrene (SBS) block copolymer [8,9] by blending with polyethylene or polypropylene. In these blends the polyolefin appears to form a continuous independent phase so that the structure is two continuous interpenetrating phases. The block copolymer phase apparently retains its morphology and contains both a polybutadiene and a polystyrene phase.

If the thermoplastic elastomer is not retained as a continuous phase in a blend, properties are controlled by the major continuous phase. A blend of 10 parts SBS block copolymer with 90 parts of polystyrene is, of course, a high-modulus, nonrubbery material.

The phase morphology of thermoplastic elastomers blended with other polymers can be controlled by the viscoelastic difference between the two as discussed in Volume 1, Chapter 7. The fineness of the dispersion of the minor phase in an intensive mixing process depends on both the viscous and elastic behavior of the components under mixing conditions. Shear rate and temperature are important variables since it is unlikely that both polymers in the blend will exhibit the same response. Good compatibility or solubility is not necessarily desirable for effective property modifications.

## B. Useful Blends

Thermoplastic elastomers are of considerable commercial importance owing to their combination of processing ease and unique properties. Certain properties, however, are not easily achieved by direct synthesis of a thermoplastic elastomer. For this reason it is useful to tailor physical properties by blending polymers.

### 1. Styrene–Butadiene–Styrene Block Polymers

The possibilities offered by compounding of the triblock copolymer can be accomplished in four basic ways [8]:

1. Extension of the rubbery phase with polymers or plasticizers soluble in this phase.
2. Inclusion in the elastomer phase of a third discrete discontinuous phase, for example, mineral fillers.
3. Extension of the polystyrene phase with compatible polymers.
4. Blending with a high modulus plastic to produce an additional continuous phase, for example, a polyolefin.

Extension of the rubbery phase with processing oils is desirable for the decrease in viscosity. Paraffinic and naphthenic oils are preferred over aromatic ones. The latter also solvates the polystyrene phase and lowers its glass transition temperature $(T_g)$ with a resultant loss in properties. The incorporation of processing oils results in a decrease in the modulus.

Mineral fillers are blended into the block copolymers by intensive mixing. These types of fillers result in a relatively small increase in modulus. Their function is to extend the system and thereby reduce cost. Reinforcing fillers such as carbon black are not required to obtain high ultimate tensile properties as they are in noncrystallizing single-phase elastomer systems, such as conventional styrene–butadiene rubber (SBR) vulcanizates. Adequate tensile properties result from having discrete polystyrene domains in excess of 0.2 weight fraction [10]. Additionally, entanglements, provided the molecular weight of the rubbery chains is several times greater than the molecular weight between entanglements, function similar to cross-links in a conventional vulcanized rubber. Since the entanglements can slip under stress, the rubbery chains become highly elongated and oriented. The orientation tends to strengthen the elastomer in addition to the filler effect.

The addition of homopolystyrene to the block copolymers produces an increase in modulus that is often used to offset the loss on modulus due to oil addition. Tensile properties are largely maintained, and by selection of the proper molecular weight polystyrene, rheological properties can be improved for injection molding and other processing operations. In fact, unblended styrene–butadiene–styrene (SBS) polymers are quite elastic in the melt at typical processing conditions owing to a retention of the two-phase structure above the $T_g$ of the polystyrene block [11]. As the temperature is increased, the polystyrene blocks and polybutadiene appear to become soluble, and the copolymer exhibits Newtonian flow behavior. The wide range of modulus, tensile, and rheological properties obtained by blends of SBS, polystyrene, and oil accounts for its use in a very large number of compositions for molded and extended articles.

The polystyrene phase can be modified with other polymers to increase the $T_g$, thereby improving the range of temperatures at which the block copolymers are useful. For example, poly(phenylene oxide) or substituted poly(phenylene oxides) are combined with SBS in $CHCl_3$ solution. An 80% SBS–20% poly(2,6-dimethyl-1,4-phenylene oxide) blend has a tensile strength of 20 $kg/cm^2$ at 300°C as compared with the unblended polymer, which has the same tensile strength at 85°C [12].

Ternary blends of AB-type styrene–isoprene block copolymer with polystyrene and polyisoprene cast from a common solvent have been studied [13] by light and electron microscopy. The results suggest a limitation of relative molecular weights of the homopolymers to those of the corresponding block of the copolymer for solubilization of the homopolymers into the respective block domains. If the fraction of block copolymer is kept large, the inherent manner of domain formation of the block copolymer from its solution appears to be maintained for the ternary or binary systems. These results should prove applicable to systems involving blends of ABA types of thermoelastic block copolymers and suggests the types of modifications that are possible by blending techniques.

Blending SBS with a high-modulus plastic to produce an additional continuous phase has proved very useful to enhance the properties of thermoplastic elastomers [8,9]. For example, an 80/20 blend of SBS with ethylene vinyl acetate copolymer produces a rubbery composite with exceptional ozone resistance. Blends with polypropylene, polyethylene, and other poly-α-olefins are mentioned in the patent literature (see Table II). Blends of about equal parts of polypropylene with SBS combine flexibility with good high- and low-temperature characteristics. The blends appear to retain a morphology consisting of two continuous phases over a wide blend ratio. This has been shown by extraction of the SBS phase with a nonsolvent for polypropylene followed by electron-scanning microscopy.

## 2. Thermoplastic Polyester Elastomers

Thermoplastic elastomers based on segmented copolyesters have only recently become commercially important. The copolymers are typically

Table II

Thermoplastic Elastomer Blend Patents: Modified Thermoelastic Elastomers

| Patent no. | File date | Inventor(s) | Subject |
|---|---|---|---|
| U.S. 3,839,501 | 10/26/72 | Y. Wei, E. G. Kent | Blends of styrene–butadiene block copolymer with crystalline alternating copolymer of isoprene and acrylonitrile |
| U.S. 3,641,205 | 3/25/68 | R. T. La Flair, J. F. Henderson | Blends of styrene–butadiene–styrene block copolymer with a resin of higher softening point, for example, polyacenaphthylene or poly($\alpha$-methylstyrene) |
| Japan 50082-162 | 11/21/73 | — | Block copolyesters blended with styrene–butadiene–styrene block copolymers |
| Japan 49000–344 | 4/17/72 | — | Blend of SBS, polyethylene, and silicone oil |
| Japan 7,242,728 | | — | Blends of SBS with poly($\alpha$-olefin), plasticizers, and fillers |
| U.S. 3,801,529 | 12/2/70 | R. C. Potter | Plasticized vinyl chloride resin compositions containing olefin block copolymer of improved processability |
| U.S. 3,589,036 | 1/16/70 | W. R. Hendricks, R. L. Danforth | Butadiene–styrene block copolymers in shoe sole compositions |
| U.S. 3,506,740 | 10/12/66 | L. T. Dempsey, L. R. Babcock | Polybutylene lubricant for high-impact polystyrenes (blend contains styrene–butadiene block copolymer) |
| U.S. 3,865,776 | 2/11/74 | W. P. Gergen | Alkenylarene block copolymer composition containing mineral white oil and polypropylene, forming kink-resistant tubing, rods, etc. |
| U.S. 3,830,767 | 5/2/73 | N. J. Condon | Block copolymer compositions containing a petroleum–hydrocarbon resin to reduce extending oil bleedout (block copolymer, e.g., polystyrene-hydrogenated polybutadiene-polystyrene) |
| U.S. 3,827,999 | 11/9/73 | R. K. Crossland | Stable elastomeric polymer–oil compositions containing selectively hydrogenated block copolymer and nonaromatic hydrocarbon oil |

(*continued*)

Table II (*continued*)

| Patent no. | File date | Inventor(s) | Subject |
|---|---|---|---|
| West Germany 2,353,314 | 10/26/72 | S. Davidson, M. Wales | Thermoplastic elastomer composition prepared by mixing selectively hydrogenated conjugated diene block copolymer with ABS resin |
| U.S. 3,792,124 | 2/24/72 | S. Davidson | Polymer compositions of block copolymer and ionic copolymer |
| West Germany 2,312,752 | 3/16/72 | S. Davidson, M. Wales | Modified block copolymers comprising hydrogenated block copolymers modified with maleated polypropylene, etc. |
| Canada 880,945 | British 8/7/68 | A. W. van Breen, M. Vlig | Thermoplastic elastomeric composition (e.g., polystyrene–polybutadiene–polystyrene plus aromatic asphaltic bitumen) |
| U.S. 3,562,355 | 3/20/68 | G. Holden | Block copolymer blends with certain polyurethanes or ethylene-unsaturated ester copolymers. |
| Britain 1,160,198 | 7/19/67 | W. R. Hendricks, B. Higginbottom, G. Holden | Elastomeric composition of styrene–butadiene block copolymer and mineral oil extruder |
| U.S. 3,562,204 | Netherlands 2/11/66 | A. W. van Breen | Thermoplastic elastomeric composition comprising block copolymers (of vinyl aromatics and conjugated dienes) and random copolymers |
| U.S. 3,562,356 | 9/24/65 11/19/68 | D. D. Nyberg W. R. Hendricks | Block copolymer (vinyl aromatic conjugated diene) blends with certain ethylene-unsaturated ester copolymers |
| U.S. 3,639,163 | 11/5/65 11/26/69 | E. T. Bishop W. R. Haefele, W. R. Hendricks | Block polymer (e.g., hydrogenated polystyrene–polyisoprene–polystyrene) insulation for electric conductors |
| U.S. 3,459,831 | 9/24/65 | M. A. Luftglass, W. R. Hendricks | Block copolymer (styrene–butadiene–styrene)–polyethylene films |
| Canada 761,526 | 2/5/65 | W. R. Hendricks, D. W. Johnson | Blends of low molecular weight poly(vinyl arenes) and self-vulcanizing block copolymers |
| Canada 761,525 | 1/6/65 | W. R. Hendricks, E. T. Bishop | Blends of poly(vinyl arenes) and self-vulcanizing block copolymers (styrene–butadiene–styrene) |
| U.S. 3,239,478 | 6/26/63 | J. T. Harlan, Jr. | Block copolymer (vinyl aromatic–conjugated diene) adhesive compositions and articles prepared therefrom (formulations contain tackifying resin and extender oil) |

Table II (*continued*)

| Patent no. | File date | Inventor(s) | Subject |
|---|---|---|---|
| Canada 761,530 | 5/22/63 | G. Holden, R. Milkovich | Blends of diene elastomers (e.g., polyisoprene) with certain block copolymers (polyisoprene–polystyrene) |
| Canada 762,544 | 6/17/63 | R. Milkovich, G. Holden, E. T. Bishop, W. R. Hendricks | Thermoplastic elastomeric block copolymer (vinyl aromatic–conjugated diene) and processable compositions (plus extending oil) |
| U.S. 3,963,802 | 6/20/75 | C. Shih | Blends of thermoplastic polyester elastomers with ethylene copolymer elastomers |
| Netherlands 2,502,638 | 1/28/74 | — | Blend of polyester thermoplastic polyurethane and PVC or PVC copolymers |
| Netherlands 2,130,938 | 6/22/70 | W. K. Fisher | Blends of polyether thermoplastic polyurethane, PVC and butadiene–acrylonitrile copolymers |
| U.S. 3,974,240 | 11/18/74 | R. D. Lundberg, R. R. Phillips, H. S. Makowski | Blends of ionic ethylene propylene elastomer with crystalline polyolefin |
| U.S. 3,974,241 | 11/18/74 | R. D. Lundberg, R. R. Phillips, L. Westerman, J. Boch | Blends of low-viscosity ionic ethylene propylene elastomer with crystalline polyolefin |

derived from terephthalic acid, poly[(tetramethylene ether) glycol], and 1,4-butanediol [14]. This results in a copolymer with crystallizable tetra-methylene terephthalate segments and amorphous poly[(tetramethylene ether) glycol terephthalate] segments randomly distributed along the poly-mer chain. The morphology of the copolymers [5] consists of crystalline lamellar domains about 100 Å in width and several thousand angstroms in length. The crystalline regions appear to be continuous and highly inter-connected. This is in contrast to the discrete polystyrene domains found in typical SBS block copolymers. Stress–strain behavior of the segmented poly-esters is consistent with this morphology. There is high initial modulus resulting from the deformation of the continuous crystalline matrix. Strain in this region is highly recoverable. At higher strains, the material draws in an irreversible manner with a disruption of the crystalline matrix and extensive orientation of the crystallites. Once the crystallites are oriented and draw no further, the stress is transmitted primarily by the continuous amorphous phase.

Modification of thermoplastic polyesters can be accomplished by blending with other polymers, fillers, and plasticizers similar to the modification carried out on triblock copolymers. The differences in the morphologies of the two systems must be taken into account, however. Blends of SBS and thermoplastic polyesters are disclosed in the patent literature (see Table II) to give enhanced property combinations. Blends of thermoplastic polyesters with ethylene–propylene elastomers are useful in lowering hardness [15]. Semicrystalline ethylene copolymers with melting points 35–65°C are preferred and the blends are prepared by conventional intensive mixing. High tensile strengths and elongations are retained over a wide composition in the blends. Blend composites consisting of thermoplastic polyester, ethylene copolymer, and fillers, such as hydrated silica and carbon black, are demonstrated as well. Given the differences in the polarity of ethylene copolymers and thermoplastic polyesters, it is somewhat surprising that useful blends could be obtained. However, once the blend morphology is established, the crystallinity in both phases should prevent separation.

### 3. Thermoplastic Polyurethane Elastomers

Like the thermoplastic polyester elastomers, thermoplastic polyurethanes are a relatively recent development. Thermoplastic polyurethane (TPU) is usually synthesized by a prepolymer route by attaching a long-chain diol and a short chain diol with a diisocyanate. The preferred diisocyanate is 4,4'- phenylmethane diisocyanate although toluene diisocyanate is sometimes used. The long-chain diols are linear terminally functional dihydroxy polyester or polyether with molecular weights between 800 and 2500, such as poly[($\epsilon$-caprolactone)diol] and poly[(ethylene glycol) adipate]. The low molecular weight diol chain extenders are typically ethylene glycol, 1,4-bis(2-hydroxyethoxy)benzene or 1,4-butane diol. The highly polar, relatively rigid blocks of diisocyanate and diol extender form hydrogen-bonded, crystalline, or glassy domains and act as tie points for the amorphous long-chain diols [16].

The stress–strain properties of TPU are characterized by exceptionally high tensile strength (up to 50 Mpascals), large hysteresis, and stress softening. With the synthesis variations possible with TPU very specific properties can be obtained. Blends, however, are increasingly important as a method of enhancing specific properties, reducing cost or both.

Polycarbodiimides blended with thermoplastic polyurethanes having polyester segments improve the hydrolysis resistance [17, 18]. A polyester urethane that would retain no measurable physical properties after immersion in 70°C water in three weeks retains 87% of its original tensile strength, 93% of its original elongation, and 75% of its original modulus under the same exposure conditions when two parts of polycarbodimide are compounded into the elastomer.

Many polymers may be blended with TPU to enhance physical properties. These polymers include acetal, PVC, PVC copolymers, acrylate–butadiene–styrene (ABS) resins, styrene–acrylonitrile (SAN) resins, polyolefins, and various elastomers. Acetal copolymer (trioxane ethylene oxide copolymer) blended with TPU by intensive mixing causes a marked increase in modulus. Apparently the highly crystalline (2500 Mpascals flexural modulus) acetal forms a second continuous phase and contributes to the properties in this fashion. Blending TPU with ABS also increases the stiffness and these more rigid compounds find use as cable sheathing. Blends with plasticized PVC allow a broader range of hardness and reduced modulus. The softer compounds are used for shoe applications, and typically the TPU grades used have low softening points because of the heat sensitivity of PVC.

Blends of TPU based on (4,4'-diphenylmethane diisocyanate) MDI–adipic acid–1,4-butane diol poly(ester urethane) prepared by recovering from a methyl ethyl ketone solution of TPU and several polymers have remarkable properties [19], as shown in Table III. In all cases the tensile strengths are increased and the elongations are decreased. The PVC and phenoxy blends retain a high degree of elongation in view of the composition of the blend.

Table III

Blends of TPU with Thermoplastic Resins

| Polymer[a] added | Tensile strength[b] (Mpascal) | 300% modulus (Mpascal) | Elongation (%) |
|---|---|---|---|
| None | 36.5 | 3.1 | 730 |
| PVC[c] | 42.7 | 35.2 | 350 |
| Phenoxy[d] | 50.3 | 44.1 | 400 |
| Nitrocellulose[e] | 55.8 | — | <100 |

[a] 50/50 blends in methyl ethyl ketone.
[b] Cast film 3–5 mils thick.
[c] Geon 427 (B. F. Goodrich Chemical Co.).
[d] PKHH (Union Carbide Plastics Co.).
[e] RS 5–6-sec type (Hercules Powder Co.).

## III. THERMOPLASTIC ELASTOMERS BY POLYMER BLENDING

### A. Morphology Requirements

For rubbery blends to be produced from nonthermoplastic elastomer materials at least one phase must exhibit elastomeric properties and be continuous. The elastomeric properties of the continuous phase can result from a lowering of the modulus and increasing of the elastic limit of the total system due to the morphology of the blend system by several mechanisms.

The simplest mechanism is a high dilution of a high-modulus plastic with an elastomer. For example, 20 parts of isotactic polypropylene blended with 80 parts of a copolymer of ethylene and propylene by intensive mixing [20] result in a TPE. The polypropylene phase in the blend exhibits a crystalline structure similar to unblended polymer, and the blend depends mainly on having continuous fine strands.

Although not accomplished by blending, polypropylene and several other crystalline polymers can also be transformed into an elastomeric material by changes in the crystal structure [21–23]. By using conditions of high stress during crystallization from the melt, highly crystalline fibers are formed with unusually high degrees of recovery from extension (e.g., 97% recovery from 100% extension). The elastic nature is due to a specific morphology. Electron microscopy shows close-packed lamellae with their normals mainly parallel to the fiber axis, which on extension tilt and spill apart, creating voids. Consistent with a mechanism of elasticity that depends on reversible bending of lamellae, the fibers have a nonrubberlike stress–temperature response.

Thermoplastic elastomers are also produced by altering the rubbery portion of the blend to make it more elastic. This has been accomplished by making the rubbery phase semicrystalline with the small amount of crystallinity acting as pseudocross-links between largely amorphous chains. The rubbery phase can also be of very high molecular weight or gelled such that it can exhibit a highly elastic response under the time scales of testing or use.

Another route to TPEs is the addition of a soluble plasticizer for one or both phases of the blend. Poly(vinyl chloride) compounded with a plasticizer and an acrylic copolymer as a dispersed phase has elastic stress–strain properties [24]. The soluble plasticizer lowers the $T_g$ of the PVC phase below the use temperature while the material remains semicrystalline. Thus, PVC that contains some 5% crystallinity becomes an elastic phase.

## B. Useful Blends

As in the case for modified thermoplastic elastomers, the incentive for combining polymers into TPEs is to obtain useful properties economically. While many of the thermoplastic blends do not resemble a high-quality natural rubber vulcanizate in stress–strain characteristics, this is not required or even desirable for many applications. The polymers used to produce the blends are, in most instances, produced on a large scale and are relatively inexpensive.

### 1. Blends of Ethylene α-Olefin Copolymers and Crystalline Polyolefins

The most useful of the blends of ethylene α-olefin copolymers and crystalline polyolefins are prepared by intensive mixing of ethylene–propylene copolymers (EPM) or terpolymers (EDPM) and polypropylene or polyethylene. The EPM copolymers are either amorphous or semicrystalline materials used as vulcanizable rubbers on a large scale. The semicrystalline varieties exhibit some elastic properties in the unblended form due to crystalline tie points [25]. Physical properties that are useful for such applications as mechanical goods and automotive parts are obtained with blend ratios of around 40/60–85/15 EPM to polyolefin, according to the patent literature (Table IV). The actual production of a blend with the specific physical and rheological properties necessary for any given application requires a higher degree of sophistication. Ethylene–propylene copolymers and crystalline polyolefins can be varied in molecular weight, molecular weight distribution, and long-chain branching. The EPM can also be synthesized with variations in composition, composition distribution, and monomer sequence length. These molecular variables in conjunction with mixing methods control the final physical and rheological properties of the blends.

### a. Morphology of the Melt

The data presented on elastomer blends where both phases are amorphous, for example, styrene–butadiene copolymer blends with different styrene content, are applicable to ethylene α-olefin copolymers and crystalline polyolefins blends in the melt. Unfortunately there is no easy way to look at the blends by electron microscopy since the melting points are high and staining is difficult.

One indirect way of looking at a blend morphology is to examine the glass transition temperature of the blend. In homogeneous mixtures a new single $T_g$ will result. The $T_g$ of isotactic polypropylene, however, is complicated by its crystallinity which is much higher. Thus, we looked at blends

Table IV

Thermoplastic Elastomers by Polymer Blending

| Patent no. | File date | Inventor(s) | Subject |
|---|---|---|---|
| Netherlands 2229–169 | 6/15/72 | — | Blends of EPDM with polyethylene and cured rubber powder |
| U.S. 3,888,949 | 6/21/73 | C. Shih | Blends of propylene α-olefin copolymers and polypropylene |
| U.S. 3,882,197 | 6/21/73 | A. Fritz, C. Shih | Blends of EP copolymers having crystalline propylene sequences, polypropylene, and stereoregular propylene–α-olefin copolymers |
| U.S. 3,817,983 | 12/22/72 | E. K. Gladding | Blends of ethylene–vinyl acetate copolymer, mineral oil, and wax |
| U.S. 3,835,201 | 8/29/72 | W. K. Fischer | Blends of polyolefin with EPDM having a high zero-shear viscosity |
| U.S. 3,957,919 | 9/24/74 | G. A. von Bodungen, C. L. Meredith | Blend of EPDM, polyethylene, and polypropylene cross-linked during mixing |
| U.S. 3,963,647 | 8/22/74 | R. M. Straub | Conductive blends of ethylene–α-olefin copolymers, propylene, and carbon black |
| U.S. 3,758,643 | 1/30/71 | W. K. Fisher | Partially cured blends of EPDM and polypropylene |
| U.S. 3,919,358 | 8/2/74 | M. Batrick, R. M. Herman, J. C. Healy | Blends of semicrystalline EPDM and polyethylene |
| U.S. 3,862,106 | 6/4/73 | W. K. Fisher | Blends of partially cured EPDM and polyolefin resin |
| U.S. 3,806,558 | 8/12/71 | W. K. Fisher | Blends of EPDM and polyolefin dynamically cured during blending to produce a thermoplastic elastomer |
| U.S. 3,037,954 | 12/15/48 | A. M. Gessler | Blends of chlorinated butyl rubber and crystalline polypropylene dynamically cured during blending |
| U.S. 3,909,463 | 2/2/72 | P. F. Hartman | Blend of butyl rubber and polyolefin mixed and heated in the presence of a bifunctional phenolic compound |
| Japan 0999693 | 9/6/73 | — | Blend of polyethylene, ethylene acrylate ester copolymers, and ethylene–α-olefin copolymers |
| Netherlands 2,303,698 | 1/26/73 | — | Blend of ethylene–butene copolymer (95.5% butene) with non cross-linked EPR |

Table IV (*continued*)

| Patent no. | File date | Inventor(s) | Subject |
|---|---|---|---|
| Netherlands 2,261,334 | 11/15/72 | — | Blend of amorphous poly(1-butene), polybutene oil, and non cross-linked EPR |
| Japan 099692 | 9/6/73 | — | Blend of polyethylene, ethylene–α-olefin copolymer, and polypropylene |
| Japan 5 0123,752 | 3/13/74 | — | Polyethylene, acrylonitrile-butadiene, copolymer, and inorganic filler blends |
| Japan 7 5,006,873 | 2/19/72 | — | Blend of acrylonitrile–butadiene rubber, PVC, and PVC plasticizer |
| Japan 4 9,006,019 | 4/21/72 | — | Blends of nitrile rubber, PVC, dioctyl phthalate, graphite flake, and carbon black |
| Japan 7 4,024,568 | 3/2/70 | — | Blend of chlorinated propylene–butadiene alternating copolymer and PVC. |
| Netherlands 2,033,453 | 7/8/69 | — | Blend of neoprene or nitrile rubber, PVC, PVC plasticizer, and filler |
| Japan 5 0022,848 | 6/28/73 | — | Vinyl chloride–vinyl acetate copolymers blended with acrylonitrile butadiene copolymers |

of atactic polypropylene with ethylene propylene copolymer that are un-complicated by crystallinity and should resemble EP–polypropylene blends in the melt.

Differential scanning calorimetry (DSC) data for blends of atactic poly-propylene (IV = 0.9 dl/gm) and blends with EPM (50 mole % propylene, IV = 2.5 dl/gm) are given in Fig. 2. The blends show a broad second order transition region that appears to be increased in temperature over the copolymer. However, a new, sharp $T_g$ does not emerge from the blends. These data suggest that the atactic polypropylene and EPM are not homo-geneous and exist as a two-phase system. From this we infer that isotactic polypropylene and EPM (of similar composition) may be inhomogeneous above the melting point.

In two-phase polymer blends the phase in highest concentration tends to become a continuous phase when both polymers are of a similar viscosity at the temperature and shear rate of mixing. At similar concentrations

*E. N. Kresge*

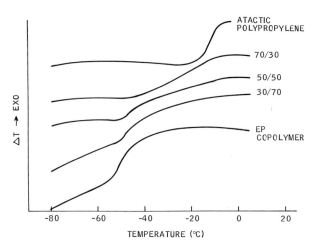

Fig. 2   Differential scanning calorimetry curves for atactic polypropylene, ethylene propylene copolymer, and blends. The numbers refer to the blend ratio.

the polymer with the lower viscosity tends to become the continuous phase [26]. Since the ethylene propylene copolymers typically used for blending are in high concentration and have viscosities five to ten times higher than typical polyethylene or polypropylene, there appears to be a driving force for both phases to be continuous in the melt.

### b.   *Morphology after Crystallization*

The structure of some EPM–polypropylene blends has been examined by electron scanning microscopy. A micrograph of a binary blend of EPM copolymer and polypropylene is shown in Fig. 3. The white portions of the micrograph are polypropylene, and the dark voids are the areas where the EPM resided prior to extraction via solvent. The polypropylene is clearly a continuous phase resembling an open-celled sponge with micron-sized porosity. Careful examination of the micrograph also shows the voids to be continuous, and therefore the rubber phase was continuous in the original sample prior to extraction. The minor phase in this blend is polypropylene (70/30 EPM/polypropylene by weight). The phase size ranges from about 2 $\mu$m to less than 0.1 $\mu$m across the polypropylene strands. The voids or EPM phase has a range of about 4 $\mu$m to 0.1 $\mu$m.

The surface of a molded part is different from the interior. Figure 4 is a scanning electron micrograph of a 70/30 EPM/polypropylene blend after extraction of the rubber phase but looking at the molded surface rather than the interior fracture surface. The surface is primarily polypropylene

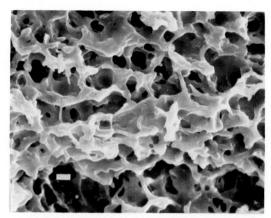

Fig. 3   Electron scanning micrograph of a binary (70/30) blend of EPM and polypropylene prepared by intensive mixing after extraction of the EPM phase with heptane. The sample was fractured under liquid nitrogen; the bar is 1 μm.

with small areas of rubber etched out. This accounts for the fact that these blends have a plastic surface "feeling" compared to an elastomer.

The range of compositions over which the interpenetrating two-phase structure could be produced was determined by making up blends of EPM and polypropylene over the entire rubber rich range in 5% intervals. For a

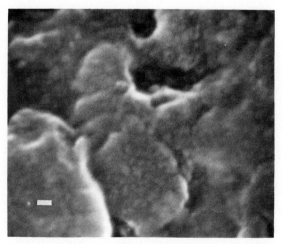

Fig. 4   Electron scanning micrograph of the surface of a binary (70/30) blend of EPM and polypropylene after injection molding. The EPM phase has been extracted with *n*-heptane; the bar is 1 μm.

binary blend, the polypropylene was a continuous phase down to 15 wt %
polypropylene in the blend.

### c.  Crystalline Structure of the Polypropylene Phase

In view of the morphological structure of the polypropylene phase as
determined by ESM, which shows elongated strands some 0.1–2 μm across,
the crystalline structure should not be spherulitic in nature. Spherulites are
typically some 10–100 times the size of the polypropylene strands. Optical
micrographs under polarized light confirm this and are shown in Fig. 5.
These micrographs were made on thin sections and the mold surface is
shown at one edge. No large spherulites are observed and there is little
difference between the apparent surface or skin structure and the interior
regions. Injection moldings of neat crystalline polypropylene show three
regions of spherulite type with the surface having a much different structure
than the interior.

Of the three forms of crystalline, isotactic polypropylene, which are (1)
smectic–nonspherulitic, (2) monoclinic–nonspherulitic, and (3)
monoclinic–spherulitic, the phase size seems to preclude the latter. The so-
called smectic type of polypropylene has a mesomorphic structure intermediate between amorphous and fully crystalline. The second type of
structure has a highly ordered monoclinic crystal structure, but like the
smectic material it does not have observable large scale spherulites. X-ray

Fig. 5   Polarized-light optical micrograph of a binary (70/30) blend of EPM and polypropylene thin section of an injection-molded bar. No spherulitic structure is found
(265 × magnification).

diffraction patterns obtained on injection molded bars of a 60/40 EPM/polypropylene binary blend show the well-defined x-ray diffraction pattern of monoclinic polypropylene with no broadening, indicating that the morphology is monoclinic and nonspherulitic in nature.

### d. Physical Properties

The stress–strain properties depend strongly on the type of crystalline polyolefin and ethylene propylene copolymer used in the blends. Properties listed in Table V for amorphous highly elastic EPDM blended with various ratios of polyethylene and polypropylene show rubberlike characteristics [27]. As expected, increasing the crystalline polyolefin phase increases the tensile strength and lowers ultimate elongation. The blends show a surprisingly low set at break (20–35%). This indicates that the polyolefin phase is not undergoing the typical drawing mechanism found when the unblended polymer is extended since drawing results in very little recovery after elongation. Polypropylene blends tend to give higher tensile values than polyethylene blends but somewhat lower elongations at failure.

Three component blends [28] of amorphous EPDM, polypropylene, and propylene–hexene copolymer (36 wt % hexene prepared with a $\gamma$-TiCl$_3$ catalyst to give isotactic polypropylene sequences) result in elastomeric properties. For example, a blend prepared by intensive mixing of equal amounts of the three polymers has a tensile strength at failure of 14 Mpascal, an elongation at break of 660% and a compression set of 74% after 22 hr at 70°C.

Table V

Properties of EPDM–Polyolefin Blends

| Blend[a] | | | | | | | |
|---|---|---|---|---|---|---|---|
| EPDM,[b] parts: | 80 | 70 | 60 | 80 | 60 | 80 | 60 |
| Polypropylene,[c] parts: | 20 | 30 | 40 | | | | |
| Low-density polyethylene,[d] parts: | | | | 20 | 40 | | |
| High-density polyethylene,[e] parts: | | | | | | 20 | 40 |
| | | | | | | | |
| Physical properties | | | | | | | |
| Tensile strength (Mpascal) | 8.3 | 10.5 | 13.9 | 5.8 | 8.0 | 8.5 | 10.2 |
| Elongation at break (%) | 220 | 150 | 80 | 290 | 190 | 210 | 130 |
| Elongation set at break (%) | 28 | 30 | 30 | 35 | 30 | 25 | 33 |

[a] Banbury mixer, about 7 min, max. temperature about 200°C.
[b] Amorphous, high molecular weight ethylene, propylene, dicyclopentadiene ($\sim 5$ wt %) terpolymer.
[c] $\rho = 0.903$ gm/cm$^3$, melt index = 4.0 gm/per 10 min at 230°C.
[d] $\rho = 0.919$ gm/cm$^3$, melt index = 2.0 gm per 10 min at 190°C.
[e] $\rho = 0.956$ gm/cm$^3$, melt index = 0.3 gm per 10 min at 190°C.

If the ethylene–propylene copolymer phase is semicrystalline due to ethylene sequences, there are profound effects on the mechanical behavior of the elastomer and its blends as well. Blends of semicrystalline EPDM with low- and high-density polyethylene are given in Table VI. Not only is crystallinity in the unstrained state important, but the increase in crystallinity induced by deformation [29–32] no doubt has a major effect on the elastomer phase. Relatively low molecular weight, gel-free, uncross-linked, semicrystalline copolymers can be extended without flow failure. Cross-linking of the samples leads to a decrease in the strength at failure of the unreinforced elastomer since it appears partially to inhibit crystallization. The improved stress–strain properties of the elastomer phase due to crystallinity improve the properties of composite.

Both polyethylene and ethylene propylene copolymers have an ortho-rhombic unit cell [33] and on cooling from the melt, blends of the two copolymers have the possibility of cocrystallizing. Moreover, at the rapid rates of crystallization normally encountered, rejection from the crystalline zones is primarily kinetically controlled, increasing the chance of crystalline regions containing segments of both polyethylene and ethylene–propylene copolymer. This could greatly influence stress–strain properties.

The chemical composition of the polyolefin blends results in good resistance to many solvents. They exhibit resistance to ozone. For good outdoor aging and autoxidation resistance ultraviolet stabilizers and inhibitors are required. Autoxidation takes place in the amorphous phase of polyolefins and, since the blends have a higher fraction of noncrystalline polymer than polyethylene or polypropylene, higher concentrations of inhibitors are required.

The combination of the spectrum of stress–strain properties, chemical

Table VI

Properties of Semicrystalline EDM–Polyolefin Blends

| Blend[a] | | | | |
|---|---|---|---|---|
| EPDM, parts: | 80 | 80 | 80 | 80 |
| Crystallinity (wt %) | 12.9 | 2.7 | 12.9 | 2.7 |
| High-density polyethylene,[b] parts: | 20 | 20 | | |
| Low-density polyethylene,[c] parts: | | | 20 | 20 |
| | | | | |
| Physical properties | | | | |
| Tensile strength (Mpascal) | 15.0 | 5.4 | 14.5 | 7.6 |
| Elongation at break (%) | 730 | 940 | 720 | 880 |

[a] Mill mixed at 150°C.
[b] $\rho = 0.95$ gm/cm$^3$.
[c] $\rho = 0.92$ gm/cm$^3$.

resistance and aging properties allows the blends to be used in many applications where elastomeric properties are desirable.

## 2. Cross-Linked Amorphous-Phase Blends

Thermoplastic elastomers have been prepared that depend on carrying out both blending and a cross-linking reaction at the same time in an internal mixer. Such materials are composed of a crystalline polyolefin and an elastomer, with the cross-linking reaction predominating in the elastomeric phase. Fisher's work [34, 35] has led to TPR™ thermoplastic rubber (Uniroyal Chemical) and is based on a "cross-linked" blend of EPDM and crystalline polypropylene. Hartman [36, 37] developed "ET" polymers (Allied Chemical Corp.) based on butyl rubber and polyethylene. The materials are melt processable by injection molding, extrusion, and the like, and on crystallization give rubbery characteristics.

It is somewhat remarkable that conducting a cross-linking reaction during mixing of a blend results in a material that has rheological properties suitable for the usual thermoplastic processing techniques. If a cross-linking reaction is carried out on a single-phase polymer system during mixing, the end result is an intractable, infusable, and unusable fine crumb. Gessler [38] noted that butyl rubber could be blended in a Banbury mixer with polypropylene having a curative (phenolic resin-cure) present for the butyl rubber. After the blending and curing operation the blend was thermoplastic in nature. Gessler called the process "dynamic curing."

### a. Morphology

The morphology of dynamically cured blends of elastomer and plastic has not been well defined. The morphological details would be expected to depend on both the nature of the blend before cross-linking takes place and the mechanism of cross-linking. At the elastomer/plastic ratios of around 75/25 used to produce rubbery blends both phases appear to be continuous during blending, as discussed in Section III.A. As the mixing continues, the temperature increases and the cross-linking reaction commences. If the cross-linking is confined to the continuous elastomer phase, for example, by using a sulfur cure through the unsaturation in an EPDM, as the polymer approaches the gel point, chain cleavage must occur for mixing to continue.

The cross-linking also partly controls the morphology of the melt since viscosity at the conditions of mixing tends to dictate which phase is continuous or dispersed. As branch points are formed, the viscosity of the phase undergoing reaction increases. This tends to make the phase that is undergoing fewer cross-linking reactions a continuous phase independent of the starting viscosities of the polymers in the blend.

The net result of cross-linking and chain cleavage during the mixing and blending could be the small gel particles of elastomer similar in size to the original morphology of the blend or around 0.5–5 $\mu$m. A phase-contrast microphotograph of an 80/20 dynamically cured blend of amorphous EPDM and polypropylene is shown in Fig. 6. The EPDM phase appears to have particles up to 10 $\mu$m in size. The polypropylene phase appears to be 2 $\mu$m or less in cross section. It is difficult to determine phase continuity since a thin section was used, and no extraction of the rubber phase was possible owing to its gelled nature. Stress–strain properties do, however, suggest a continuous polypropylene phase.

There is the potential for some grafting during dynamic vulcanization. Peroxide curing of an EPDM in the presence of polypropylene could result in an occasional cross-linking reaction between the two polymer molecules. Similarly, in the dynamic curing of butyl rubber and polyethylene blends with a phenolic resin, the small amount of unsaturation in the polyethylene could participate in grafting reactions. In addition, the rate of cross-linking is low in the polypropylene due to chain cleavage at tertiary carbons and in the polyethylene due to a low level of unsaturation. Fractionation with a solvent for the polyolefin phase suggests little grafting between the two phases.

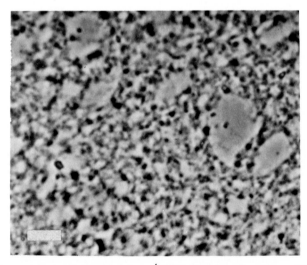

Fig. 6   Phase-contrast micrographs of a cross-linked rubber phase binary (80/20) blend of EPDM and polypropylene; the bar is 10 $\mu$m.

### b. Thermoplastic Rubber (TPR)

The physical properties of TPR [39, 40] depend on the actual grade, while other properties are inherent to polyolefins. They have good chemical resistance, good electrical properties, low specific gravity, and good colorability via color concentrates. The polymers remain flexible down to $-50°C$ and, due to the relatively high melting point of the polypropylene crystallinity (about 165°C), exhibit high retention of physical properties at elevated temperatures. Properly stabilized olefin polymers have good environmental and aging resistance.

Typical mechanical properties for materials with a range of hardness are listed in Table VII. The hardness range available is from 65 to 92 Shore A, and intermediate hardnesses and corresponding properties can be obtained by blending various grades. The softer grades, particularly, exhibit a rubber feel, traction, resilience, flexibility, and set properties. The harder grades are most comparable to the rubbery plastic characteristics of thermoplastic polyurethanes.

The rheological properties of TPR are consistent with their apparent structure. In general, the viscosity is decreased as the polypropylene content is increased, and all grades exhibit pronounced shear thinning. At higher shear rates typical of normal thermoplastic processing operations ($10^2$–$10^3$ $sec^{-1}$) the viscosity is around $5 \times 10^3$ poise, which is similar to a typical resin such as ABS. At lower shear rates the viscosity is high compared to typical resins, and there is no sign that the viscosity will become Newtonian at lower shear rates. This phenomenon is due to the high gel fraction in the blend.

### c. Butyl Rubber–Polyethylene Blends

Blends with useful physical properties are prepared with both low and high density polyethylene and butyl rubber at ratios of 3/1 through 1/1

Table VII

Typical Properties of Thermoplastic Rubber[a]

| Property[b] | TPR 1700 | TPR 2800 | TPR 1900 |
|---|---|---|---|
| Hardness (Shore A) | 74 | 87 | 92 |
| Tensile strength (Mpascal) | 6.9 | 9.4 | 13.3 |
| Elongation at break (%) | 200 | 180 | 250 |
| Set at break (%) | 20 | 30 | 50 |
| Aging at 150°C (hr to failure) | 300 | 400 | 300 |

[a] Uniroyal Chemical TPR®.
[b] From Morris [39].

polyethylene to butyl rubber [41] by intensive mixing with a phenolic resin curative. Stress–strain properties are in the rubbery plastic range, with the higher levels of polyethylene producing a slight yield during elongation. This results in considerably more tensile set than the blends containing lower amounts of polyethylene. Plastic deformation is also more in evidence for high-density than low-density polyethylene blends. Typical physical properties are listed in Table VIII. The polymer exhibits good electrical properties and can be loaded efficiently with fillers.

The rheological properties of ET polymers allow processing by the usual thermoplastic methods. The viscosity is around $10^4$ poise at shear rates between $10^2$ and $10^3$ sec$^{-1}$ at 230°C. Like the EPDM–polypropylene cross-linked blends there is a marked reduction in viscosity with increasing shear rate, and there is no indication of a lower Newtonian flow region at shear rates of 1 sec$^{-1}$. Higher rubber levels increase the viscosity and makes the material more difficult to process. Melt instability also occurs at lower shear rates with increased rubber level.

Butyl rubber blends with polyethylene and polypropylene over the entire compositional range have been studied by Deanin *et al.* [42] and their physical properties presented. The blends were prepared by milling and no curatives were present. The morphology of the blends were not investigated.

### d.  Silicone–Polyethylene Blends

Cross-linked elastomer phase blends of poly(dimethylsiloxane) and polyethylene have been prepared using mechanical shear to induce formation of free radicals during the mixing procedure [43]. Either polyethylene or ethylene–vinyl acetate copolymer was used. The poly(dimethylsiloxane) was a copolymer containing 2.5–4 mole % unsaturation by incorporation of methylvinylsiloxane. During the mixing free radicals are generated and these

Table VIII

Physical Properties of Butyl Rubber—Polyethylene ET Polymers[a]

| Property | Blend ratio | | | |
| --- | --- | --- | --- | --- |
|  | 3/1 HDPE | 1/1 HDPE | 3/1 LDPE | 1/1 LDPE |
| Hardness (Shore D) | 55 | 40 | 38 | 30 |
| Tensile strength (MPascal) | 24.4 | 16.7 | 16.7 | 14.0 |
| Elongation at break (%) | 500 | 440 | 450 | 500 |
| Tensile set (% after 200% elongation) | 125 | 72 | 56 | 31 |

[a] From P. F. Hartman [37].

could lead to both grafting and cross-linking. The grafting reaction would require a polyethylene macroradical to add across the double bond in the silicone phase, or coupling would have to take place between radicals on the polyethylene and on the silicone.

Extraction studies show cross-linking of the silicone phase has occurred and elemental analysis after extraction suggests some grafting. However, occlusion of nongrafted polymer during extraction could not be ruled out. The extent of reaction is controlled by the unsaturation level in the rubber phase and the polyethylene phase contains many fewer branch points. Electron micrographs show distinct silicone particles about 1–5 $\mu$m in diameter dispersed in a continuous polyethylene matrix at a 10% silicone level. The polyethylene remains continuous up to 75% silicone (the highest level shown). The extraction and microscopy data suggest that the silicone is dispersed as a microgel phase in a continuous uncrosslinked polyethylene phase. This type of morphology appears to be consistent with all of the cross-linked blends containing an elastomer and a polyolefin phase.

Mechanical and rheological properties are consistent with the morphological picture. The blends can be extruded and injection molded and the stress–strain properties are in the rubber–plastic range.

## REFERENCES

1. D. C. Allport and W. H. Janes, "Block Copolymers," Wiley, New York, 1973.
2. J. C. Saam, A. H. Ward, and F. W. Fearon, *Amer. Chem. Soc. Polym. Preprints* **13**(1), 525 (1972).
3. A. Noshay and M. Metzner, *Amer. Chem. Soc. Polym. Preprints* **13**(1), 292 (1972).
4. M. Kojima and J. H. Magill, *J. Appl. Phys.* **15**(10), 4159 (1974).
5. R. Cella, *J. Polym. Sci. Symp. No. 42*, p. 727 (1973).
6. R. W. Seymore, A. E. Allegrazza, Jr., and S. L. Cooper, *Macromol.* **6**(6), 896 (1973).
7. M. Pineri and C. Meyer, *J. Polym. Sci. Polym. Phys. Ed.* **12**, 115 (1974).
8. K. Van Henter, *Meeting IISRP 13th* (June 19, 1972).
9. A. L. Bull and G. Holden, Rubber Div., Amer. Chem. Soc. (April 1976).
10. T. L. Smith, *J. Polym. Sci. Polym. Phys. Ed.* **12**, 1825 (1974).
11. C. I. Chung and J. C. Gale, *J. Polym. Sci. Polym. Phys. Ed.* **14**, 1149 (1976).
12. Asaki, Chem. Ind. Co., Japanese Patent, 8097-945 (1972).
13. T. Inoue, T. Soen, T. Hashimoto, and H. Kawai, *Macromolecules* **13**(1), 87 (1970).
14. M. Brown, *Rubber Ind. (London)* **9**(3), 102 (1975).
15. C. K. Shih, U.S. Patent, 3,963,802 (June 15, 1976).
16. R. Bonart, *J. Macromol. Sci. Phys.* **B2**(1), 115 (1968).
17. W. Neumann, J. Peter, H. Holtschmidt, and W. Kallert, *Preprints, Rubber Technol. Conf., 4th, London* (May 22–25, 1962).
18. U.S. Patent, 3,193,522 (1965).
19. C. S. Schollenberger, *in* "Polyurethane Technology" (P. F. Bruins, ed.), p. 197. Wiley (Interscience), New York, 1969.
20. W. K. Fisher, U.S. Patent, 3,835,201 (September 10, 1974).

21.  R. G. Quynn, *J. Macromol. Sci. Phys.* **B4**, 953 (1970).
22.  R. G. Quynn and H. Brody, *J. Macromol. Sci. Phys.* **B5**, 721 (1971).
23.  B. Cayrol and J. Petermann, *J. Polym. Sci. Polym. Phys. Ed.* **12**, 2169 (1974).
24.  J. T. Lutz, Jr., *Plast. Eng.* **9**, 40 (1974).
25.  D. Puett, K. J. Smith, Jr., and A. Ciferri, *J. Phys. Chem.* **69**(1), 141 (1965).
26.  J. L. Work, *Polym. Eng. Sci.* **13**, 46 (1973).
27.  W. K. Fisher, U.S. Patent, 3,835,201 (September 10, 1974).
28.  A. Fritz and C. Shih, U.S. Patent, 3,882,196 (May 6, 1975).
29.  G. Ver Strate and Z. W. Wilchinsky, *Bull. Amer. Phys. Soc.* **15**, 332 (1970).
30.  I. W. Bassi, P. Corradini, G. Fagherazzi, and A. Valvassori, *Eur. Polym. J.* **6**, 709 (1970).
31.  E. N. Kresge, J. T. Kehn, J. P. Daugherty, and G. Ver Strate, *Synthetic Rubber Symp. No. 2*, p. 58 (1969).
32.  F. P. Baldwin and G. Ver Strate, *Rubber Chem. Technol.* **45**(3), 709 (1972).
33.  P. Zugenmaier and H. J. Cantow, *Kolloid Z. Z. Polym.* **230**, 229 (1968).
34.  Rubber World p. 49 (February 1973).
35.  W. K. Fisher, U.S. Patent, 3,806,558 (April 23, 1974).
36.  Rubber World p. 59 (October 1970).
37.  P. F. Hartman, U.S. Patent, 3,909,463 (September 30, 1975).
38.  A. M. Gessler and W. H. Haslett, U.S. Patent, 3,037,954 (June 5, 1962).
39.  H. L. Morris, *J. Elastomers Plast.* **6**, 121 (1974).
40.  P. Morin and J. R. Johnson, *Plast. Technol.* p. 49 (September 1975).
41.  P. F. Hartman, C. L. Eddy, and G. P. Koo, *SPE J.* **26**, 62 (1970).
42.  R. D. Deanin, R. O. Normandin, and C. P. Kannankeril, *Amer. Chem. Soc. Org. Coatings Plast. Preprints* **35**(1), 259 (1975).
43.  J. R. Falender, S. E. Lindsey, and J. C. Saam, *Polym. Eng. Sci.* **16**(1), 54 (1976).

# Chapter 21

# Polyolefin Blends: Rheology, Melt Mixing, and Applications[†]

## A. P. Plochocki

*Industrial Chemistry Institute*
*Plastics Division*
*Warsaw, Poland*

## I. SCOPE

In this chapter I first outline reasons why polyolefins are blended, modified rheologically, or melt mixed. This is followed by a short exposition of the

[†] In addition to its discussion of polyolefins, this chapter was included to serve as an introduction to the vast Eastern European and Soviet polymer blend literature and technology, which is developing somewhat differently than that in Western countries. Some ideas are only briefly introduced since neither space nor logistics permit their full explanation or development here. [ —EDITORS]

pertinent engineering rheology of the pure components and their blends. The rheological characteristics of molten polyolefins are the primary factors in forming the supermolecular structure of their blends, and thereby govern processability and performance characteristics. Also the melt-mixing processes, both dispersive and laminar, as well as the available measures of goodness of mixing, merit brief presentation. Naturally enough, the technical aspects of processing techniques for polyolefin blends, as well as their end-use properties, are discussed. An overview of the analytical and testing methods employed for these blends is given in the last section. The list of references forms a reasonably complete survey of polyolefin blend research carried out in the Soviet Union and Eastern European countries (some of which is presented for the first time for the reader in the United States). Its importance lies in both the high quality of work done in the Soviet Union within "schools" headed by Kuleznev, Lipatov, Vinogradov, and others and the fact that the majority of research on polyolefin (also with other polymers) blends conducted in the United States, Japan, and Western Europe is considered proprietary. Hence, the results of research done in the West, except for studies conducted in academic institutions, are available mainly through the necessarily awkward form of patent applications.

## II.  REASONS FOR BLENDING POLYOLEFINS

### A.  General

Polyolefins form the largest group of commercial thermoplastics, and before enumerating the trends in their composition development a review of the salient features of polyolefins is pertinent [1–7]. They constitute a group of polymers of complex macromolecular structure: wide molecular weight distributions that are diverse in shape, substantial short and long-chain branching [8–16], and high crystallinity [5, 17], coupled with a multitude of spherulitic forms and orientability [6, 18]. Their density is the lowest among polymers and strongly influences performance characteristics [3, 19].

The characteristics pertinent for melt mixing (blending) are [1, 4, 6, 18, 20–21]: excellent dielectric properties, water repellence, nonpolarity, high melt viscosities that respond weakly to temperature changes but strongly to shear rate or stress changes [6, 25, 26, 26a] coupled with inherently high melt elasticity [16, 16a, 27–30] and melt strength [31, 32]. Polyolefin glass transition temperatures, with the notable exception of polypropylene, are low [34, 35, 277].

Polyolefin blends display the characteristic features [2–4, 6, 11, 13, 15, 17, 18, 20, 34–49] of macromolecular incompatibility [50–52] with a texture consisting of either spherulites "swollen" with the lower-crystallinity component [53–55] or droplets-in-a-matrix alternative with fibrils-in-a-matrix "composites" [56]. There is evidence that the latter supermolecular structures persist in the molten state [5, 10, 57], rendering feasible the existence of so-called "rheologically particular" compositions (RPC)[†] [59, 177, 184]. These RPCs typically show extrema in plots of the melt viscoelasticity or end-use properties versus composition [28, 59, 60–67]. By and large, the properties of polyolefin blends are nonlinear functions of composition and the component properties [60, 68, 69, 71–74].

The melt mixing of polyolefins should be distinguished from blending [75–77]. Blending in the language of polyethylene manufacturing processes means the granulate (or pellet) blending operation. Melt mixing is carried out in two stages [78–82], dispersive followed by laminar mixing, and it is aimed typically at the production of new grades of polyolefins tailored to specific applications often unaccessible or uneconomic by an adjustment of the polymerization process and/or chemical modification [7, 8, 46, 47, 83–86]. If an increase in toughness of a polyolefin is required, blends with rubber and polyisobutylene answer this need [1, 2, 6, 34, 42, 46, 87–96]. For applications demanding a combination of toughness with high processing efficiency, blends are prepared with ethylene–vinyl acetate copolymer (EVA) [97], thermoplastic rubbers, ethylene–propylene rubber (EPR) [98–105], or styrene–butadiene–styrene copolymer (SBS) [70, 71, 106–111]: a third component generating melt heterogeneity, for example, isotactic polypropylene (iPP) or poly(dimethylsiloxane) (PMDS), is added [46, 47, 99, 112]. Polyolefins are melt mixed with other polymer types for three main reasons: manufacture of readily fibrillized plastics (e.g., polyolefin–polystyrene compositions [83]), fiber grade plastics for self-texturization (e.g., poly(ethylene terephthalate) (PET)–polyolefin [113]) and controlled permeability and water adsorption resistant plastics (e.g., polyamide– and polycarbonate–polyolefin systems [114–117].

Typical examples of commercial polyolefin blends are: impact-resistant polypropylene; superthin, highly orientable film grades of polyolefins; polyolefins for high-speed cable covering and blow molding of environmental stress cracking (ESC) resistant containers [67, 68, 73, 84, 85, 102, 103, 113, 114, 115–133].

---

[†] Evidently this term refers to the compositions in which maxima or minima occur when certain rheological or physical properties are plotted versus blend composition. The literature in English never seems to recognize these compositions (e.g., IV.2 in Section II of this chapter and [177, 184, 184a]) through any specific terminology. [—EDITORS]

Strong pressure for environmental protection as well as an ever increasing plastics waste accumulation has initiated numerous investigations of blending as a solution to the plastics scrap reclamation problem (e.g., [113, 134]). Extensive organizational and equipment design activities have been reported in Japan [134]. The property–composition problem posed by a typical municipal plastic waste (i.e., three- or more-component systems rich in PVC and polystyrene as well as polyolefins) was systematically studied by Paul and co-workers with a comprehensive evaluation of graft copolymers and chlorinated polyethylenes (CPE) as compatibilizers [135–138] (see Chapter 12).

## B.   Review of Selected Patents

Within the last 20 years, well over 100 patents were granted on processes and/or compositions for manufacturing rheologically modified (i.e., melt-blended) polyolefins. A comprehensive selection of polyolefin blend patents are presented in Table I. These are divided into four groups, designated by Roman numerals, according to type and use. The largest number of these patents (about 44%) forward claims for polypropylene–polyethylene (group I) systems. The majority of the patents within Group I result from trial and error industrial research directed toward an increase in toughness and lowering the brittleness temperature of polypropylene or toward improvements in the processability of "ultra-high molecular weight", linear polyethylene. Some of these claims rest on thus far unknown physical or rheological phenomena. The most significant among the former are: the enormous orientability of the iPP–LDPE (low-density polyethylene) systems (cf., e.g., I.25, I.36, I.48; the notation employed here specifies the group of the patents in Table I, and the consecutive number within the group, respectively); the molecular weight dependence of the brittleness temperature lowering effect (I.9); and the effectiveness of small amounts of iPP in controlling the development of spherulitic structure in blends (I.17, I.36). The trend in the Group I systems is toward an increase in the number of blend components, with a particular emphasis on block and, to a lesser extent, on random polyolefin copolymers or atactic polypropylene (aPP); hence, the claims in novel patents typically specify three or more components. The high processability of the iPP–PE blends is inherently associated with the structural effects of the blend melt flow: the composition and component dependent generation of the microheterogeneous melt structure (I.20, I.41, I.42). Further development or exploitation of this rheological effect for blends is possible by introduction of minute amounts of organosilicone polymers (I.36, II.8 [5]), which amplify the melt viscosity supression and melt elasticity-controlling phenomena due to the melt

Table I

Review of Patents Involving Polyolefins in Blends[a]

| Patent number | Country and year | Components Base | Components Complementary | Merits and applications | Comments |
|---|---|---|---|---|---|
| | | **GROUP I: POLYPROPYLENE–POLYETHYLENE POLYOLEFIN BLENDS** | | | |
| 893,540 | UK 1962 | PP | HDPE | Low temps, resistance | |
| 934,640 | UK 1963 | PP | HDPE | Blow molding | |
| 952,089 | UK 1964 | PP | HDPE | High impact, low-temperature resistant, molding | |
| 978,633 | UK 1964 | HDPE | PP, EVA | Wire coating, blow molding | High CSR (i.e. $\dot{\gamma}_{cr}$) |
| 982,753 | UK 1965 | LDPE | PC, PP | High impact, ESC resistant, blow molding | Low (MMI) |
| 3,153,681 | U.S. 1964 | PP | None; size distribution | Low-temperature resistant, molding | |
| 992,388 | UK 1965 | PP | LDPE | Coating, extrusion | |
| 3,192,288 | U.S. 1965 | PP | PIB | Molding, impact resistant | Mixing process specified |
| 1,026,254 | UK 1966 | PP | HDPE, EP (amorph.) | High impact, low-temperature resistance, molding | |
| 1,209,736 | West Germany 1966 | PP | HDPE | High-impact, low-temperature resistance which increases with av. mol. wt. (e.g. soln. viscosity VN) of PE | |
| 3,256,367 | U.S. 1966 | PP | EP | Improved impact, molding | |
| 3,265,771 | U.S. 1966 | PP | HDPE*) | Improved impact, molding | *)Controlled branching |
| 3,281,501 | U.S. 1967 | PP | HDPE*) | Improved impact, molding | *)Controlled VN |
| 1,195,489 | West Germany 1964 | LDPE, aPP, EP | PIB, BR | Cable jackets, blow molding | ESC resistant |
| 1,204,820 | West Germany 1967 | HDPE | EP | High impact, molding | CaCO$_3$ (filler) |
| 1,065,568 | UK 1967 | PP | EP, HDPE or other poly-(α olefin) | High impact, molding | Composition range claim |
| 1,077,217 | UK 1967 | HDPE | None | High-impact, molding | Controlled molecular weight distribution |
| 1,095,427 | UK 1967 | LDPE | EP, EB, TPX etc.–controllable crystallization polyolefins | Cable jackets of controlled structure | Small spherulites |
| 1,562,860 | France 1967 | PP | PE | Extruded, drawn products, film | |
| 5871 | Japan 1967 | PE | aPP | Tough, injection molding | |

*(continued)*

Table I (*continued*)

| No. | Patent number | Country and year | Base | Complementary | Merits and applications | Comments |
|-----|---------------|------------------|------|---------------|-------------------------|----------|
| | | | | **Components** | | |
| 20 | 3,355,520 | U.S. 1967 | LDPE | EVA or HDPE | Wire coating | Equal melt viscosities |
| 21 | 3,340,123 | U.S. 1967 | PP | LDPE | Paper, film coating | |
| 22 | 1,522,255 | France 1968 | PP | LDPE | Coatings (cellophane) | LDPE of MMI ~ 1.5 |
| 23 | 1,526,733 | France 1968 | HDPE | EP, EB-1, PB-1 | Impact resistant, molding | |
| 24 | 1,546,825 | France 1968 | PP | PE, oxidized | Impact resistant, molding | |
| 25 | 1,562,860 | France 1968 | PP | LDPE | Films, high strength | |
| 26 | 3,372,049 | U.S. 1968 | PP | aPP, HDPE, PB-1 | Biaxally oriented film for adhesives | |
| 27 | 1,114,589 | UK 1968 | PP | EP, LDPE | High-impact, molding | |
| 28 | 1,144,853 | UK 1969 | PP | EP, nucleating agent | Higher-impact, molding | PE as an EP substitute |
| 29 | 1,154,447 | UK 1969 | PP | EP-block, HDPE | High-impact, molding | |
| 30 | 11,025 | Japan 1969 | PP | HDPE | Film of high transparency and low friction coefficient | Properties as PP film |
| 31 | 1,810,829 | West Germany 1969 | PP | PE containing 30% (acetylene) carbon black | Cable jackets, conductive (electrically) | |
| 32 | 1,560,566 | France 1969 | PP | HDPE or EP | Impact resistant, moldings | |
| 33 | 3,420,916 | U.S. 1969 | HDPE | Degraded PO | Blow molding (controlled melt recovery i.e. MMI) | Process claim |
| 34 | 3,668,281 | U.S. 1972 | PP | ClPE, antypirens | Pipe, blow-molding compound | Low flammability |
| 35 | 86,534 | Poland 1973 | HDPE* | PP | Processability increase, Pipe, profile | *)Ultra high molecular weight PE |
| 36 | 97,112 | Poland 1973 | LDPE | PP, EP, PB-1, EB-1, TPX, etc., and PDMS | Stiff, impact-resistant molding, and extrusion: films | Microheterogeneit of melt |
| 37 | 3,755,502 | U.S. 1973 | PP | LDPE and EB-1, isopentene–dibutene | Oriented films, highly sealable | |
| 37a | 801,435 | Belgium 1973 | LDPE | iPP, aPP | Foamed sheet leatherlike | |
| 38 | 7,431,929 | Japan 1973 | LDPE | SBR, aPP | High-impact, foamed moldings | |
| 39 | J4,8,096,688 | Japan 1973 | PP | EP, aPP | foamed moldings | |
| 40 | 2,306,893 | West Germany 1973 | PP | HDPE, aPP | Molding, high flow, weld strength | Both iPP, aPP the same MWD |
| 41 | 2,306,892 | West Germany 1974 | PP | HDPE, aPP | Molding, high flow, high weld strength | |
| 42 | 2,187,830 | France 1974 | HDPE*), PP | PP, EVA, EP, EA copolymer | Adhesive tape, tearing-strength film Biaxial orientation | *)HDPE advantageous |
| 43 | 3,849,520 | U.S. 1974 | PP | Mixture of PB-1 and EB-1 | Oriented "shrink" films : transparent, nonblocking | PB-1 improves sealability |
| 44 | 3,914,342 | U.S. 1974 | HDPE or E copolymer | PP, EP | Cable jackets, blow molding, high resistance against ESC high processability, films | Process claim |

Table I (*continued*)

| No. | Patent number | Country and year | Components Base | Components Complementary | Merits and applications | Comments |
|---|---|---|---|---|---|---|
| 45 | J7.4,015,044 | Japan 1974 | HDPE | EP or BR | High impact, molding | |
| 46 | J7.4,030,264 | Japan 1974 | HDPE | EP | High impact, molding | |
| 47 | J4.9,114,678 | Japan 1974 | PP | LDPE | Biaxially oriented films, 90°C shrinkage temperature | |
| 48 | J4.9,114,679 | Japan 1974 | PP | EP, PE | Transparent, biaxially oriented films of 80°C shrinkage temperature | |
| 49 | J5.0,076,157 | Japan 1975 | aPP | LMWPE, PB-1 | Copper adhering cable insulation | |
| 50 | J5.0,008,848 | Japan 1975 | PP | LDPE | Paper coating, good necking | Shear degraded LDPE of MFI ≈ 50 |
| 51 | J7.4,048,470 | Japan 1975 | PP | EP, terpene | Film, heavy-duty shrink packing | Terpene = Picolite (TM) |
| 52 | J4.9,112,946 | Japan 1975 | PP or BEP | EPR | High-impact molding | |

GROUP II: BLENDS OF POLYOLEFINS WITH OTHER POLYMERS

| No. | Patent number | Country and year | Base | Complementary | Merits and applications | Comments |
|---|---|---|---|---|---|---|
| 1 | 1,193,104 | France 1959 | LDPE | Graft copolymer ES, EVA | Extrusion and molding | Apparently the first use of compatibilizers |
| 2 | 897,643 | UK 1962 | EP | PP | High-impact molding | — |
| 3 | 705,481 705,482 | Canada 1965 | PE | PC, EP, EB-1, BS, EVA | High: extrudability ESC resistance, gloss and low: MMI, $T_b$. | See UK 982,758 |
| 4 | 990,273 | UK 1965 | PE | PP–PS, PMM, TPX, BE | Films, coatings, fiber | Compatibilizer introduced |
| 5 | 3,725,372 | U.S. 1969 | PE | Cross-linked PE | — | — |
| 6 | 3,668,281 | U.S. 1970 | PP | — | — | — |
| 7 | 3,674,757 | U.S. 1970 | PO | — | Extrusion grade | — |
| 8 | 3,723,402 | U.S. 1970 | PP | Organosilicon compound | — | — |
| 9 | 3,755,502 | U.S. 1971 | PP, EB-1 | Dipentene–isobutene copolymer*) | Films of low sealability temperature Good optics, tough | *)(commercial grade, 1:1) |
| 10 | 3,767,638 | U.S. 1971 | PO | — | — | — |
| 11 | 97,228 | Poland 1973 | PE | PA, PC, PET, poly-hydroxyester, PPO copolymer A, copolymer PO and organosilicon compounds | Molding, extrusion blowing grades Stiffness, controlled permeability, good processability | High processability due to micro-heterogeneous melt |
| 12 | J4.8,018,333 | Japan 1973 | PP | PS, PEwax | High-speed molding | — |
| 13 | J4.8,031,241 | Japan 1973 | PP | PS, HIPS | Luster effect profiles (Decorative) | — |
| 14 | J7.4,031,551 | Japan 1974 | PP | Terpene resin (Picolite) | Oriented film. High sealability | — |
| 15 | J4.9,013,267 | Japan 1974 | PO | SBS, foaming agent | Sheet extruded, foamed | — |

(*continued*)

Table I (*continued*)

| No. | Patent number | Country and year | Components Base | Components Complementary | Merits and applications | Comments |
|-----|---------------|------------------|------|---------------|-------------------------|----------|
| 16 | J4.9,072,339 | Japan 1974 | LDPE | EPR | Lasting electric strength cable insulation | ESC, corona discharge resistant |
| 17 | J4.9,126,903 | Japan 1974 | PVC | PE Soaps | Synthetic paper, non-woven textiles | Fibrillation |
| 18 | 3,793,283 | U.S. 1974 | PP or block EP | SBS | Flexible, high-impact foamed, moldings | — |
| 19 | 3,808,304 | U.S. 1974 | PP | PB-1 | Heat*) shrinkable film | *)100°C |
| 20 | 3,821,332 | U.S. 1974 | PE | PIB | Impact high, Blow, injection molding | Particulate |
| 21 | 814,477 | Belgium 1974 | PP | PB-1, high II | Highly transparent film, high shrinkage, sealability | 0.85:0.15 (optimal PP/PO ratio) |
| 22 | 816,311 | Belgium 1974 | PE or PP | PS | Sheet-paper substitute, | Biaxial orientation |
| 23 | 7,309,337 | Republic of South Africa 1974 | EP with hexene-1 | E–methacrylic acid copolymer | Highly sealable films | — |
| 24 | 2,254,440 | West Germany 1974 | LDPE | PB-1, TPX, PP, HDPE | Film:. Shrinkable, high tear strength | — |
| 25 | 006,396 | UK 1974 | PET | PP | Opaque, translucent film | Printing, drawing |
| 26 | 097,014 | Poland 1975 | LDPE or PP | HDPE or EP elastomers SBS and PMDS or HMW PO | High-impact film or molding grades Good processability | High processability due to micro-heterogeneous melt |
| 27 | 3,865,903 | U.S. 1975 | PP | Pentadiene polymer | High-sealability films | — |
| 28 | 3,887,640 | U.S. 1975 | EP | LDPE | Coating of high adhesion and cuff resistance | MMI $\geqslant$ 50 |
| 29 | 3,919,358 | U.S. 1975 | EP | LDPE | Moldings, liners, tubing | — |
| 30 | 818,455 | Belgium 1975 | PO, PE, PP, TPX | Silicone rubber | Cable insulation with low-temperature flexibility | Mixing at $\dot{\gamma} = 10 \sec^{-1}$ |
| 31 | 822,619 | Belgium 1975 | EP | PB-1 | Films: shrinkable, of low blocking and high tear resistance and sealability | Poultry packing |
| 32 | 455,124 | USSR 1975 | PP | Silicone rubber | Cable sheathing of $T_b = -80°C$ | — |
| 33 | 2,433,005 | West Germany 1975 | EPDM | PE, PP | Heat shrinkable elastomers | Vulcanizable |
| 34 | 7,504,141 | Netherlands 1975 | HDPE | ES, E–MM, PS PMM, PVC | Molding, High strength | — |
| 35 | J.4,853,025 | Japan 1974 | PP | Terpene resin (PETROSIN) TM | Oriented film of good bending set | See II.14 |
| 36 | J.4,843,031 | Japan 1973 | PO | SBS, PS or other "polyaromatics" | High-impact molding | Impact increase |
| 37 | 2,012,193* | West Germany | PB1 | PP, LDEP or LDEB (copolymer) | Tear-resistant film, tubes | — |

GROUP III: POLYOLEFIN–POLYAMIDE BLENDS

| No. | Patent number | Country and year | Base | Complementary | Merits and applications | Comments |
|-----|---------------|------------------|------|---------------|-------------------------|----------|
| 1 | 3,373,222 | U.S. 1967 | PA | HDPE and carboxylated HDPE | Blow molding, containers | — |

Table I (*continued*)

| No. | Patent number | Country and year | Components | | Merits and applications | Comments |
|-----|---------------|------------------|------|----|-----|----|
| | | | Base | Complementary | | |
| 2 | 3,373,223 | U.S. 1968 | PA, PO | Graft AE, EMM | Low-permeability bottles | PO = PP, HDPE |
| 3 | 1,122,696 | West Germany 1968 | — | — | — | — |
| 4 | 1.301,064 | France 1968 | — | — | — | — |
| 5 | 919,098 | UK 1968 | — | — | — | — |
| 6 | 656,484 | Italy 1968 | — | — | — | — |
| 7 | 606,944 | Belgium 1968 | — | — | — | b) |
| 8 | 267,917 | Netherlands 1968 | — | — | — | — |
| 9 | 236,128 | Austria 1968 | — | — | — | |
| 10 | 3,645,932 | U.S. 1972 | PA6, PA66 | aPP, TPX, PC etc.*) | Molding, high stiffness | *)PP or PE alternatively −modifiers for PA crystallizability |
| 11 | 3,668,278 | U.S. 1972 | PP | Tinctorial modifiers | Molding, high flow, fibers | — |
| 12 | 097,228 | Poland 1973 | Polyethylene | AE copolymer, PMDS | Blow and injection molding | Controlled permeability and good processability |
| 13 | J7.4,018,782 | Japan 1974 | PA | HDPE | Molding, extrusion, (Abrasion resistant) | — |
| 14 | J7.4,049,179 | Japan 1974 | PA | PB-1, TPX, HDPE, PP | Printable, packing films | Metaxylene ring containing PA |
| 15 | 2,426,671 | West Germany 1975 | PA6 | Graft EMM-A | Impact-resistant, molding | PE of MFI = 15 |
| 16 | J4.9,117,574 | Japan 1974 | PA6 (contg 0.3 SURLYN− LDPE blend) | Ionomer (SURLYN) | Double ply films of good adhesion | Coextrusion of ionomer melt blend PD and PA |
| 17 | 1,403,794 | UK 1975 | PA 66, PP or EP | Graft PP–PA copolymer | Sheets, films, molding (automotive) | Graft ppd. in an extruder |

GROUP IV: LINEAR–BRANCHED POLYETHYLENE SYSTEMS

| No. | Patent number | Country and year | Base | Complementary | Merits and applications | Comments |
|-----|---------------|------------------|------|----|-----|----|
| 1 | 1,109,864 | France 1956 | LDPE | HDPE | Extrusion | — |
| 2 | 2,862,762 | U.S. 1959 | LDPE | LDPE of melt viscosity ratio 1:10⁵ | Extrusion and molding grades of excellent processability | Apparently the first description of blending. Only LDPE |
| 3 | 827,363 | UK 1960 | HDPE | LDPE | "HDPE-like," high stiffness, | — |
| 4 | 843,697 | UK 1960 | HDPE | LDPE | impact-resistant moldings | — |
| 5 | 871,586 | UK 1961 | LDPE or HDPE | — | High stiffness, molding | Mechanochemical treatment of irradiated PE |
| 6 | 875,040 | UK 1961 | HDPE | HDPE | High ESC resistance, blow molding | Components differing in av. molecular weight. |
| 7 | 904,985 | UK 1962 | PE*) | Cross-linked PE*) | Molding, Extrusion | *)Same PE type: control of MMI, FR feasible |
| 8 | 933,194 | UK 1963 | LDPE | HDPE*) | Low haze films | *)VN$_{HDPE}$ ≃ 3 VN$_{LDPE}$ |
| 9 | 942,369 | UK 1963 | MDPE | LD or HDPE | Extrusion coating | MD-medium density grade PE 3 component blends |

(*continued*)

Table I (*continued*)

| No. | Patent number | Country and year | Components Base | Components Complementary | Merits and applications | Comments |
|-----|---------------|------------------|------|--------------|-------------------------|----------|
| 10 | 1,483,603 | France 1967 | LDPE | HDPE | Extrusion coating, High shrink stability | Photographic products |
| 11 | 3,379,303 | U.S. 1968 | LDPE | HDPE | Wire insulations | Surface smoothnes (high CSR, $\gamma_{cr}$) |
| 12 | J5.0,056,448 | Japan 1974 | PE (of VN ⩾ 7) | PE (of VN ⩽ 5) | High-speed extrusion | Critical shear increase with addition of a low mol. wt. PE |
| 13 | J4.9,064,667 | Japan 1974 | HDPE*) | LDPE | Tough, blow film | *)of FR > 3 |
| 14 | 3,888,409 | U.S. 1975 | LDPE*) | LDDE*) | Cable through filling comps. | *)MFI = 0.2–25 (†)FR 50.2 ⩽ 14 |
| 15 | 820,878 | Belgium 1975 | LDPE | HDPE | High weld strength, strong weld bags | Shrinkable pallet coats |
| 16 | J7.5,006,022 | Japan 1975 | LDPE | HDPE | Blown film, Excellent drawdown | HDPE of 5–20 $CH_3/1000\,C$ |
| 17 | P190,436 | Poland 1976 | LDPE | HDPE, PDMS | Large blow moldings, injection molding, pipe | Improved processability |
| 18 | 2,304,998 | West Germany 1973 | HDPE | LDPE | Transparent, tough films | PE might be substituted with PB-1 or EB-1, EPB-1 |

*a* Abbreviations used in table:

BR = butadiene rubber
EA = ethylene–acrylate copolymer
EB = ethylene–butene-1 copolymer
EMM = ethylene–methyl methacrylate copolymer
EP = ethylene–propylene copolymer
EPB-1 = ethylene–propylene–butene-1 terpolymer
ES = ethylene–styrene copolymer

EVA = ethylene–vinyl acetate copolymer
PC = polycarbonate
PH-1 = polyhexene-1
PIB = polyisobutylene
PO = polyolefin
PDMS = poly(dimethylsiloxane)
TPX = poly(4-methylpentene-1)

*b* Data from Kunstst. Gummi [117].

*Nomenclature*

CSR = critical shear rate (or $\dot{\gamma}_{cr}$), $sec^{-1}$
ESC = environmental stress cracking, hrs
FR = flow rate, a standardized measure of melt fluidity similar to melt index (gm/10 min) [ASTMD1238
   e.g. FR.230.2A = FR at 230° C, 2-kG load using die A.
MFI = melt flow index, or FR at 190°C and under 2.16 kG, die "A" for LDPE.
MMI = melt memory index (e.g. [25]) or
MR = melt recovery (e.g., extrudate swelling)
VN = viscosity (of a soln.) number [ASTM D1601]
   II = Isotacticity Index [2, 6a]
$T_b$ = brittleness temperature ([3, 6] etc.)

structure-regulating action of isotactic polypropylene as well as the phase transitions at the critical blend concentration (RPC).

The second largest group of patents encompasses blends of polyolefins with other polymers: elastomers and plastics (Group II in Table I, about 29% of the patents). This group includes blends of polypropylene or polyethylene with olefin copolymers and polymers of higher olefins. In the latter case, the main effect is an increase of intercomponent adhesion, resulting in enhanced sealability; lower shrink temperatures in films (II.21, II.24, II.31); and stronger weld lines in moldings. The general trend in all Group II blends is to compound the components with graft copolymers of the blend components to act as a compatibilizer (II.1, II.4, II.13, II.23, II.34) (see Chapter 12). Blends with polystyrene and poly(vinyl chloride) (PVC) are used in manufacturing substitute paper (II.17, II.22) or for production of decorative profiles (II.13). Blends with polycarbonates are used in the production of controlled permeability containers and hydrophobic engineering molding parts (II.3, II.11). Elastomers are incorporated into polyolefins in order to increase the toughness of moldings and to produce more pliable films (II.15, II.18, II.26) as well as to enhance the flexibility of formed sheets and other shapes. Silicone rubber is used in production of cable sheaths of longer lasting electric strength that are flexible at low temperatures (II.30, II.32). Again the trend toward multicomponent blends is evident, with the main effort in this case directed toward substantial increases in processability through exploitation of the melt structure heterogeneization (II.11, II.26) effect.

Although polyolefin–polyamide blends belong to the group discussed above, they were separated into Group III because of their number and basically similar effect evoked, viz. reduction of permeability (with respect to that of polyolefins) and water absorption (compared to polyamides). This group comprises about 14% of all the patents on polyolefin blends. In order of application frequency, Group III blends are used in the manufacture of low-permeability containers (III.1, III.2, III.13) (see Volume 1, Chapter 10), sheets and films with good printability and improved adhesion to other polymers (III.14, III.16, III.17), and high-flow injection-molding grades of plastics for engineering applications (III.3–10, III.13, III.15, III.17). The base polyolefin employed in either polyethylene or polypropylene.

The last, but important, and historically the first, group of polyolefin blends consist of high- and low-density polyethylenes (HDPE and LDPE). Interestingly, the first patent (IV.2) describing blends of LDPEs is apparently the first record of a critical concentration effect (RPC) in blend melt flow—the composition of 1/3-component ratio is claimed to be the best processable mixture. This effect was rediscovered much later for other blends. Technical reasons for the manufacture of polyethylene blends have been to increase the stiffness of LDPE and the toughness and flexibility of HDPE (IV.3–5). With increasing numbers of HDPE grades, more sophisticated systems were introduced for which the molecular structure parameters: average molecular weight (IV.6, IV.8, IV.14) and branching (IV.16) were specified for the components. The applications have progressed from injection molding to blow molding (IV.6, IV.17), then to high-processability extrusion grades (IV.11, IV.12), and the present trend is the production of tough, transparent films (IV.13, IV.15, IV.18).

## III. ENGINEERING RHEOLOGY OF MOLTEN POLYOLEFIN BLENDS AND THE MELT BLENDING PROCESS

Successful melt blending of polyolefins to obtain selected end-use properties and/or processability is the outcome of skillful employment of their rheological characteristics [6, 25, 26, 107, 139]. The ratio of the component rheological characteristics determines the results of dispersive mixing, that is, the fineness of the dispersion of the minor component [112, 155]. The elastic characteristics of the components determine, in turn, the formation of droplet-in-matrix or layers-in-matrix texture upon completing the laminar melt mixing stage of the blending process [140]. Reliable knowledge of the rheological characteristics over a wide range of shear rate or stress is of paramount importance in predicting or modeling the characteristics of a blend and/or the course of the blending process. The complementary physical characteristics required for these applications are surface tensions [140] of molten polyolefins and a knowledge of the interphase layer behavior [39, 141] (see Volume 1, Chapter 6). In this section, I would like to outline the salient features of polyolefin rheological characteristics with suggestions as to the way in which their rate or stress dependencies should be accounted for in engineering applications. A short summary on "blending laws" applicable for industrial melt mixing operations follows. Next, a technique for rheotechnical selection of polyolefin-blend compositions is described. In the last subsection I discuss the so-called "rheologically particular" compositions (RPCs) of polyolefin blends, the super-molecular structure of molten blends, interphase layer effects, and aggregation phenomena in flowing melts.

### A. Rheological Characteristics

The main rheological characteristics of molten polyolefins [6, 25, 26, 26a] consist of rate- or stress-dependent viscosity and elasticity (Fig. 1). The

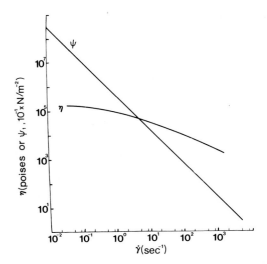

Fig. 1 Shear rate dependence of melt viscosity $\eta$ and elasticity, $\psi_1 = N_1/\dot{\gamma}^2$, observed for Malen ® J 400 polypropylene at 463°K. (From Plochocki *et al.* [53].)

latter is typically represented by the normal stress difference $N_1$ (see Volume 1, Chapter 7). Little is known about the tensile melt viscosity of polyolefins in spite of the importance this characteristic has in the development of blow molding grades of blends and in the formation of fibrillar blend texture in convergent flows (Fig. 18 and [156, 156a,b].

The viscosity of molten polyolefins [6, 9, 10, 16, 21, 22, 25, 26a, 32, 142] and the polymers blended with them [27, 28, 62, 63, 86, 92, 93, 143] have been studied extensively [90, 91, 102, 144–147], including a comprehensive interlaboratory program of investigation [25], using routine capillary or rotational viscometers and laboratory mixers [148]. However, the form of presentation, with few exceptions [18, 31, 59, 62, 64], does not render them applicable for engineering use. The most appropriate form for this application seems to be the logarithmic parabola approximation for the viscosity curve:

$$\log \eta = B_0 + B_1 \log \dot{\gamma} + B_2 \log^2 \dot{\gamma} \qquad (1)$$

where $\eta$ is melt viscosity in poises and $\dot{\gamma}$ is shear rate in reciprocal seconds. The term $B_0$ is related to a standardized viscosity at $\dot{\gamma}^0 = 1 \ \mathrm{sec}^{-1}$, that is, $\eta^0 = 10^{B_0}$, and $B_1$ is a measure of the shear rate dependence of the viscosity corresponding to the power law exponent when $B_2 = 0$. The approximation is usually carried out by applying the least-squares regression technique to the experimental data on any engineering calculator. One may easily extend the approximation to include the effects of temperature, hydrostatic pressure,

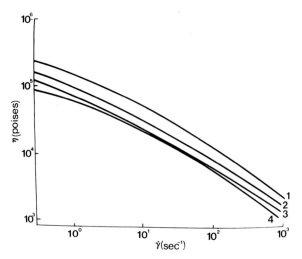

Fig. 2   Composition dependence of the melt viscosity curves for polypropylene–LDPE blends at 463°K : (1) LDPE; (2) blow molding; (3) injection molding grades of blends; (4) polypropylene. (From Plochocki *et al.* [53].)

and even composition by introduction of additional terms accounting for these effects and their interactions [117a]. An example of the rate and composition dependence of the melt viscosity for polypropylene–LDPE blends is presented in Fig. 2. Similar data are available for polyamide–HDPE (Fig. 3), polyethylene (or polypropylene)–polystyrene [149, 150], polyethylene–EVA [4, 102, 151], and other blends of polyolefins [16, 27, 28, 62, 63,

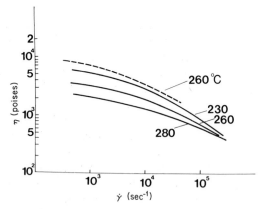

Fig. 3   Temperature dependence of the viscosity curves for ASE 9010 (———) and BSE 9010 (———) polyamide–polyethylene blends. (Courtesy *Kunstst. Gummi* [117].)

90, 93, 103, 143–147]. From the viewpoint of engineering applications (both composition selection and scaleup of melt-mixing processes) the literature data suffer from two drawbacks: (1) they are in the form of graphs, which show the magnitude and the shape of the $\eta(\dot{\gamma})$ function but these depend nonlinearly on the composition of a blend [59, 64]; (2) the data published thus far do not account for the melt flow behavior of some polyolefins blends characterized by two or more flow mechanisms (Fig. 4). The solution to the problem, in part, appears to be with the use of the logarithmic parabola approximation parameters, for example, $B_0$, $B_1$ which represent the melt viscosity, $\eta^0$ and its rate dependence, $n$ resp. (there is $B_1 \simeq n - 1$). Since the confidence limits for the parameters can be readily estimated,[†] an unambiguous comparison of the "rheology"[‡] of polyolefins and their blends is

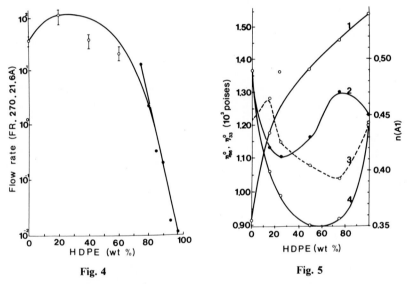

Fig. 4                                    Fig. 5

Fig. 4   Composition dependence of melt fluidity (expressed as the flow rate FR at 543°K. ASTMD 1238, where FR = 0.015 exp 0.26 wt% PP [229]) for polypropylene–HDPE (ultra high molecular weight) blends. Note the two regions of the composition dependence. (FR is essentially a "general" melt flow index.—EDITORS) (Courtesy *Trans. Soc. Rheol.* [65].)

Fig. 5   Composition dependence of various melt viscosity parameters at 463°K, for polypropylene–HDPE blends: (1, 2) parameters for shear rate dependence of the melt viscosity, which is the logarithmic parabola parameter $A_1$ and the power law exponent $n$ respectively; (3, 4) the standardized melt viscosities 'as measures with long, L/D ~ 66, and shorter, L/D ~ 33, capillaries respectively. (Courtesy *Trans. Soc. Rheol.* [65].)

[†] This constitutes a significant advantage over presentations using either a modified power law [63, 152], Ellis [62] or Bueche-Harding [3, 153] equation constants.

[‡] That is, melt viscoelasticity within the engineering applications range.

therewith feasible. Also the composition-dependence evaluation executed with the approximation parameters is consistent for different viscometers and various viscosity measures employed (Fig. 5). The treatment suggested here for the composition dependence of shear viscosity $\eta$ might also be applied to that of the tensile melt viscosity $\lambda$ [154], in spite of the unusual form of its stress-dependence curve as reported by Cogswell for HDPE–LDPE and polyamide–polypropylene blend systems (Fig. 6) [156, 156a]. It seems that the third term of the analogous approximation equation whose coefficient, $B_2{}^*$, which we may distinguish with an asterisk from $B_2$ of Eq. (1), should be treated with closer attention than $B_2$ in the case of shear viscosity. This is inferred from data on (estimated) tensile viscosity-rate dependence observed for various LDPEs differing in homogeneity [157]. The effect of temperature on polyolefin blend melt viscosity, and how this changes with composition has been evaluated for only a few systems. Very strange Arrhenius plots have been reported for polyamide–polypropylene systems [158]. Complex forms of the composition dependence of the activation energies for polyethylene–rubber have been reported [159]. Because of these, I propose systematic research into the temperature dependence of the logarithmic parabola approximation parameters $B_i$ ($i = 0, 1, 2$). Until the results of such studies are available, one has to resort to qualitative data scattered throughout the literature, and, with notable exception of Semjonov's monograph [26a], unevaluated.

Within the range of shear rates or stresses pertinent for engineering applications, the principal melt elasticity [6, 9, 22, 25, 30, 53, 65, 144, 149, 153, 160–163] measure is extrudate swell or puff-up (Barus effect), $\beta$ [16, 16a, 164, 165]. Encouraging are Han's [160] data on the end pressure measurement for capillary (slit) melt flow: elasticity evaluation seems to be simpler using this

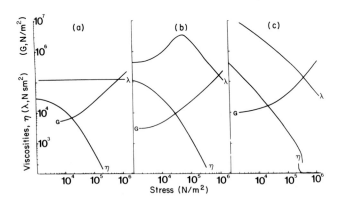

Fig. 6   The effect of blending on shear ($\eta$) and tensile ($\lambda$) viscosity and apparent modulus $G$: (a) LDPE; (b) HDPE–LDPE 38/62 blend; (c) HDPE. Similar effect was observed for a polyamide–polypropylene blend. (Courtesy N. Cogswell [ICI Ltd., 1971].)

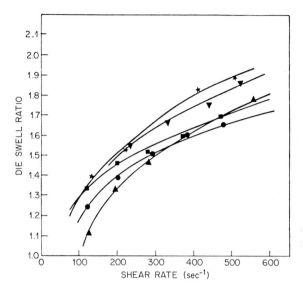

Fig. 7 Shear rate and composition dependence of the extrudate swelling $\beta$ for polyethylene–polystyrene blends at 473° K. (●) PE; (▲) PS; (■) PS/PE = 20/80; (▼) PS/PE = 50/50; (★) PS/PE = 80/20, $L/D = 4$. (Courtesy *A.I.Ch.E.J.* [164].)

technique [146]. Composition and shear stress dependence of this swelling $\beta$ is illustrated for HDPE–polystyrene blends in Fig. 7 and for polypropylene–HDPE blends in Fig. 8. Problems involving engineering applications of the composition dependence of melt elasticity require careful selection of the rheological characteristic used to measure it–direct application of the viscoelasticity measures, which are adequate for polymers within the linear range, lead to somewhat inconclusive results [151, 153]. The basic step here is the selection of a "sound" relationship between $\beta$ and the recoverable strain $\gamma$. This point is illustrated in Fig. 9, where the extrudate swelling $\beta$ from two capillaries ($L/D = 33$ and 66) is used to calculate $\gamma_e$ using both the empirical relationship of Rigbi (see ref. 31 in [65]) and the more rigorous treatment of Tanner (subscripts R and T, respectively [166]). The elastic strain measure selected has a dramatic influence on the rate dependence of the apparent shear modulus $G$. There is consistent experimental evidence that Tanner's treatment yields the most reliable $\gamma_e$ values. An example of this is the overlap in recoverable shear strain measured directly by rheogoniometry and that evaluated from extrudate swelling data (cf. Utrackie *et al.* in [65]). Majority of research on molten blends elasticity in the past was directed toward an elucidation of the composition dependence of the relaxation spectra $H_i(\tau)$ ($i$th degree of approximation, $\tau$ is relaxation time), yet these results are of

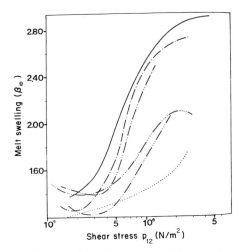

Fig. 8   Shear stress and composition dependence of the extrudate swelling $\beta$ for polypropylene–HDPE blends at 469° K. Designation based on the HDPE content (wt%) in blends: (——) 0 wt% HDPE and 100 wt% iPP; (—·—) 15/85 HDPE/iPP; (—··—) 25/75 HDPE/iPP; (—···—) 50/50 HDPE/iPP; (······) 75/25 HDPE/iPP; (···) 100 wt% HDPE. (Courtesy *Trans. Soc. Rheol.* [65].)

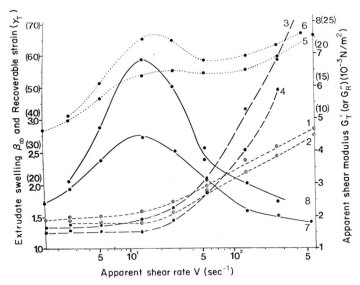

Fig. 9   Shear rate dependence of several melt elasticity measures derived from the extrudate swelling $\beta$ measured for polypropylene at 463°K using $L/D = 33$ and 66 capillaries using the Rigbi (R) and Tanner (T) $\beta$–elastic strain $\gamma$ relationships: (1, 2) extrudate swelling; (3, 4) Tanner's elastic strain, $\gamma_T$; (5, 6) $G_R$ moduli estimated from Rigbi's strain $\gamma_R$; (7, 8) $G_T$ calculated from $\gamma_T$.

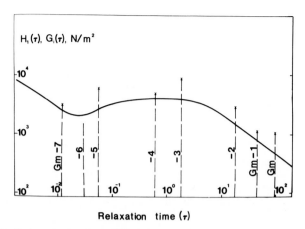

**Relaxation time (τ)**

Fig. 10 The first approximation relaxation spectrum, $H_1(\tau)$, for a polypropylene–HDPE blend (containing 75% of the latter) at 463°K from extrudate swelling and dynamic measurements. The discrete spectrum $G_i(\tau)$ established via Procedure X is shown for comparison. (Courtesy W. Wachmiller, 1973.)

limited value for engineering applications. By way of illustration a spectrum for a polypropylene–HDPE blend, derived from combined dynamic shear and steady-flow extrudate swelling data is presented in Fig. 10. These results were derived using the first approximation, continuous spectrum $H_1(\tau)$ and the discrete spectrum $G_i(\tau)$ established via "Procedure X."[†] The two display satisfactory coincidence. An illustration of the composition dependence of the spectra $H_1(\tau)$ is presented in Fig. 11. For practical purposes, the spectra offer little assistance even if they were approximated from reasonably high shear data (as those presented here were). This is the consequence of the involved relationships between the spectra and the "engineering" measures of molten-blend elasticity. A more convenient approach employs the first normal stresses difference, $N_1 = p_{11} - p_{22}$, and its shear rate-dependent coefficient $\psi_1$. The term $N_1$ is directly measurable in viscometric flows (using, e.g., a rheogoniometer or Rheometrics instrument) or may be evaluated from extrudate swelling via the recoverable strain $\gamma$ [65, 86, 100]. We have the following relationships for the normal stresses coefficient and its shear rate dependence:

$$\psi_1 \equiv N_1/\dot{\gamma}^2 \tag{2}$$

$$\log \psi_1 = C_0 + C_1 \log \dot{\gamma} \tag{3}$$

[†] Procedure X is a method for successive graphical evaluation of discrete relaxation times from viscoelastic data and is described by Tobolsky [167]. [—EDITORS]

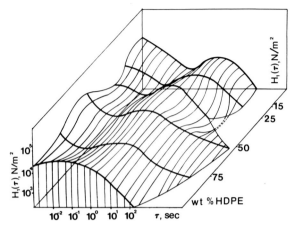

Fig. 11 Composition dependence of the relaxation spectra $H_1(\tau)$ at 463°K for polypropylene–HDPE blends. (Courtesy M. Ostrowski, 1972.)

Assuming a power law dependence of $N_1$ on the shear rate (which means $m = C_1 + 2$),

$$N_1 = N_1{}^0 \dot\gamma^m \qquad (4)$$

The elasticity coefficient $\psi_1$ offers an expedient measure of molten-blend elasticity. Indeed, for many polyolefins and their blends quite satisfactory correlation was established for Eq. (3) within six decades of shear rate, that

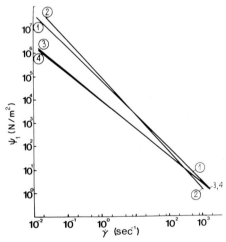

Fig. 12 Composition dependence of melt elasticity, $\psi_1(\dot\gamma)$, curves for polypropylene–LDPE blends (see Fig. 2 for description). (From Plochocki et al. [53].)

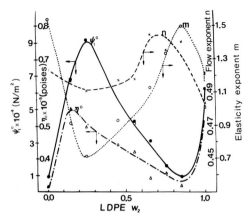

Fig. 13 Composition dependence of the melt viscoelasticity characteristics for polypropylene–LDPE blends at 463°K: standardized viscosity $\eta^0$, elasticity $\psi_1^0$, and the parameters for their rate dependence: $n$ and $m$, respectively. (From Plochocki and White [59].)

is, both rheogoniometric and capillary rheometry data [59]. Hence, by an analogy with the melt viscosity power law, one has $\psi_1^0 = 10^{C_0}$ as the standardized melt elasticity and $m$ as the measure of elasticity shear rate dependence. Melt elasticity curves for polypropylene–LDPE blends and their pure components are presented in Fig. 12. The blends are injection (2) and blow molding (3) grades. A comparison of the composition dependence of the engineering measures of the melt viscosity $\eta^0$ and elasticity, $\psi_1^0$ as well as of their rate dependences is attempted in Fig. 13 for the same blends. The technique outlined, therefore, allows for unambiguous determination of the particular compositions (RPC) for the polyolefin blends, and by referring to the rate-dependence exponents, $n$ and $m$, an assessment of the persistence of these particular features (e.g., high fluidity, low elasticity) at high shear rates or stresses can be made.

## B.   Blending Laws in Engineering Rheology Practice [170, 171a]

The standardized measures of melt viscosity and elasticity are obviously convenient for predicting the rheological characteristics of a blend of selected composition from known characteristics of the components. In order to perform such a feat some reliable blending law has to be available. The attempts undertaken in this direction were aimed at refinements of empirical mixing rules, due to Arrhenius ([1887] see ref. 4 in [65]): [152], [176], Lees ([1900]—e.g. [178]): [177, 177a] and Bingham [1922]: [169a, 171, 173a, 184a].

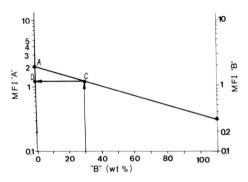

Fig. 14   Technical mixing rule employed in LDPE blending to obtain an AB composition of selected MFI value. (From *USI Chem. Appl.* [181].)

In spite of many [9, 15, 27, 28, 62, 63, 75, 77, 158, 168–170] publications on blending rules for polyethylenes and polyethylene–wax blends only simple linear or logarithmic mixing rules are available for use with polyolefins blends over the range of viscoelastic characteristics of interest to process and product engineers [10–13, 142, 143, 152, 171–181]. The logarithmic mixing rule, introduced in the late 1950s for guidance in blending LDPEs differing in (MFI), is illustrated in Fig. 14 [181]. A kind of composition superposition [13, 175] approach, together with blending laws developed by Bogue *et al.* [171] for the relaxation spectra [65, 168, 169, 174, 180], although more quantitative and "sophisticated" rheologically, are of little help for the practitioner since they are only applicable to low molecular weight polyolefins [9, 10, 13, 174] and/or in the linear viscoelasticity range. Apparently, the only "engineering" type blending law available is that developed by Kasajima and Mori for polypropylene–HDPE blends [176]. Their approach is based on the

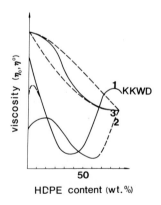

Fig. 15   Patterns in the composition dependence of melt viscosities of HDPE blends: (1) (KKWD) iPP–HDPE (Phillips type), $\eta_0$; (2) iPP–HDPE (Ziegler type), $\eta_0$; (3) polybutene-1–HDPE, $\eta_0$. The dashed lines illustrate Takayanagi's model prediction [64, 169, 177].

nineteenth century Arrhenius mixing rule. It has been successfully applied, however, to blends of components differing substantially in melt viscosities and the fit of calculated and experimental data is quite satisfactory.

Further development in polymer models may be expected along the lines of Kerner's approach as reformulated by Takayanagi (Fig. 15) with appropriate considerations of the rate-dependent rheological characteristics. An alternative lies with an adaptation of suspension models, which are quite successful in predicting the sigmoid shape of the property–composition dependencies for multimodal component suspensions [172a] and filled rubbers [173a, 184], respectively.

### C. Rheotechnical Selection of Polyolefin Blend Compositions

Rheometry of molten polyolefins yields characteristics of steady flow as well as information on the flow behavior above the critical shear rate (or stress). Both were employed extensively in a semiempirical composition selection, which resulted in blends that are outstanding with respect to high fluidity [62, 63, 90, 91, 93, 102, 164] and/or high critical shear rate [161]. The routine is to make a plot of the characteristic in question versus composition, thereby localizing the composition range(s) in which the characteristic attains the level desired [21, 22, 28, 29, 56, 64, 65, 112, 182]. Newer approaches employ statistical optimization techniques [183]. These approaches have been used successfully to select compositions of iPP–LDPE blends for film-blowing and blow-molding grades [53, 84].

### D. The "Rheologically Particular" Compositions (RPC)

There is substantial experimental evidence for the generality of the sigmoid shape [28, 59, 62, 63, 162, 178, 184] of the property–composition graphs for polyolefin blends. To a first approximation this is readily explained by the packing density concept derived from knowledge about flow behavior of concentrated suspensions and/or the effects associated with the interchange of the matrix-suspended particle roles [172, 185] between the micro-heterogeneous blend components. Since at the extrema-range compositions the properties display more radical changes than those foretold by either concept, and, furthermore, since these changes are often in the favorable direction, a number of research efforts were undertaken to elucidate the intuitively obvious role of the interphase layer, separating components of a microheterogeneous blend [39, 140a, 141, 186–189]. Indeed, even routine measurements of the composition dependence of the specific melt volume,

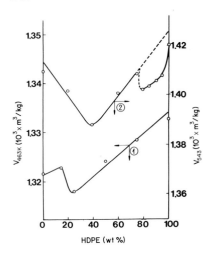

Fig. 16 Composition dependence of the specific melt volume, $v_T$, for polypropylene–HDPE blends: (1) at 463°K for blends consisting of similar average molecular weight components; (2) at 543°K for components widely differing in molecular weight. (Courtesy *Trans. Soc. Rheol.* [65].)

performed on the flowing melt [65] (Fig. 16), rendered support to the early findings of Letz [190, 191] (comprehensively developed by Lipatov's school [39, 126, 128] and also confirmed by Kryszewski [54]) that at the particular concentration range the density of the melt at the interphase significantly exceeds that of the components [65, 141, 191]. Thus, the flow mechanisms at these compositions obviously differ from the typically bicomponent polymer melt [22, 145, 150, 173, 192] system since the mutual "wetting" [59, 65, 66, 105, 193–196] of the components depends both on their molecular structure [5, 9, 10, 11, 13, 21, 31, 42, 142, 197–201] as well as on the thermal and shear history of the system [112, 174, 202]. The differences among superficially quite similar systems (LDPE–polypropylene and HDPE–polypropylene blends)

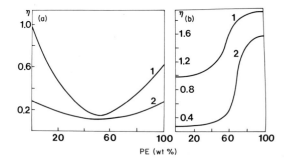

Fig. 17 Composition and thermal treatment dependence of the melt viscosity of (a) polypropylene–LDPE and (b) polypropylene–HDPE blends, where 1 = high and 2 = moderate cooling rates. (Courtesy *Vysokomol. Soedin.*, Eds. [195].)

are astounding (Fig. 17). One of the main causes for the differences is the extent and size of branching in LDPE, which preclude effective diffusion of polypropylene [195, 203]. Practical consequences of the interphase layer effects are twofold. First, a substantial manipulation with the "particular concentration" blends is feasible [29] by introducing appropriate microhetero-genizing agents [6, 56, 57] at the dispersive mixing stage in the blend manufacture. This aids the development of the surface of the interphase layer and reduces the size of the component grains [150]. In turn, a change in the flow mechanism results, and particulate [9, 10, 31, 86, 92, 139, 140, 142, 143, 146, 147] flow is initiated. It is characterized by a lower melt viscosity and a substantially lower (especially at higher shear rates) melt elasticity [149, 160]. Second, fibrillar texture formation in the flowing melt via the mechanism shown in Fig. 18 will be enhanced [56, 56a, 100, 156a], and the resulting blend will be tougher and more orientable than a "straight" blend [48].

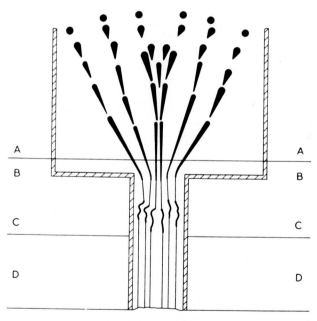

Fig. 18   Empirical model for fibrillar texture formation in a molten polyolefin blend while under tensile stress resulting from convergent flow: (A) reservoir; (B) primary fibrillation zone with (C) "snap back" (elastic recovery) effects evident; the fixing of fibrillar texture; (D) melt flow with fibers orientation. (Courtesy G. V. Vinogradov, 1975—by permission of the publishers IPC Business Press Ltd ©.)

## IV. MELT-MIXING PROCESS ENGINEERING

The large scale production of polyolefin blends is carried out in two stages: dispersive mixing (typically involving melting and kneading) [23, 81] followed by laminar mixing [112, 204, 205]. The spatial distribution and the initial component grain size depend on the dispersive mixing efficiency. Hence, the effectiveness of laminar mixing in the production of a blend characterized by the preferable fibrillar or finally stratified texture depends to a large degree on the dispersive mixing stage. The latter is typically carried out with high-speed mixers [1, 2, 78, 79, 81, 204], which operate on the principle of generating high-intensity shear fields within the slit separating the rotors, as shown in Fig. 19 [81]. The dispersive mixing process variables are: mixer wall temperature, $T_b$; rotors speeds (or, if the speed of rotation is constant, the time of mixing $t$); and the pressure pushing the material against the rotors $p_T$. The performance of a dispersive mixer for a given set of mixing process variables is illustrated with Fig. 20. The inherent complexity of the process [33, 148, 163, 206–212, 311] limits the available solutions to the problem of the operational variables–blend quality relationship to that of a statistical approach based on factorial design. From the minimum number of experiments run on extreme levels of $T_b$, $p_T$, and $t$, paralleled with measurements of the properties of the

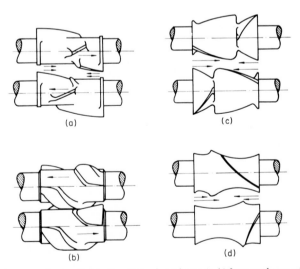

Fig. 19   Basic types of rotor designs for dispersive mixers: (a, b) four- and two-wing Banbury; (c) four-wing Werner–Pfleiderer; (d) three-wing Shaw Intermix. Arrows indicate the flow direction of the sheared melt. (Courtesy *European Rubber J.* [81].)

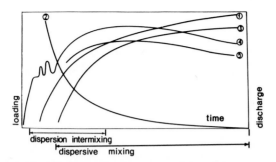

Fig. 20 The dispersive melt mixing process. Time dependence for: (1) blend temperature; (2) the amount of unincorporated minor component; (3) degree of mixing; (4) power consumption; (5) viscosity of the blend. (Courtesy *European Rubber J.* [81].)

blends produced in these experiments, a set of property–process variables equations can be generated [183, 213]. The equations are subsequently employed in generating operating parameter surfaces, which outline the permissible combinations of the mixing process variables for which the required levels of performance characteristics (properties) are attainable. A formally similar technique of statistical optimization of the processing of polymer blends is discussed by Kartashova *et al.* [183].

Recently a new type of dispersive mixer has been developed that is known as a "motionless mixer" [80, 80a, 146–7, 154, 214–215]. It operates on the principle of alternately separating and joining the molten polymer flow by a set of appropriately shaped flow restrictions. So far, applications of motionless mixers are largely limited to small-scale blending operations done during processing. The reader is referred to the literature for further details [80, 146, 160, 210, 214, 215]. There are also extruder screws designed to combine dispersive and laminar melt mixing [24, 217, 222, 228, 311]. Similar "combined mixing" action is claimed for double-screw extruders [218].

The dispersive mixing is typically followed by *laminar mixing*, which is illustrated conceptually in Fig. 21 [112]. The net result of this process depends on the extent of the shear strain $\bar{\gamma}$, which, in turn, is a function of the pressure-to-drag flow ratio $\Phi$. This ratio depends on the extruder geometry, the flow behavior of the melt, the extrusion process throughput, $Q$, and the back pressure $\Delta p$. For a Newtonian fluid it is given by

$$\Phi = h^2 \Delta p [\sin \phi/(6\eta_h \pi D N \cos \phi L)] \qquad (5)$$

where $h$ is screw channel depth, $\eta_h$ melt viscosity in the screw channel, $D$ barrel diameter, $N$ screw rotation speed, $\phi$ screw helix angle, and $L$ barrel length (see, e.g., [204–205, 225]).

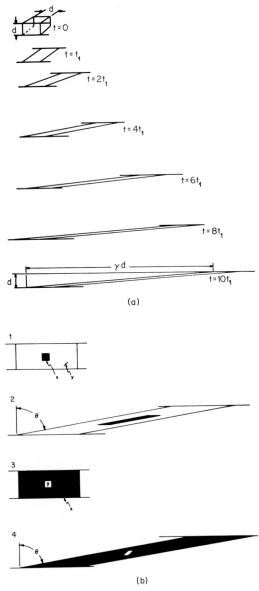

(a)

(b)

Fig. 21 Models for constant shear rate laminar mixing. (a) Strain of a molten polymer "cube" as a function of the deformation time. (b) Strain in the element of a blend: (1, 2) components of similar melt viscosity and elasticity; and (3, 4) blend containing minor component of higher viscosity and elasticity. Data for each part of (b) are as follows: (1) reference state at $t = 0$, $\gamma_x = \gamma_y = 0$; (2) if $\eta_x(\dot{\gamma}) \leqslant \eta_y(\dot{\gamma})$ and $[p_{11} - 2p_{22} + p_{33}]_x(\dot{\gamma}) \leqslant [p_{11} - 2p_{22} + p_{33}]_y(\dot{\gamma})$, then at $t = t_0$, $\gamma_x = \gamma_y = \tan \Theta$; (3) reference state at $t = 0$, $\gamma_x = \gamma_y = 0$; (4) if $\eta_x(\dot{\gamma}) < \eta_y(\dot{\gamma})$ and $[p_{11} - 2p_{22} + p_{33}]_x(\dot{\gamma}) < [p_{11} - 2p_{22} + p_{33}]_y(\dot{\gamma})$, then at $t = t_0$, $\gamma_y = [\eta_x(\dot{\gamma})_x/\eta_y(\dot{\gamma})_y]\gamma_x = \tan \Theta$. (Courtesy *Trans. Soc. Rheol.* [112].)

346

The engineering approach to the *laminar mixing* process is based on quantifying the strain $\bar{\gamma}$, which obviously affects the goodness of mixing and the polyolefin blend texture (supermolecular structure) [4, 219, 220, 223, 226, 236]. From an estimate of the particle size and its distribution resulting from the dispersive mixing, one could evaluate the size reduction resulting from laminar mixing and then attempt application of van Oene's treatment (see Volume 1, Chapter 7) to predict the texture of a blend. There are at least three techniques for evaluating the goodness of mixing.

First, we have the qualitative measure developed by Squires and Wolf, known as the quality index, $N_Q$, which is based on the thermal uniformity of the extruded melt [221]. It is built on the observation that the lower the amplitude of the temperature oscillation during the extrusion process, the higher is the uniformity of the melt [82, 205, 222]. If we designate the length of a functional zone $i$ of the extruder screw as $L_i$ and the depth of the screw channel there $h_i$, and assume further that melt extruded at a flow rate $Q$ is characterized by specific volume $v_T$, specific heat $c_p$, and thermal conductivity $k$, then for $N_Q$

$$N_Q = (kv_T/c_p)(D/Q)\sum (L_i/h_i) \tag{6}$$

where $D$ is barrel diameter.

Schenkel's mixing degree $M$ allows a straightforward quantitative estimate of the shear strain imposed on the melt in the metering zone of the extruder:

$$M = 3.72(L_m/h_m)(1.64 + \Phi)/(1 - \Phi) \tag{7}$$

where the index $m$ refers to the metering zone, and the numerical constants result from assuming the most typical helix angle $\phi = 17° 41'$ [24].

A somewhat more realistic strain measure is the one suggested by Pinto and Tadmor in the form of the weight-average total strain (WATS), since it accounts both for the distribution of residence times [205, 217, 223, 224] and the two shear regions in the channel. Evaluation of WATS is rather involved because it requires two numerical integrations (the averaging procedure) using judiciously selected integration limits. However, it can be performed by even a small computer within a reasonable time, yielding results of high accuracy. Basically

$$\text{WATS} = \int_{t_0}^{\infty} \bar{\gamma} f(t)\, dt =$$
$$= F(L/h, \phi, \Phi) \tag{8}$$

where $f(t)$ is the residence time $(t)$ distribution (RTD) function [225, 226].

The sensitivity of WATS toward the laminar melt-mixing process parameters (i.e., screw speed $N$ and the extrusion flow ratio $\Phi$) is illustrated in Fig.

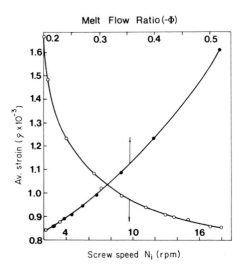

Fig. 22   Computer simulation of the laminar (melt) mixing process for a blow-molding grade polypropylene–LDPE blend for an industrial extruder showing the effects of screw rotation speed, *N* and melt flows ratio Φ on the degree of mixing $\bar{\gamma}$ that'is, average strain (WATS). (Courtesy Mrs. E. Zakroczymska, 1973.)

22 for the isothermal mixing process in a large (15 in. dia.) extruder and for the blow-molding grade of a polypropylene–LDPE blend. The dependence of the WATS–Φ relationship on the extruder screw type is illustrated for a $2\frac{1}{2}$-in. extruder [82] and high molecular weight LDPE and HDPE with Fig. 23. For the polyethylenes and smaller extruder the relationship seems to be linear and independent of the melt flow characteristics.

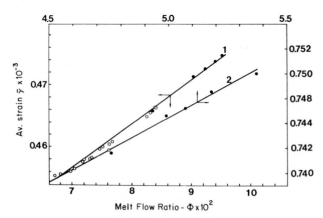

Fig. 23   The laminar mixing in the $2\frac{1}{2}$-in. diameter instrumented extruder with ( ● ) HDPE and ( ○ ) LDPE: the influence of the flows ratio, Φ, on the degree of laminar mixing $\bar{\gamma}$ for two screws: (1) short (SO25); (2) long metering zone (SO22) screw. (From Ruminski and Plochocki [82].)

In addition to the significance of the strain measures discussed above (WATS in particular) for the evaluation of the particle size reduction in the laminar mixing and hence the texture of a blend, these might also be used to check the effectiveness of the (computer-simulated) mixing process [217]. While the goodness of mixing in the dispersive mixing process might be controlled with microscopic [206] or other optical techniques [227, 228], as well as with subsequent measurements of the blend properties (e.g., the critical shear rate level, Fig. 24 [214–216]), the laminar-mixing quality measures discussed above allow an additional check. For this purpose, as illustrated in Fig. 25, one of the most effective characteristics is the quality index $N_Q$. On the other hand, the melt memory index, MMI [25, 229] is the most sensitive and expedient "off-line" mixing quality estimate [230]. It is worth emphasizing that the sensitivity of the melt elasticity to the homogeneity of the blend is still more effective if adapted to steady, viscometric melt flow [165].

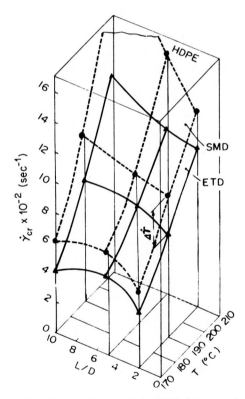

Fig. 24   Static (dispersive) mixer performance for HDPE: influence of capillary length and temperature on the mixing induced increase in critical shear rate $\dot\gamma_{cr}$. Model experiments with a capillary viscometer with (SMD) and without (EMD) the static mixer in the reservoir. (Courtesy *Japan. Soc. Rheol. Jr.* [215].)

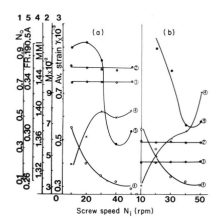

Fig. 25   Comparison of the sensitivities to screw speed for the various goodness of mixing estimates, as calculated: (1) quality index $N_Q$ [221]; (2) mixing degree $M$ [24]; (3) weight average total strain WATS [225] and measured; (4) melt memory index MMI (extrudate swelling measured according to ASTM D1238); (5) flow rate FR (190° C, 5 kG). Instrumented extruder ($D = 2\frac{1}{2}$ in.) with a long metering zone screw (a) and a long compression zone screw (b), using LDPE. (Courtesy Dabrowski, Galas a. Podedworny Ltd., 1974.)

Dependence of polyolefin-blend quality on the mixing process and component characteristics has been studied extensively [3, 9, 23, 24, 31, 33, 39, 56, 75, 144, 160, 163, 170, 173, 192, 197, 206–208, 211, 212, 215, 216, 220, 231–236] and exploited for the development of new techniques in polyolefin manufacture [216, 238]. Most of the research listed above was performed on polyolefins (especially LDPE[†]) [23, 230, 231, 236, 238, 286] and to a lesser extent on polyolefin blends. From among the latter group, the most prominent research has been on polyethylene–polyisobutylene (PIB) [211] and iPP–HDPE [197] systems. This research revealed that the thermal treatment applied to polyolefin blends affects their melt flow behavior almost to the same extent as does the laminar mixing [195, 234].

For some polyolefin blends, the significant influence of mechanochemical processes, during the melt mixing operation, on the blend properties was demonstrated [58, 198–200]. It is worth realizing that the long-term (several years) field service of items manufactured from polyolefin blends more often than not will be affected by diffusion of the components within the blend, which is due to the thermodynamic instability of microheterogeneous blends [203].

## V.   BEHAVIOR OF POLYOLEFIN BLENDS IN PROCESSING

Most commercial polyolefin blends are sold under polyolefin trade names, and their processing is covered by pertinent catalogues and technical

---

[†] The most widely employed laminar melt mixing process is that of homogenization of LDPE in film- and cable-grades production [23, 227, 230, 231, 236, 237].

literature, supplied by the manufacturers (e.g., [85, 121, 133]). Here I would like to point out some of the factors influencing the processability of the blends and the effects of processing on the structure of the blends. Processability is a result of blend structure, which in turn, depends on the microstructure (average molecular weight, $\overline{M}_w$, $\overline{M}_z$ [8a, 160, 239], molecular weight distribution [11, 14, 77, 143, 239], and branching [16, 16a] and macrostructure (powder grain size [76, 83, 142] and/or uniformity [75, 173, 175]) of the components.

The behavior of a blend during processing and the properties of its products are obviously influenced by mechanochemical processes [58, 97, 198–200, 232] and the melt shearing in the processing equipment [29, 98, 112, 142, 214, 231, 236]. While the former might be reasonably controlled with adequate stabilizers, the shearing history effects are obscured by an interaction with the thermorheological phenomena [66, 94, 104, 189, 195, 197, 240–242].

In attempts to evaluate structure–property relationships and their dependence on processing [6, 60, 201], one should realize that blends manufactured from commercial polyolefins pose additional obstacles. Property-molecular weight relationships (e.g., tensile strength–solution viscosity, VN, see Fig. 26) depend significantly on the polyolefin type: linear polyethylenes that are nominally the same but originate from different polymerization processes (e.g., Ziegler or Phillips [1, 7]) when blended with a LDPE, yield two "families" of blends [60, 131].

On the other hand, LDPE used in the manufacture of blends are usually delivered in granulate form, which is prepared by mixing granulates from various batches. Typical gross nonuniformity of such granulates is shown

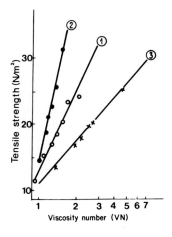

Fig. 26 The influence of linear polythylene type (1, 3 Ziegler; and 2, Phillips) on the tensile strength–average molecular weight (viscosity number VN, T 125—see, e.g., ref. 33–35 in [64]) relationship for HDPE–LDPE blends.

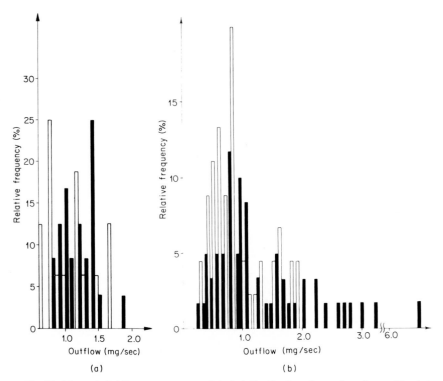

Fig. 27   The melt fluidity (or apparent melt index) distribution of granulates from a blended, extrusion-grade LDPE (filled bars) of (a) satisfactory and (b) poor processibility. The open bars represent a reference granulate obtained from blended LDPE via melt mixing followed by granulation. (Courtesy *Polimery* (*Warsaw*) [75].)

with Fig. 27. A polyolefin blend manufactured from LDPE similar to that presented in Fig. 27(b) would display, during processing, excessively high extrudate swelling and low critical shear rate.

Processing leads to transformations in the supermolecular structure of the blends. These transformations might be divided into three main types:

1.   Structure changes due to orientation effects, which may be further influenced by the thermal history of the melt. The differences in product performance characteristics due to orientation effects are illustrated in Fig. 28 by tensile stress–strain curves for samples injection and blow molded from the same blend. Combined effects of composition and rheothermal treatment enforced during processing upon the elastic properties of polyethylene–rubber blends (a, b) and on the anisotropy of the tensile strength of film extruded from polypropylene–polyethylene blends (c) is shown in Fig. 29.

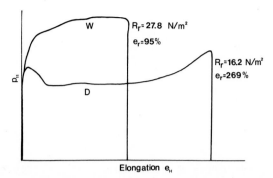

Fig. 28 Influence of processing on the tensile stress–strain behavior of a blow-molding grade, ternary polypropylene–HDPE–LDPE blend: (D) sample machined from barrel wall; (W) injection molded. Elongation rate $= 1.7 \times 10^{-3} \, \text{msec}^{-1}$. (Courtesy Mr. J. Bojarski, 1976.)

2. Structural changes due to generation of fibrillar and/or lamellar forms [56, 56a, 112].

3. Redistribution of the blend components due to shearing flow [243, 244] is illustrated in Fig. 30. This type of structural change leads to difficulties in processing by weakening the weld lines in moldings [53].

Obviously these transformations often act in parallel and are accompanied by a homogenization effect [238, 245, 246], leading to a narrowing of blend

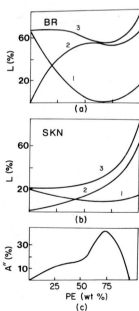

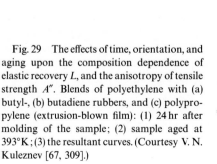

Fig. 29 The effects of time, orientation, and aging upon the composition dependence of elastic recovery $L$, and the anisotropy of tensile strength $A''$. Blends of polyethylene with (a) butyl-, (b) butadiene rubbers, and (c) polypropylene (extrusion-blown film): (1) 24 hr after molding of the sample; (2) sample aged at 393°K; (3) the resultant curves. (Courtesy V. N. Kuleznev [67, 309].)

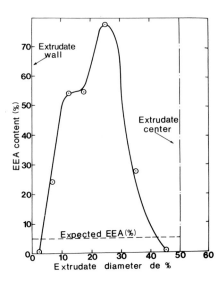

Fig. 30 Solution DTA-determined redistribution of ethylene–ethyl acrylate copolymer (solid line) in an EEA–LDPE blend induced by melt flow. Compare with Fig. 33. (Courtesy *J. Appl. Polym. Sci.* [244].)

molecular weight distribution and generation of branched structure [247]. There have been a number of studies aimed at establishing common features governing the processing of the blends [6, 20, 36, 84, 106, 110, 122, 133, 245–249, 309] and optimizing processing parameters [183, 250].

Other developments in the processing of polyolefin blends include: polymer–lubricant-type blends, in which low molecular weight polyethylene [245–246] or PMDS [6] is the processing aid; extrusion of thin, tough and high orientability films [48, 84, 240, 251]; injection molding [263]; processing of blends containing particulate fillers [6, 38, 124, 154, 252, 253]; and optimization of the processing (operation) parameters [254].

## VI.  END-USE PROPERTIES AND APPLICATIONS OF POLYOLEFIN BLENDS

Properties of microheterogeneous polyolefin blends display several characteristic features: nonlinear dependence on composition (nonadditivity); strong dependence on the degree of orientation and rheothermal treatment; and substantial flexibility for change to suit particular applications using, for example, fillers and compatibilizers [38, 129, 130, 158, 247, 253]. Thus far, a comprehensive survey is available only for polypropylene blends [6]. There have been a number of competent contributions describing properties of commercial pilot-plant size or semicommercial blends, however: HDPE–LDPE [14, 15, 131, 251, 256]; polyamide–polypropylene [113, 115, 158];

polyamide–polyethylene [83, 117, 117a, 257]; and polyethylene–polystyrene [48, 89, 129, 130]. Most of the literature discusses the properties of blends as used in production of films [14, 48, 84, 103, 108, 113, 122, 126, 193, 240, 251, 257, 258, 259, 260, 279], profiles and containers [84, 121, 249, 261, 262], moldings [71, 106, 116, 121, 125, 127, 129, 130, 250, 256, 262, 263, 264], insulation and sheaths (jackets) for cables [19, 102, 183], and miscellaneous products such as fibers (e.g., [37], see also Chapter 16), etc. [48, 154, 253, 265].

A substantial amount of research has been devoted to the mechanical properties of polyolefin blends [104, 211, 238, 246, 256, 266–268, 277]. Typical temperature dependence of the shear modulus $G$ for a polyamide–polyethylene blend is shown in Fig. 31, in which the blend is contrasted with the components.

The trend in evaluation of blend performance characteristics and composition optimization [183, 250, 263, 269] has been to use statistically designed measurements performed with limited number of three-component blends such as the PS–PIB–PE system [269].

The evaluation and control of polyolefin blend orientation was also thoroughly investigated [37, 111, 250, 258, 260, 265, 270], particularly in the case of film grade (iPP–LDPE) blends, which offer possibilities for substantial material savings [53]. A number of papers on blend properties cover comparative evaluation of performance characteristics for blends and corresponding block or graft copolymers [266, 267, 271], evaluation of thermal

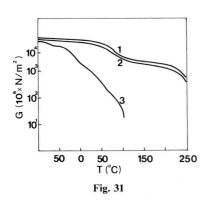

Fig. 31

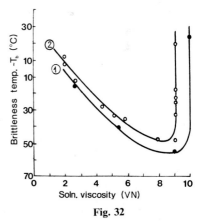

Fig. 32

Fig. 31   The temperature dependence of the shear modulus $G$ (ASTM D2236-70) for "ASE 9010" (polyamide–polyethylene) blend (2) and for the components: (1) polyethylene; (3) polyamide. (Courtesy *Kunstst. u. Gummi* [117].)

Fig. 32   The influence of the average molecular weight of polyethylene (viscosity number) on the brittleness temperature $T_b$ of polypropylene–HDPE blends containing ca. 40 wt % HDPE. The blends were prepared by: (1) coprecipitation and (2) melt-mixing (see Table I, I.9.)

properties [71, 257], performance characteristics [14, 15, 110, 113, 240, 251, 254, 269], and structure [67, 271a].

For the main group of polyolefin blends, viz. polypropylene–polyethylene systems, the principal characteristic is that of the low-temperature toughness. In Fig. 32 the brittleness temperature is plotted as a function of the average molecular weight (expressed as solution viscosity) of HDPE for the constant composition of 40 wt %. The $T_b$ suppression is great for the less homogeneous (powder-mixed) sample, and above a limiting molecular weight the component interaction changes character.

## VII. ANALYSIS OF POLYOLEFIN BLENDS: STRUCTURE, COMPATIBILITY, AND COMPOSITION EVALUATION

In this closing section, an attempt is made to review various analytical techniques that have been used not only for polyolefins but for other systems as well.

Frequently, rather complex techniques that encompass several methods are used [7, 8, 34, 35, 37, 98, 115, 123, 131, 187, 195, 254, 256, 263, 265, 269, 271a–276, 295]. These were summarized and discussed from the viewpoint of industrial applications [72, 133, 260] as well as usefulness for compatibility evaluation [132, 266, 278] and optimization of blend composition [103]. Many of these techniques are based on the chromatography of products evolved during pyrolysis of a blend sample [186, 280–283].

Techniques of thermal analysis exploit either observations of melting (or crystallization) behavior [62, 276, 284–286] or measurements of thermal properties [287, 288] via evaluation of traces recorded in differential thermal analysis (DTA), solution DTA [244], differential scanning calorimetry, thermogravimetric analysis, etc. [37, 50, 51, 52, 55, 108, 196, 241, 265, 266, 274, 275, 287, 289–292]. Composition and structure of blends are also evaluated with parallel investigations on corresponding block copolymers [271, 293] using thermal analysis techniques. These were found to be effective in an assessment of blend-component compatibility [294, 295]. To illustrate applications of DTA and solution DTA (SDTA) for polyolefin blends typical traces for (a) polypropylene–HDPE blends, and (b) polypropylene rubber (a E–P copolymer)–LDPE respectively, are presented in Fig. 33.

Compatibility and the supermolecular structure of polyolefin blends, which has been the object of numerous studies [36, 39, 40, 42, 50, 51, 287], require a more precise view of some differences in usage of the term "compatibility" between people working with solutions and those working

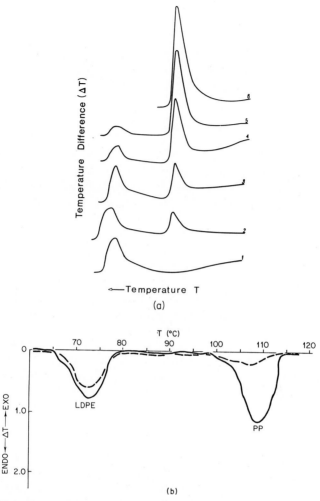

Fig. 33   DTA traces for polyolefins blends: (a) polypropylene–HDPE; (b) ethylene–propylene copolymer (rubber "PPR")–LDPE (from solution DTA (SDTA) measurements.) Solid line—the components, dash line "85/15"—LDPE/PPR blend. (1) Polypropylene; (2) HDPE; (3) 15% wt% HDPE; (4) 25% wt% HDPE; (5) 50 wt% HDPE; (6) 75 wt% HDPE. (Courtesy (a) *Kolloid Z.* [295] and (b) *J. Appl. Polym. Sci.* [187], respectively.)

with bulk blends. One may postulate a "technical compatibility" concept characterized by the depth of interpenetration at the interphase boundary, as illustrated in Fig. 34 [39, 141, 202]. With such a definition of compatibility, one has some guidance for selecting the scale to be used for texture evaluation. In the case of polyolefin blends it would amount to the medium range

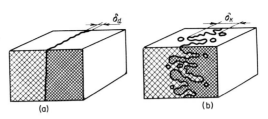

Fig. 34    Model of the interphase layer in a polyolefin blend: (a) after dispersive mixing; (b) after laminar mixing and/or heat treatment. (Courtesy Yu. S. Lipatov [141].)

magnification in optical microscopy. Electron microscopy of the fracture surfaces would have to be employed for the blends containing a compatibilizer [129, 130] and/or component(s) of distinct (i.e., spherulitic) texture (e.g. [5, 62, 195]). Obviously, for studies on the influence of thermorheological factors on the texture and compatibility of a blend, the technique of grain size evaluation would be useful. An evaluation of compatibility from measurements with solutions (linearity of the soln. viscosity-composition plot—see e.g. [64] for reference) might, in technical applications, be substituted with the check based on solution viscosity (e.g., VN) melt flow rate (FR) plot [60].

The compatibility of polyolefin blends has been investigated in various aspects: in comparison with corresponding copolymers [296–298]; in order to optimize the composition [103, 299]; its effects on blend orientation [67, 111, 258–260, 265, 278, 270, 271a] and the effectiveness of compatibilizers [158, 271]. A number of studies were devoted to compatibility itself [17, 37, 43, 49–52, 55, 66, 72, 87, 88, 96, 105, 113, 126, 185, 186, 187, 193–196, 197, 203, 212, 252, 274, 275, 281, 288, 289, 290, 291, 292, 294, 295, 300–303]. Almost equally numerous were studies into polyolefin blend structure [54, 95, 104, 105, 112, 128, 141, 159, 188–191, 197–200, 201, 233, 234, 242–244, 266, 276, 286, 292, 305].

Relaxation spectrometry of polyolefin blends, comprehensively summarized by Schmieder [34] in the early 1960s is based on the assumption of a one-to-one correspondence between mechanical energy dispersion peaks and the macromolecular relaxation processes in a polymer. It is now an almost routine technique for localizing the molecular and supermolecular motions on the energetic (temperature and/or frequency) scale [22, 35, 37, 49, 52, 65, 87, 95, 137, 171, 172, 175, 180, 185, 186, 194, 253, 265, 268, 271a, 275, 277, 301, 302, 303, 306, 307]. A typical plot of temperature–composition dependence of the components $G'$ and $G''$ of the complex modulus is presented in Fig. 35 for polypropylene–HDPE blends [295]. Persistence of the relaxation processes, characteristic for the microheterogeneous blends components, as well as the differences in sensitivities of various viscoelasticity measures toward temperature changes, is evident (see also Fig. 9).

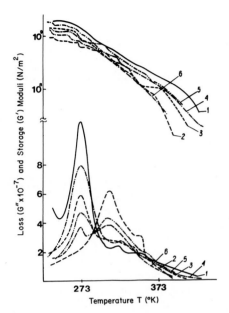

Fig. 35 Temperature and composition dependence of the complex shear modulus $G^*$ components, storage $G'$ and loss $G''$ moduli: (1) polypropylene; (2) HDPE blends containing (3) 15; (4) 25; (5) 50; (6) 75 wt% HDPE. (Courtesy *Kolloid Z.*, Eds. [295].)

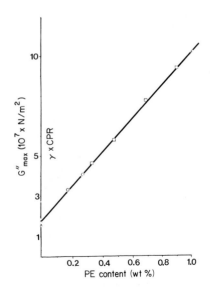

Fig. 36 Composition dependence of the loss modulus $G''_{max}$ for the "$\gamma$" characteristic relaxation process (CPR) for polypropylene–polyethylene blends: ($\square$) iPP–HDPE and ($\bigcirc$) HDPE–aPP. (Courtesy *J. Polym. Sci. C* [35].)

The low-temperature "$\gamma$" relaxation process observed for polyolefins is characterized by good repeatability and persistence, also after the melt mixing operation. It was selected, therefore, for presenting the composition dependence of blend viscoelasticity in Fig. 36.

Relaxation spectrometry of polyolefin blends was employed in:

1. studying compatibility [94, 105, 169, 196, 197, 266, 295];

2. approximation of the relaxation spectra [29, 65, 73, 174, 184, 299, 303, 304] for blends and comparing these with spectra for pertinent copolymers [73, 74, 267, 271, 297, 298, 303, 308]; and rarely,

3. for studying blend structure [34, 35].

Spectroscopy and spectrophotometry is employed for polyolefin blends [37, 72, 104, 105, 132, 195, 260, 271a, 274, 278, 300–302] either for routine composition evaluations, or in more specific applications: (1) orientation [111, 270]; (2) composition optimization [103]; (3) compatibility [96]; (4) structure [305]; and (5) degradation [283].

Recently a new technique employing photoactive tracers effective in studying the interface layer phenomena and spatial distribution of components in a blend is finding increasing application [39, 132, 258, 309].

## REFERENCES‡

1.   A. Renfrew and P. Morgan (eds.), "Polythene," p. 376. Wiley (Interscience), New York, 1960.

2.   T. O. J. Kresser, *in* "Crystalline Olefin Polymers" (R. A. V. Raff and K. W. Doak, eds.), Vol. 2, p. 404. Wiley (Interscience), New York, 1964.

*3.   H. V. Boenig, "Polyolefins: Structure and Properties," p. 248. Elsevier, Amsterdam, 1966.

4.   R. Vieweg et al. (eds.), "Polyolefine," pp. 214, 283, 495. Hanser, Munchen, 1969.

*5.   G. P. Andryanova, "Fizikokhimya Poliolefinov." Khimya, Moskva, 1974.

*6.   D. V. Ivanyukov and M. L. Fridman, "Polipropilen-Svoistva, Primenenie," p. 107. Khimya, Moskva, 1974.

*6a.   H. P. Frank, *Polypropylene*, p. 96 ff., McDonald, London, 1968.

†7.   G. H. Burke, *Classification of Polyethylene Compounds*, ISO TC61/WG 5 Circular 11.14 (1973).

†8.   C. Strazielle and H. Benoit, *Pure Appl. Chem. J.* **26**, 451 (1971); **42**, 615 (1975).

†8a.   F. M. Mirabella and J. F. Johnson, *J. Macromol. Sci.* **12C**, 81 (1975).

9.   W. L. Wagner and K. F. Wissbrun, *SPE Tech. Papers* **8**, 3–2 (1962).

10.   W. F. Busse and R. Longworth, *J. Polym. Sci.* **58**, 49 (1962).

11.   R. Z. Naar and R. F. Heitmiller, Techniques in Polymer Science, *Soc. Chem. Ind. London Monogr.* No. 17 (1965).

12.   M. Nagano, N. Kawada, and T. Yamamoto, *Nagoya Kogyo Daigaku Gakuho* **23**, 173 (1971).

13.   N. Nakajima et al., *Trans. Soc. Rheol.* **15**, 759 (1971).

14.   V. J. Smith, SPE RETEC on Polyolefins Papers, Houston, Texas, p. 38 (March 1975).

15.   H. H. Zabusky and R. F. Heitmiller, *SPE Trans.* **4**(1), 17 (1964).

‡ Designation; *monographs, †review or summary (basic, conceptual) paper.

16.  R. A. Mendelson and F. L. Finger, *J. Appl. Polym. Sci.* **19**, 1061 (1965).
16a.  M. Fleissner, *Angew. Makromol. Chemie* **33**, 75 (1973).
†17.  M. G. Gubler and A. J. Kovacs, *J. Polym. Sci.* **34**, 551 (1959).
*18.  A. G. Sirota, "Modifikatsya Struktury i Svoistv Poliolefinov," p. 145. Khimya, Leningrad, 1974.
19.  V. N. Kotrelev, K. I. Zernovo, I. D. Avtokratova, and T. I. Yurkevich, *Plast. Massy No. 7*, p. 18 (1968).
*20.  Y. Ito, "Polyethylene: Problems/Solutions in Processing and Manufacturing," p. 21. Plastic Age, Tokyo, 1968.
21.  G. V. Vinogradov, I. V. Kurbatova, M. A. Martynov, and N. A. Sibiryakova, *Kolloid Zh.* **29**, 186 (1967).
22.  G. V. Vinogradov, L. I. Ivanova, Yu. G. Yanovskii, and E. I. Frenkin, *Vysokomol. Soedin.* **10B**, 726 (1968).
23.  R. S. Adams, *SPE Tech. Papers* **21**, 383 (1975).
†24.  G. Schenkel, *Kunststoffe* **60**, 196 (1970).
†25.  J. Meissner, *Kunststoffe* **57**, 397; 702 (1967) and *Pure Appl. Chem. J.* **45**, 551 (1975).
†26.  A. P. Plochocki and G. Miller, *Polimery (Warsaw)* **21**, 128 (1976).
†26a.  V. Semjonov, *Adv. Polymer Sci.* **5**, 387 (1968).
27.  N. Nakajima and P. S. L. Wong, *Trans. Soc. Rheol.* **9**, 3 (1968).
28.  A. P. Plochocki and J. Baranówna, *Polimery (Warsaw)* **16**, 223 (1971).
29.  S. Hirata, H. Hasegawa, and A. Kishimoto, *J. Appl. Polym. Sci.* **14**, 2025 (1970).
30.  C. D. Han, T. C. Yu, and K. U. Kim, *J. Appl. Polym. Sci.* **15**, 1149 (1971).
31.  M. Shida and L. V. Cancio, *J. Appl. Polym. Sci.* **14**, 3083 (1970).
32.  R. L. Tusch, *SPE Tech. Papers* **11**, 7 (1965).
33.  E. Dolling, *Plastverarbeiter* **21**, 448 (1970).
†34.  K. Schmieder, in "Kunststoffe..." (J. Nitsche and K. A. Wolf, Eds.), vol. 1, p. 791, Springer, Berlin, 1963.
†35.  C. A. F. Tuijnman, *J. Polym. Sci. Part C* **16**, 2379 (1967).
†35a.  R. F. Boyer, *J. Polymer Sci. Symposia* **50**, 189 (1975).
*36.  K. A. Korniev (ed.), "Modifikatsya Svoistv Polimerov i Polimernykh Materialov." Naukova Dumka, Kiev, 1965.
*37.  A. B. Pakshver (ed.), "Karbotsepnye Volokna," pp. 82, 181. Khimya, Moskva, 1966.
*38.  "Modifitsirovannye i Napolnennye Termoplasticheskie Materialy. (Properties, Processing and Application Trends.)" LDNTP, Leningrad, 1968.
*39.  Yu. S. Lipatov (ed.), "Makromolekuly na Granitse Rozdela Faz." Naukovaya Dumka, Kiev, 1971.
*40.  Y. M. Tarnopolskii (ed.), "Modifikatsya Polimernykh Materialov." Zinatne, Riga, Vol. 2, 1969; Vol. 3, 1972; Vol. 4, 1974.
41.  V. E. Gul and V. N. Kulezney, "Struktura i Mekhanicheskie Svoistva Polimerov," p. 290. Vyssha Shkola Izd., Moskva, 1972.
*42.  A. G. Shwarts and B. N. Dinzburg, "Sovmeshanie Kauchukov s Plastikami i Sinteticheskimi Smolami." Khimya, Moskva, 1972.
†42a.  Th. Laus, *Angew. Makromol. Chemie* **60/61**, 87 (1977).
†43.  P. Dubois, *Ind. Plast. Mod. Paris* **17**, 101 (1965).
†44.  K. Thinius, *Plaste Kautschuk (Leipzig)* **15**, 164 (1965).
†45.  D. Feldman and R. Rusu, *Mater. Plast. (Bucharest)* **6**, 8 (1969); **7**, 106 (1970).
†46.  P. Arnaud, *Off. Plast. Caoutch. Paris* **18**, 780, 788, 792 (1971).
†47.  J. E. Hauck, *Mater. Design. Eng.* **62**, 101 (1965).
†48.  W. H. Skoroszewski, *Plast. Polym. London* **40**(147), 142 (1972).
*49.  M. Takayanagi and S. Manabe, "Polymer Blends: from Research to Application," pp. 12, 36. Newspaper (Japan) Publ. (June 14–15, 1968).

362                                                                    A. P. Plochocki

†50. P. J. Flory et al., Macromolecules 1, 287 (1968).
51. M. Hirami, Polymer 8, 482 (1967).
52. G. L. Slonimskii, I. N. Musaelyn, and V. A. Kazantseva, Vysokomol. Soedin. 6, 219, 818, 823 (1964).
53. A. P. Plochocki, L. Grabiec, and T. Bek, Opakowanie (Warsaw) 164(10), 22; 165(11), 25 (1976).
†54. M. Kryszewski, Plaste Kautschuk 20, 743 (1973).
†54a. F. P. Warner, W. J. MacKnight and R. S. Stein, J. Polymer Sci. (Phys. Ed.), 15, 2113 (1977).
55. M. Inoue, J. Polym. Sci. Part A 1, 3427 (1963).
56. O. V. Romankevich, A. V. Yudin, S. E. Zabello, and A. N. Gopkalo, Kolloid Zh. 35, 1083 (1973).
†56a. M. Yakob, M. V. Tsebrenko, A. V. Yudin, and G. V. Vinogradov, Int. J. Polym. Metals 3, 99 (1974); Faserforsch. Textiltech. 27, 333 (1976).
57. M. S. Akutin and G. M. Ozerov, Plast. Massy No. 11, p. 49 (1966).
†58. A. Casale and R. S. Porter, Adv. Polymer Sci. 17, 1 (1975).
*58a. N. K. Baramboym, "Mechanochemistry of Polymers", RAPRA, MacLaren, London, 1964.
59. A. P. Plochocki and J. L. White, Preprints, Int. Congr. Rheol., 7th, Gothenburg, August (1976), 512 and Polymer Eng. Sci., in print.
60. A. P. Plochocki, Plaste Kautschuk (Leipzig) 13, 71 (1966).
61. T. Kamiya, Jpn. Plast. Age 3(8), 34 (1965).
62. O. F. Noel, Jr., and J. F. Carley, Polym. Eng. Sci. 15, 117 (1975).
63. K. Maciejewski and R. G. Griskey, ANTEC SPE-T21, 32nd, Montreal, Canada, p. 23 (1974).
64. A. P. Plochocki, J. Appl. Polym. Sci. 16, 987 (1972).
65. A. P. Plochocki, Trans. Soc. Rheol. 20, 287 (1976).
66. V. E. Gul, E. A. Penskaya, N. A. Zanemonets, and T. A. Zanina, Vysokomol. Soedin. 14A, 291 (1972).
67. V. N. Kuleznev, Yu. V. Yevreinov, V. D. Klykova, and M. I. Shaposhnikova, Kolloid Zh. 35, 281 (1973).
68. W. A. Mack, Mod. Plast. 48(8), 62 (1971).
69. A. P. Plochocki, Polimery (Warsaw) 9, 23 (1963).
†70. G. Holden, E. T. Bishop, and N. R. Legge, J. Polym. Sci. Part C 26, 37 (1969).
71. N. A. Sibiryakova and A. I. Tsvetkova, Plast. Massy No. 8, p. 58 (1974).
72. J. P. Tordella and P. F. Dunion, J. Polym. Sci. Part A-2 8, 81 (1970).
73. J. R. Richards, R. G. Mancke, and J. D. Ferry, J. Polym. Sci. Part B 2, 197 (1964).
74. A. Piloz, J. Y. Decroix, and J. F. May, Angew. Makromol. Chem. 44, 77 (1975).
75. B. Misterek, Polimery (Warsaw) 15, 30 (1970), A. Plochocki, and L. Cxarnecki, idem 16, 34 and 83 (1970).
76. M. Dimitrov and R. Hegele, Kunststoffe 61, 815 (1971).
77. E. T. Hill, Plast. Paint Rubber. No. 11–12, p. 71 (1962).
†78. W. Mack, SPE J. 26(2), 31 (1970); D. Loehr, Kunststoffe 63, 738 (1973).
†79. Mixing Equipment (anonymous), Plast. Technical. 19(6), 141, 143 (1973).
†80. Static Mixers, Plast. Technol. 19(10), 14 (1973); Pat. Application, Poland: P-187, 368.
80a. "N". Static Mixers, Plast. Technology 19(10), 14 (1973) and Pat. Application (Poland) no. P-187, 368.
†81. H. Palmgren, Eur. Rubber J. 156, 30, 70 (1974).
82. W. Ruminski and A. Plochocki, Polimery (Warsaw) 19, 98, 159 (1974).
83. E. V. Minkin, N. K. Baramboim, and A. G. Ankudimova, Plast. Massy No. 4, p. 8 (1967).
84. N. Zdyb, M. Rajkiewicz and W. Budzynski, Polimery (Warsaw) 20, 149 (1975).
85. "TENITE" Polyallomers, Eastman Chem. Int. Bull. MB16D (October 1973).

86. N. Minoshima and Sh. Kobayashi, *Kobunshi Kagaku* **18**, 953 (1971).
87. Ye. I. Abramova, V. I. Andreev, and V. A. Voskresenskii, *Zh. Prikl. Khim* **42**, 1614 (1969).
88. Y. Ya. Goldman, Yu. S. Polyakov, I. V. Kurbatova, and N. A. Sibiryakova, *Plast. Massy* No. 7, p. 17 (1972).
89. D. V. Ivanyukov, V. V. Amerik, E. V. Zhiganova, N. N. Gorodetskaya, and V. F. Petrova, *Plast. Massy* No. 3, p. 32 (1972).
90. J. Beniska and D. Krasnov, *Plast. Hmoty Kauc.* **5**, 358 (1968).
91. B. Krasnov and J. Beniska, *Plast. Hmoty Kauc.* **5**, 328 (1968).
†92. T. S. Lee, *Proc. Int. Congr. Rheol.*, *5th* **4**, 421 (1970).
93. M. A. Natov and E. Kh. Dzhagarova, *Vysokomol. Soedin.* **8**, 1835 (1966).
94. Kh. K. Kardanov, R. B. Chakakhov, and Yu. V. Zelenev, *Plaste Kautschuk (Leipzig)* **20**, 613 (1973).
95. J. Micek, *Plast. Hmoty Kauc.* **10**, 199, 236 (1973).
96. S. M. Ohlberg, R. A. V. Raff, and S. S. Fenstermaker, *J. Polym. Sci.* **35**, 531 (1959).
97. B. H. Krevsky and P. R. Junhaus, *Plast. Technol.* **9**(5), 34 (1963).
98. Z. I. Salina, L. Yu. Zlatkevich, B. V. Andryanov, and L. E. Vlaksina, *Plast. Massy* No. 2, p. 23 (1971).
99. T. Boiangiu, *Mater. Plast. (Bucharest)* **8**, 544 (1971).
100. E. I. Frenkin, G. A. Yermilova, Yu. G. Yanovskii, and G. V. Vinogradov, *Plast. Massy* No. 10, p. 32 (1970).
*100a. G. V. Vinogradov and A. Ya. Malkin, "Reologlya Polimerov," p. 323ff. Khimya Eds., Moskva, 1977.
*101. G. V. Vinogradov and A. Ya. Malkin, "Reologlya Polimerov," p. 323, Khimya Eds., Moskva, 1977
102. E. Laue and F. Woods, *Kautsch. Gumm. Kunst.* **26**, 155 (1973).
103. L. F. Benkova, V. I. Bukhgalter, and V. P. Emelyanova, *Plast. Massy* No. 7, p. 27 (1972).
104. M. Kryszewski, A. Galeski, T. Pakula, J. Grebowicz, and P. Milczarek, *J. Appl. Polym. Sci.* **15**, 1139 (1971).
105. S. Onogi, T. Asada, and A. Tanaka, *J. Polym. Sci. Part A-2* **7**, 171 (1969).
106. M. S. Akutin, Z. I. Salina, V. P. Menshutin, and B. V. Andryanov, *Plast. Massy* No. 2, p. 47 (1970).
107. K. van Henten, *Plastica 25*, 143 (1972).
108. I. G. Kokoulina, L. G. Dorofeeva, S. N. Ilin, and T. S. Yanvarova, *Plast. Massy* No. 4, p. 46 (1972).
109. B. G. Mudzhiri, *Plast. Massy* No. 10, p. 79 (1973).
110. W. S. Smith, Jr., *ANTEC SPE-TP21, 33rd, Atlanta, Georgia*, p. 394 (1975).
111. M. S. Akutin, A. N. Shabadash, Z. I. Salina, V. A. Galubev, and N. P. Besonova, *Vysokomol. Soedin.* **14B**, 769 (1972).
†112. J. M. Starita, *Trans. Soc. Rheol.* **16**, 339 (1972).
113. V. G. Raevskii and M. N. Tolmachew, *Plast. Massy* No. 9, p. 6 (1971).
114. G. C. Travers, *Rubber Plast. Age* **24**(9), 111 (1962).
115. F. Ide and A. Hasegawa, *J. Appl. Polym. Sci.* **18**, 963 (1974).
116. "Panlite E", *Jpn. Plast. Age* **10**(1), 41 (1972).
117. G. Illing, *Kunststoffe Gummi* **7**, 275 (1968).
117a. D. Braun and U. Eisenlohr, *Angew. Makromol. Chemie* **58/59**, 227 (1977).
118. H. Kato and T. Chino (eds.), "Japan Plastics Industry Annual," p. 39, 93. Plastics Age Co., Tokyo, 1974/1975.
119. K. Takagi and S. Yoshioka, *Jpn. Plast. Age* **3**(9), 11; (12), 19 (1965).
120. J. R. Falender, S. E. Lindsay, and J. C. Saam, *SPE Tech. Papers* **21**, 543 (1975).

†121.  Anonymous, *Mod. Plast. Int.* **6**(8), 40 (1976).
122.  L. L. Fomina, N. N. Lopanina, D. F. Kagan, E. A. Penskaya, V. V. Ananev, M. N. Tolmacheva, and V. E. Gul, *Plast. Massy* No. 7, 72 (1975).
123.  F. Ide, T. Kodama, and A. Hasegawa, *Kobunshi Kagaku* **29**, 259, 265 (1972).
124.  F. Ide and I. Sasaki, *Kobunshi Kagaku* **30**, 641 (1973).
125.  G. M. Otopkov, A. V. Skachek, and A. A. Natetkov, *Kozh. Obuv. Promyshlennost* **11**(10), 28 (1969).
126.  S. V. Vlasov, G. V. Sagalaev, Yu. N. Diligenskii, and L. I. Kurakin, *Plast. Massy* No. 2, p. 34 (1973).
127.  S. Yamada, S. Sakajiri, and R. Hirako, preprint, *IUPAC Int. Congr. Macromol., 8th, Tokyo*, 176 (1966).
128.  E. V. Lebedev, Yu. S. Lipatov, and V. P. Privalko, *Vysokomol. Soedin.* **17A**, 148 (1975).
129.  V. M. Barentsen and D. Heikens, *Polymer* **14**, 579 (1973).
130.  J. R. Stell, D. R. Paul, and J. W. Barlow, *Polym. Eng. Sci.* **16**, 496 (1976).
131.  J. Bojarski and A. P. Plochocki, *Polimery* (*Warsaw*) **6**, 188, 230 (1961).
132.  Yu. S. Lipatov, V. A. Vonsyatskii, E. G. Mamuniya, and G. Ya. Boyarskii, *Vysokomol. Soedin.* **16B**, 838 (1974); *Plasteu. Kautschuk* (*Leipzig*) **20**, (1973).
133.  "MOPLEN," Tech. Bull. Montedison S.p.a. (September 1976) (courtesy Dr. F. Ranalli).
134.  S. Hamaya, *Jpn. Plast. Ind. Annu.* **17**, 43 (1974).
135.  C. E. Locke and D. R. Paul, *J. Appl. Polym. Sci.* **17**, 2597, 2791 (1973).
136.  C. E. Locke, C. E. Vinson, and D. R. Paul, *Polym. Eng. Sci.* **12**, 157 (1972).
137.  C. F. Locke and D. R. Paul, *Polym. Eng. Sci.* **13**, 202, 308 (1972).
138.  R. E. Robertson and D. R. Paul, *J. Appl. Polym. Sci.* **17**, 2579 (1973).
139.  Z. K. Walczak, *J. Appl. Polym. Sci.* **17**, 153, 169, 177 (1973).
140.  H. Van Oene, *J. Colloid Interface Sci.* **40**, 448 (1972).
†140a.  T. Nose, *Polymer J. Jpn.* **8**, 96 (1976).
141.  Yu. S. Lipatov, L. I. Bezruk, and E. V. Lebedev, *Kolloid Zh.* **37**, 481 (1975); *Vysokomol. Soedin.* **18B**, 77 (1976).
142.  G. Doering and G. Leugering, *Kunststoffe* **53**, 11 (1963).
143.  R. N. Haward, B. Wright, G. R. Williamson, and G. Thackray, *J. Polym. Sci. Part A-2* **8**, 2977 (1964).
144.  D. F. Berezhnaya, D. F. Kagan, and L. I. Zakharchuk, *Proc. All-Sov. Symp. Rheol., 8th, Homel, USSR*, **1**, 108, (1975) [*Chem. Abstr.* **84**(20), 136161].
145.  C. D. Han, *J. Appl. Polym. Sci.* **17**, 1289 (1973).
146.  C. D. Han, Y. W. Kim, and S. J. Chen, *J. Appl. Polym. Sci.* **19**, 2831 (1975).
147.  C. D. Han and Y. W. Kim, *Trans. Soc. Rheol.* **19**(9), 245 (1975).
148.  G. C. N. Lee and J. R. Purdon, *Polym. Eng. Sci.* **9**, 360 (1969).
149.  C. D. Han, *J. Appl. Polym. Sci.* **15**, 2579 (1971).
150.  J. L. White, R. C. Ufford, K. R. Dharod, and R. L. Price, *J. Appl. Polym. Sci.* **16**, 1313 (1972).
151.  T. Fujimura and K. Iwakura, *Konbunshi Ronbunshu* **31**, 617 (1974).
152.  K. Hayashida, J. Takahashi, and M. Matsui, *Proc. Int. Congr. Rheol., 5th, Kyoto* **4**, 525 (1970).
153.  T. Fujimura and K. Iwakura, *Kobunshi Kagaku* **27**, 323 (1970).
154.  C. D. Han and Y. W. Kim, *J. Appl. Polym. Sci.* **18**, 2589 (1974).
†155.  R. B. Bird, *Int. Conf. Mixing Proc. Polym., Preprints, Delft*, p. 47 (June 1976).
156.  F. N. Cogswell, *Trans. Soc. Rheol.* **16**, 383 (1972).
†156a.  F. N. Cogswell, "Converging Flow a. Stretching Flow: a Compilation," BSR Conference, Edinburgh, Sept. 1977.
157.  L. Czarnecki, A. Plochocki, and J. L. White, *Proc. BSR Conf.* "Polymer Rheol. a. Plastics Processing", 288 (September 1976) and *Polimery* (*Warsaw*) **22**, 214 (1977).

158. F. Komatsu and A. Kaeriyama, *Muroran Kogyo Daigaku* 7, 719 (1972).
159. M. A. Natov, L. Peeva, and E. Djagarova, *J. Polym. Sci. Part C* 16, 4197 (1968).
160. C. D. Han and T. C. Yu, *J. Appl. Polym. Sci.* 15, 1163 (1971).
161. C. D. Han and R. R. Lamonte, *Polym. Eng. Sci.* 12, 77 (1972).
162. C. D. Han and T. C. Yu, *Polym. Eng. Sci.* 12, 81 (1972).
163. T. Fujimura and K. Iwakura, *Kogyo Kagaku Zasshi* 73, 1641 (1970).
164. C. D. Han and T. C. Yu, *AIChE J.* 17, 1512 (1971).
165. L. Czarnecki and A. P. Plochocki, "Evaluation of Thermoplastics Processability," (to Warsaw Polytechnic, Poland) Patent Appl. P-177 200 (January 8, 1975).
166. R. I. Tanner, *J. Polym. Sci. Polym. Phys. Ed.* 8, 2067 (1970), and Z. Rigbi, *SPEJ*, No. 9, p. 22 (1953).
*167. A. V. Tobolsky, "Properties and Structure of Polymers," pp. 188–193. Wiley, New York, 1960.
168. S. Onogi, S. Ueki, and H. Kato, *Zairyoshiken* 15, 371 (1966).
†169. S. Uemura and M. Takayanagi, *J. Appl. Polym. Sci.* 10, 113 (1966).
169a. A. N. Dunlop and H. L. Williams, *J. Appl. Polym. Sci.* 17, 2945 (1973).
170. Y. Shoji, M. Sato, T. Yahata, and F. Komatsu, *Res. Rep. Muroren Inst. Technol.* 7(3) (1972).
†171. D. C. Bogue, T. Masuda, Y. Einaga, and S. Onogi, *Polymer Jpn.* 1, 563 (1970).
*171a. T. Masuda, Ph.D.Diss., Dept. Polymer Chem., Univ. Kyoto, March 1973.
172. J. A. Faucher, *J. Polym. Sci. Part A-2* 12, 2153 (1974).
†172a. R. J. Farris, *Trans. Soc. Rheol.* 12, 291 (1968).
173. R. F. Heitmiller, R. Z. Naar, and H. H. Zabusky, *J. Appl. Polym. Sci.* 8, 873 (1964).
†173a. G. C. Derringer, *J. Appl. Polym. Sci.* 18, 1083 (1974) and *Rubber Chem. Technology* 47, 825 (1974).
†174. A. S. Hill and B. Maxwell, *Polym. Eng. Sci.* 10, 289 (1970).
175. M. Horio, T. Fuji, and S. Onogi, *J. Phys. Chem.* 68, 778 (1964).
†176. M. Kasajima and Y. Mori, *Kobunshi Kagaku* 37, 915 (1973).
177. V. N. Kuleznev, I. V. Konytikh, G. V. Vinogradov, and I. P. Dmitrieva, *Kolloid Zh.* 27, 540 (1965).
177a. W. M. Prest and R. S. Porter, *Polym. J. Jpn.* 4, 163 (1973).
178. B. L. Lee and J. L. White, *Trans. Soc. Rheol.* 19, 481 (1975).
179. R. Longworth and W. F. Busse, *Trans. Soc. Rheol.* 6, 179 (1962).
180. A. Ya. Malkin, N. K. Blinova, G. V. Vinogradov, M. P. Zabugina, O. Yu. Sabsai, V. C. Shigalova, I. Yu. Kircherskaya, and V. P. Shatov, *Eur. Polym. J.* 10, 445 (1974).
181. USI Chem. Appl. Note, *Mod. Plast.* 35(7), 229 (1958); see also [143] p. 3001.
182. T. Boiangiu, *Mater. Plast. (Bucharest)* 3, 140 (1966).
†183. T. M. Kartashova *et al.*, *Plast Massy* No. 9, p. 29 (1969).
†184. J. C. Halpin and J. L. Kardos, *J. Appl. Phys.* 43, 2235 (1972).
†184a. K. Ninomiya, *J. Colloid Sci.* 18, 421 (1963) and *J. Macromol. Sci.* 3, 237 (1969).
†185. T. Soen, T. Horino, Y. Ogawa, K. Kyuma, and H. Kawai, *J. Appl. Polym. Sci.* 10, 1499 (1966).
186. J. Pavlinec and N. I. Kalaforov, *Eur. Polym. J.* 7, 1445 (1971).
187. H. P. Schreiber, *J. Appl. Polym. Sci.* 16, 539 (1972).
188. K. Baba, *Rep. Progr. Polym. Phys. Jpn.* 16, 253 (1973).
189. J. Barton and J. Rak, *J. Appl. Polym. Sci.* 11, 499 (1967).
190. J. Letz, *J. Polym. Sci. Part A-2* 7, 1987 (1969).
191. J. Letz, *Kolloid Z.* 236, 38 (1970).
192. T. C. Yu and C. D. Han, *J. Appl. Polym. Sci.* 17, 1203 (1973).
193. I. E. Chalykh, L. Yu. Zlatkevich, and V. G. Raevskii, *Vysokomol. Soedin*, 11B, 120 (1969).
194. E. I. Frenkin, Yu. G. Yanowskii, and G. V. Vinogradov, *Mekh. Polim.* No. 6, p. 895 (1966).

†195. M. Kryszewski, T. Pakula, and G. Grebowicz, *Vysokomol. Soedin.* **16A**, 1569 (1974).
†196. G. V. Vinogradov, Yu. G. Yanovskii, Y. N. Kuleznev, and I. V. Kurbatova, *Kolloid Zh.* **28**, 640 (1966).
197. G. V. Vinogradov, A. Ya. Malkin, V. N. Kuleznev, and V. F. Larionov, *Kolloid. Zh.* **28**, 809 (1966).
198. N. K. Baramboym and V. F. Rakityanskii, *Plast. Massy* N. 11, p. 34 (1971).
199. N. K. Baramboym and V. F. Rakityanskii, *Vysokomol. Soedin.* **13B**, 662 (1971).
200. N. K. Baramboym and V. F. Rakityanskii, *Kolloid. Zh.* **31**, 129 (1974).
201. R. E. Smelkov, V. V. Petryaev, and I. I. Senichkina, *Plast. Massy* No. 1, p. 32 (1966).
†202. M. Kryszewski, A. Galeski, T. Pakula, and J. Grebowicz, *J. Colloid Interface Sci.* **44**, 85 (1973).
203. V. G. Raevskii, V. B. Zamyslov, L. Yu. Zlatkevich, and V. E. Gul, *Vysokomol. Soedin.* **8**, 1145 (1966).
*204. J. M. McKelvey, "Polymer Processing," p. 299. Wiley, New York, 1962.
*205. Z. Tadmor and I. Klein, "Engineering Principles of Plasticating Extrusion," p. 332. Van Nostrand-Reinhold, Princeton, New Jersey, 1970.
206. H. Barth, *Plastverarbeiter* **21**, 560, 633 (1970).
207. V. V. Bogdanov, A. M. Voskresenskii, E. V. Korotyshev, R. G. Mirzoev, and V. N. Krasovskii, *Proc. All-Sov. Symp. Rheol., 8th* p. 258. Homel, USSR (May 1975).
208. E. Dolling and R. Reutenbach, *Plastverarbeiter* **22**, 859 (1971).
209. J. Huebner, *Kunststoffe Rundsch.* **20**, 504 (1973).
210. J. Shimizu, *Jpn. Plast. Age* **12**(8), 30 (1974).
211. N. S. Mayzel and O. B. Vasilev, *Plast. Massy* No. 3, p. 50 (1967).
212. M. Shundo, M. Imoto, and Y. Minoura, *J. Appl. Polym. Sci.* **10**, 939 (1966).
†213. A. P. Plochocki and G. Miller, *Przemysl Chem.* **55**, 539 (1976).
214. Y. Oyanagi, *Kobunshi Ronbunshu* **32**, 382 (1975).
215. Y. Oyanagi, *J. Soc. Rheol. Jpn* **4**, 10 (1976).
216. I. K. Yartsev *et al.*, USSR Patent, 312760-S (1971).
217. Z. Tadmor and I. Klein, *Polym. Eng. Sci.* **13**, 382 (1973).
218. D. B. Todd, *Polym. Eng. Sci.* **15**, 437 (1975).
219. K. F. Jerchel, *Chem. Eng. Tech.* **44**, 552 (1972).
220. G. M. Konovalov, V. P. Krivoshev, and A. A. Grafov, *Plast. Massy* No. 9, p. 30 (1972).
†221. P. H. Squires and C. F. W. Wolf, *SPE J.* **27**, 68 (1971).
222. B. H. Maddock, *SPE J.* **23**(7), 23 (1967).
223. D. M. Bigg and S. Middleman, *IEC Fundamentals* **13**, 66, 184 (1974).
224. D. M. Bigg, *Polym. Eng. Sci.* **15**, 684 (1975).
†225. G. Pinto and Z. Tadmor, *Polym. Eng. Sci.* **10**, 279 (1970).
226. R. W. Torner, M. S. Akutin, and A. V. Melik-Kasumov, *Plast. Massy* No. 8, p. 31 (1969).
227. V. I. Giegerich, *Plastverarbeiter* **23**, 807 (1972).
228. G. Menges and V. Giegerich, *Gummi Asb. Kunstst.* **25**, 818 (1972).
229. A. P. Plochocki, *Polymery (Warsaw)* **17**, 328 (1972).
230. P. Machowski, A. Plochocki, and L. Czarnecki, *Polimery (Warsaw)* **22**, 82 (1977).
231. I. V. Konoval, E. I. Evdokimov, F. G. Gilimyanov, A. I. Muzykantova, and M. I. Vysotskaya, *Plast. Massy* No. 7, p. 8 (1972).
232. V. A. Sakrevski and I. E. Korsunov, *Plaste Kautschuk (Leipzig)* **19**, 92 (1972).
233. K. Satake, *J. Appl. Polym. Sci.* **15**, 1819 (1971).
234. K. Satake, *Nippon Gomu Kyokaishi* **42**(3), 177 (1969).
235. D. Snider and T. E. Carrigan, *AIChE J.* **14**, 813 (1968).

236. V. S. Vasilenko, I. V. Konoval, E. I. Evdokina, F. G. Gilimyanov, and A. I. Muzykantova, *Plast. Massy* No. 8, p. 26 (1973).
237. T. Sakai, K. Tsuchiya, and Y. Arakida, *Kobunshi Kagaku* **28**, 868 (1971).
238. I. K. Yartsev, V. I. Pilipovskii, and G. V. Vinogradov, *Plast. Massy* No. 7, p. 10 (1968).
†239. H. G. Stabler, R. N. Haward, and B. Wright, *Soc. Chem. Ind. (London) Monogr.* No. 26, p. 327 (1967).
240. S. V. Vlasov, G. V. Sagalaev, V. N. Kuleznev, and I. G. Mazurkov, *Plast. Massy* No. 9, p. 37 (1975).
241. T. Yoshimoto and A. Miyagi, *Kogyo Kagaku Zasshi* **69**, 1771 (1966).
242. O. V. Romankevich, A. N. Gopkalo, L. V. Grzhimalovskaya, and I. V. Kurbatova, *Vysokomol. Soedin.* **16B**, 684 (1974)—see also [56a].
243. H. P. Schreiber and S. H. Storey, *J. Polym. Sci. Part B* **5**, 723 (1965).
244. H. P. Schreiber, *J. Appl. Polym. Sci.* **18**, 2501 (1974).
245. K. A. Kaufman and C. S. Imig, *Mod. Plast.* **36**(6), 137 (1959).
246. S. E. Poskvitovskaya and M. D. Medvedeva, *Plast. Massy* No. 8, p. 46 (1969).
247. Y. Kinoshita and H. Nakamura, *Kobunshi Kagaku* **28**, 864 (1971).
248. *Plast. Technol.* **13**(11), 57 (1967).
249. E. A. Tonkopii, D. F. Kagan, G. I. Shapiro, M. D. Fridman, and L. F. Danilenko, *Plast. Massy* No. 6, p. 31 (1974).
250. W. M. Listkov and G. P. Starevskaya, *Plast. Massy* No. 8, p. 21 (1974).
251. Z. G. Suleymanova, B. I. Ryzhov, and M. Sh. Kadyrov, *Plast. Massy* No. 9, p. 48 (1975).
252. L. S. Gutsu, S. Sh. Shaposhnik, and V. I. Golub, *Mekh. Polim.* No. 3, p. 532 (1971).
253. A. B. Ayvazov and V. Ya. Zhivaev, *Mekh. Polimerov* No. 1, p. 138 (1970).
254. V. I. Serenkov and I. P. Berkovich, *Plast. Massy* No. 5, p. 13 (1968).
255. J. J. Bikerman and D. V. Marshall, *J. Appl. Polym. Sci.* **7**, 1031 (1963).
256. L. Dijkema and T. Frieling, *Kunststoffe* **51**, 703 (1961).
257. V. P. Popov and V. V. Belozerov, *Plast. Massy* No. 9, p. 42 (1975).
258. V. G. Baranov and K. A. Gasparyan, *Vysokomol. Soedin.* **11B**, 809 (1969).
†259. M. Maeda, T. Oda, S. Hibi, S. Makino, M. Asano, and K. Ando, *Proc. Ann. Conf. Jpn. Soc. Plast. Eng.* **C1**, 589 (1974).
†260. M. Maeda, T. Oda, S. Hibi, S. Makino, M. Asano, and K. Ando, *Kobunshi Kagaku* **30**, 282, 288 (1973).
261. W. A. Mack, *Brit. Plast.* **35**(3), 144 (1962).
262. Anonymous, *Mod. Plast. Int.* **6**(8), 11 (1976).
263. V. L. Peglovsky, V. I. Sidorenko, and G. V. Livin, *Khim. Prom. Ukr.* No. 1, p. 12 (1970).
264. V. I. Sidorenko, V. L. Peglovskii, and G. V. Liviji, *Kozh. Obuv. Prom.* **12**(5), 27 (1970).
265. Y. Atarashi, *Kobunshi Kagaku* **21**, 210 (1964).
266. T. Horino *et al.*, *Zairyo* **16**(166), 494 (1967).
267. T. Horino, E. Iwami, K. Ban, T. Soen, and H. Kawai, *Kogyo Kagaku Zasshi* **73**, 1615 (1970).
268. A. P. Plochocki, *Polimery (Warsaw)* **14**, 433 (1969).
269. E. G. Fedoseeva, R. I. Feldman, and V. I. Filatova, *Mekh. Polim.* No. 6, p. 1133 (1968).
270. V. I. Gerasimov, I. N. Kabalinskii, G. N. Torontseva, and D. Ya. Tsvankin, *Vysokomol. Soedin.* **10A**, 2465 (1968).
271. J. Bares and M. Pegoraro, *J. Polym. Sci. Part A-2* **9**, 1271, 1287 (1971).
271a. A. Nishioka and J. Furukawa, *J. Appl. Polym. Sci.* **14**, 1183 (1970).
272. R. Longworth and D. L. Funck, *J. Appl. Polym. Sci.* **10**, 1611 (1966).
273. T. Ogawa, S. Tanaka, and T. Inaba, *J. Appl. Polym. Sci.* **18**, 1351 (1974).
274. S. Fujiwara and M. Narasaki, *J. Polym. Sci. Part B* **1**, 139 (1963).

368                                                                      A. P. Plochocki

275.  N. V. Mikhailov, E. Z. Fainberg, V. O. Gorbacheva, and Chen-Tsin-Khai, *Vysokomol. Soedin.* **4A**, 237 (1962).
276.  Z. Sobiczewski, *Makromol. Chem.* **63**, 100 (1963).
277.  A. P. Plochocki and Z. Kohman, *Polimery (Warsøw)* **11**, 403 (1966); **12**, 16 (1967).
278.  T. Asada, T. Fukao, A. Tanaka, and S. Onogi, *Zairyoshiken* **17**, 59 (1968).
279.  L. F. Benkova, A. D. Shulyak, M. A. Martinov, G. A. Zhivulin, and G. A. Patrikeev, *Plast. Massy* No. 7, p. 27 (1972).
280.  Yu. B. Shilov and E. T. Denisov, *Vysokomol. Soedin.* **14A**, 2385 (1972).
281.  E. Hagen and G. Hazkoto, *Plaste Kautsch.* (*Leipzig*) **16**, 21 (1969).
282.  G. Hazkoto and E. Hagen, *Muanyag.* (*Budapest*) **7**, 210 (1970).
283.  S. Damyanov, I. Glavchev, and E. Kashcheva, *Angew. Makromol. Chem.* **35**, 1 (1974).
284.  B. H. Clampitt, *Anal. Chem.* **35**, 577 (1963), see [291].
285.  B. H. Clampitt, *J. Polym. Sci. Part A-2* **3**, 671 (1965).
286.  T. Sato and M. Takahashi, *J. Appl. Polym. Sci.* **13**, 2665 (1969).
287.  J. L. Zakin and R. Simha, *J. Appl. Polym. Sci.* **10**, 1455 (1966).
288.  M. A. Natov, L. B. Peeva, and B. L. Serafinov, *Makromol. Chem.* **117**, 231 (1968).
289.  E. M. Barral, R. S. Porter, and J. F. Johnson, *J. Appl. Polym. Sci.* **9**, 3061 (1965).
290.  N. Nakajima and F. Hamada, *Kolloid. Z.* **205**(1), 55 (1965).
291.  M. Kunimura, T. Nagasawa, and S. Hoshino, *Kogyo Kagaku Zasshi* **73**, 1402 (1970).
292.  M. A. Natov and L. B. Peeva, *Vysokomol. Soedin.* **8**, 1846 (1966).
293.  E. M. Barral and E. J. Gallegos, *J. Polym. Sci. Part A-2* **5**, 113 (1967).
†294.  M. A. Natov and L. B. Peeva, *Angew. Makromol. Chem.* **6**, 144 (1969).
295.  A. P. Plochocki, *Kolloid. Z.* **208**, 168 (1966).
296.  T. Horino, E. Iwami, T. Soen, and H. Kawai, *Kogyo Kagaku Zasshi* **73**, 1611 (1970).
†297.  S. P. Kabin and O. T. Usyarov, *Vysokomol. Soedin.* **2**, 46 (1960).
298.  Yu. V. Selenev and G. H. Bartenev, *Plaste Kautschuk* (*Leipzig*) **17**, 731 (1970).
†299.  T. Okamoto and M. Takayanagi, *J. Polym. Sci. Part C* **23**(2), 597 (1969).
300.  M. A. Martynov, V. M. Yuzhin, A. I. Malushyn, and G. F. Tkachenko, *Plast. Massy* No. 10, p. 6 (1965).
†301.  E. L. Vinogradov, I. V. Kurbatova, M. A. Martynov, and N. A. Sibiryakova, *Plast. Massy* No. 7, p. 24 (1972).
302.  E. L. Vinogradov, M. A. Martynov, N. A. Sibiryakova, and I. V. Kurbatova, *Vysokomol. Soedin.* **14A**, 1652 (1972).
303.  J. Y. Decroix, A. Piloz, A. Douillard, J. F. May, and G. Vallet, *Eur. Polym. J.* **11**, 625 (1975).
†304.  T. Pakula, M. Kryszewski, J. Grebowicz, and A. Galeski, *Polym. Jpn. J.* **6**, 94 (1974).
305.  A. D. Yakovlev, N. Z. Evtyukov, G. T. Tkachenko, M. A. Martynov, and A. I. Marei, *Vysokomol. Soedin.* **16B**, 878 (1974).
306.  M. Jarzynska, A. Plochocki, and E. Treszczanowicz, *Polimery (Warsaw)* **14**, 505 (1969).
307.  S. Krozer and M. Wajnryb, *Polimery (Warsaw)* **7**, 367 (1962).
308.  K. Nakamura *et al.*, *Rep. Progr. Polym. Phys. Jpn.* **11**, 23 (1968).
*309.  V. N. Kuleznev, "Mnogokomponentnye Polimernye Sistemy." Khimya, Moskva, 1974.
310.  R. B. Seymour and M. L. B. McGee, *Mod. Plast.* **50**(1), 98 (1973).
311.  V. Schlueter, *Kunststoffe Plast.* **19**, 87, 90 (1972).
312.  M. L. Kerber, *Plast. Massy* No. 5, p. 59 (1971).

# Chapter 22

# Blends Containing Poly(ε-caprolactone) and Related Polymers

*J. V. Koleske*

*Union Carbide Corporation
Chemicals and Plastics
Research and Development Department
South Charleston, West Virginia*

## I. INTRODUCTION

Various polymeric materials are often known for specific or unique characteristics. Certain polymers are known for their elastomeric properties, others for their clarity, others for their toughness, and so on. Poly(ε-caprolactone), too, has its unique characteristic. Among its other attributes, it is known for its ability to blend with a wide variety of polymers. In this chapter, the miscibility of poly(ε-caprolactone), other polylactones, and copolymers containing ε-caprolactone are described.

### A. Poly(ε-caprolactone)

Poly(ε-caprolactone) (PCL) is a partially crystalline polymer that has a moderately low melting point of 60°C. It is available from Union Carbide

369

Corporation in three molecular weight grades. One, PCL-700, is a tough, extensible polymer with $\overline{M}_w$ of about 40,000. The others, PCL-300 and PCL-150, have $\overline{M}_n$ of about 10,000 and 5000, respectively. They are hard, brittle materials. Low molecular weight ($\overline{M}_n \sim 300$–3000) oligomers or polyols are also commercially available. However, these polyols have utility as intermediates for polyurethanes and are not used as blending materials. As molecular weight decreases, end groups play a role in affecting blend-ability, and it has been found that the compatibility can be markedly affected. For this reason, the main polymer of concern in this chapter is PCL-700.

## 1.  Polymerization

### a.  Linear and Branched Homopolymers

Poly($\varepsilon$-caprolactone) is prepared from $\varepsilon$-caprolactone through an addition type of mechanism [1]. The cyclic ester monomer will open in the presence of an initiator that contains an active hydrogen atom as shown in Eq. (1):

$$x\,O\!-\!CH_2\!-\!CH_2\!-\!CH_2\!-\!CH_2\!-\!CH_2\!-\!\overset{\overset{O}{\|}}{C} + ROH \xrightarrow{\ \Delta\ } RO\!\!\left[\!C\overset{\overset{O}{\|}}{}\!-\!(CH_2)_5\!-\!O\right]_{\!x}\!\!H \quad (1)$$

The resulting polymer is formed without the loss of any compound; and for this reason, the reaction is considered to be an addition type of heterocyclic polymerization. While the polymerization will proceed without a catalyst, catalysts such as stannous octanoate greatly facilitate the reaction. When a diol is used as the initiator, a dihydroxy-terminated polymer is formed. If water is used as the initiator, the resulting polymer is terminated with both carboxyl and hydroxyl functionality. Branched polymers can be prepared by using triols, tetraols, etc., as initiators. Many other examples could be given, but these few serve to illustrate the polymer's structure. The polymer PCL-700 is terminated with hydroxyl groups.

### b.  Graft Copolymers

Graft copolymers can be prepared from polymers that contain suitable functionality. A recent study [2] describes polymers that have a styrene–$\beta$-hydroxyethyl methacrylate (22.5/0.5) backbone to which 77% $\varepsilon$-caprolactone has been grafted. These "tone-grafted styrene" or TGS polymers contain long pendant chains of polymerized $\varepsilon$-caprolactone, and each pendant chain is hydroxyl terminated. It is also possible to utilize poly($\varepsilon$-caprolactone) as the backbone polymer and graft monomers such as acrylic acid, acrylonitrile, or styrene to it [3].

### c. Block Copolymers

Block copolymers of the ABA, poly(ε-caprolactone)–poly(alkylene oxide)–poly(ε-caprolactone), type have been prepared [4–6] by using a hydroxyl-terminated, low molecular weight alkylene oxide polymer as an initiator. The (AB)$_n$-type block copolymers of ε-caprolactone and ethylene oxide have also been prepared by alternatively feeding the monomers during polymerization [7]. Block copolymers of PCL with poly(ethylene terephthalate) [8], styrene–butadiene [9], and siloxane [10] are also known.

### 2. Physical Properties

### a. Crystal Structure

Bittiger *et al.* [11] investigated the crystal structure of PCL. The polymer has an extended, zigzag conformation and the unit cell is orthorhombic in character. Its crystalline structure is quite similar to that of polyethylene with the a and b dimensions of the polymers almost identical. Of course, the fiber axis or c dimension of PCL is much larger than that of polyethylene. The large difference in melting point between these two polymers (63°C for PCL and 136°C for polyethylene) was explained in terms of entropic effects. Poly(ε-caprolactone) has a greater degree of rotational freedom about the chain backbone which results in a higher entropy of fusion relative to polyethylene. This effect is strong enough to overcome the increased polarity that exists in the polylactone. In its usual state, PCL is about 50% crystalline [12].

### b. Solution Properties

Poly(ε-caprolactone) is soluble in a variety of solvents, such as benzene, toluene, methylene chloride, chloroform, carbon tetrachloride, tetrahydrofuran, cyclohexanone, dihydropyran, and 2-nitropropane. It can be solubilized in methyl ethyl ketone, dimethylformamide, acetone, and ethyl acetate by warming; however, after aging these solutions about a week at room temperature, phase separation occurs.

Hydrodynamic properties and unperturbed dimensions of fractionated PCL have been studied [13,14]. Table I contains the Mark–Houwink constants for PCL in various solvents. Since the reduced viscosity at a concentration of 0.2 gm/dl is an easily determined parameter, the following relationships have been calculated [16]:

$$(\eta_{sp/c})_{c=0.2} = 3.21 \times 10^{-5} \overline{M}_w^{0.93}, \qquad \text{benzene,} \quad 30°C \qquad (2)$$

$$(\eta_{sp/c})_{c=0.2} = 8.33 \times 10^{-6} \overline{M}_w^{1.08} \qquad \text{chloroform,} \quad 30°C \qquad (3)$$

Table I

Constants for the Expression $[\eta] = K M^a$
$[\text{Poly}(\varepsilon\text{-caprolactone})], [\eta]$ in dl/gm]$^a$

| Solvent | $K \times 10^5$ | $a$ | Type mol wt average | Ref. |
|---------|------------------|------|---------------------|------|
| Benzene | 9.94 | 0.82 | Weight | [13] |
| Benzene | 12.5 | 0.82 | Number | [15] |
| Chloroform | 4.59 | 0.91 | Weight | [16] |
| Dimethylformamide | 19.1 | 0.73 | Weight | [13] |
| Pyridine | 5.15 | 0.86 | Weight | [16] |

$^a$ All at 30°C.

These expressions may be used to yield an approximation of $\overline{M}_w$ from a one-point measurement. The solution properties of the polymer under theta conditions are described by

$$[\eta]_\Theta = 1.1 \times 10^{-3} \overline{M}_w^{0.5} \tag{4}$$

*c. Physical Properties*

As was indicated above, high molecular weight PCL is a tough polymer with a moderate melting point. Typical physical properties are given in Table II. It is interesting to note that the volume resistivity of PCL is about four orders of magnitude lower than that of most polymers. Thus, it

Table II

Physical Properties of PCL at Room Conditions$^{a,b}$

| | |
|---|---|
| 1% Secant modulus | 50,000 psi |
| Yield stress | 1600 psi |
| Elongation | ~750% |
| Tensile strength | 3500 psi |
| Melting point [11] | 63°C |
| Glass transition temperature (1 Hz) | |
|     Partially crystalline [17, 18] | 213°K |
|     Amorphous [17] | 202°K |
| Volume resistivity (23°C, 50% RH) | $4.76 \times 10^{11}$ ohm/cm |

$^a$ Reprinted from G. L. Brode and J. V. Koleske [1], *J. Macromol. Sci. Chem.* **A6**(6), 1109 (1972), copyright 1972 Marcel Dekker, Inc., by courtesy of Marcel Dekker, Inc.
$^b$ $\overline{M}_w \sim 40{,}000$ or greater.

is comparable in this property to phenol–formaldehyde polymers and nitro-cellulose. Another factor, which is of importance to the blendability of PCL, is the difference in glass transition temperature $T_g$ between the amorphous and the partially crystalline polymer. The experimental evidence describing the effect of crystallinity on the $T_g$ of PCL has been detailed [17, 18], and it is important to use the value of 202°K when calculating $T_g$ for compatible blends from relationships such as that devised by Fox [19] or Gordon and Taylor [20].

## B. Definition of Compatibility

At the beginning of this chapter, it was indicated that PCL has a unique ability to blend with a variety of other polymers over wide composition ranges. The materials that result when such blends are prepared have varied and useful properties. Investigations that have been carried out indicate that three types of blends result when PCL is blended with other polymers. These are blends with (a) true compatibility, (b) mechanical compatibility, and (c) crystalline interaction.

### 1. Compatible

The definition of compatible polymer blends has been given in Volume 1, Chapter 5. In accord with this definition, blends of PCL with other polymers are considered to be compatible when they exhibit a single glass transition temperature intermediate between that of either homopolymer. Obviously, the $T_g$ of the blend will depend on the amount of either component present.

### 2. Mechanically Compatible

Incompatible polymer blends are characterized as exhibiting more than one glass transition temperature. In such blends, a $T_g$ associated with each component of the blend is evident. Mechanically compatible blends would be considered as incompatible by this definition. However, the mechanical properties of such blends are not what would be expected of an incompatible blend. Rather than being very brittle or weak in character, mechanically compatible blends do not exhibit deleterious alteration of properties and often exhibit improvements in selected properties.

### 3. Crystalline Interaction

When polycaprolactone is blended with other partially crystalline polymers, the blends are crystalline in character. In fact, rather than disrupting crystallinity, it often appears that crystallinity is enhanced and new interactions take place.

## II. BLENDS OF POLY(ε-CAPROLACTONE)
## AND OTHER POLYMERS

### A.  Compatible Blends

It is well known that it is difficult to blend polymers and achieve compatibility. For some compatibility to occur, $\Delta G$ must be negative in the expression $\Delta G = \Delta H - T \Delta S$. Although the entropic change is positive and favors compatibility when polymers are mixed, the change is small since only relatively small numbers of molecules are involved when polymeric compounds are considered. In addition, mixing is usually an endothermic process and the enthalpic change is expected to be positive and thus opposed to compatibilization. Arguments such as these led Flory [21] to state that "incompatibility of chemically dissimilar polymers is observed to be the rule and compatibility is the exception" (p. 555). Since it will be shown that PCL is compatible with a number of dissimilar polymers, it stands out as being unique in its blending ability.

### 1.  Poly(vinyl chloride) [17]

Poly(ε-caprolactone)–poly(vinyl chloride) (PVC) blends can be readily prepared by hot compounding processes. The compositions are flexible, transparent, plasticized products (see Chapter 17). PVC effectively destroys the crystallinity of PCL. Thus, blends of two rigid polymers become soft and pliable compositions. However, at high PCL contents ($>30\%$), the blends become translucent and more rigid due to a crystallization of the PCL as time elapses. This effect is dependent on blend composition. The curves shown in Fig. 1 describe the increase in modulus that takes place with time in PCL–PVC blends. It is readily apparent that several weeks pass before the change is noticeable in blends containing $40\%$ PCL, and little or no change is apparent at lower levels of PCL. As the PCL content increases above $40\%$, the crystallization takes place more rapidly. In fact, when blends containing $80\%$ or more PCL are examined, quenching in liquid nitrogen is required to obtain clear blends. As is described by the data given in Fig. 2, such blends are compatible only for the time scale of the measurement. After the temperature is increased above $T_g$, the PCL rapidly crystallizes as indicated by the modulus increase for the 80/20 PCL/PVC blend. Results similar to these were also found by Shur and Rånby [22], who investigated PCL–PVC blends by a gas permeation technique.

Extrapolation of the $T_g$ values for these blends by the Gordon–Taylor [20] method resulted in a $T_g$ for PCL of 202°K when the polymer is in an amorphous state. This value is 16 degrees lower than is obtained for the polymer when it is room temperature aged and 11 degrees lower than is

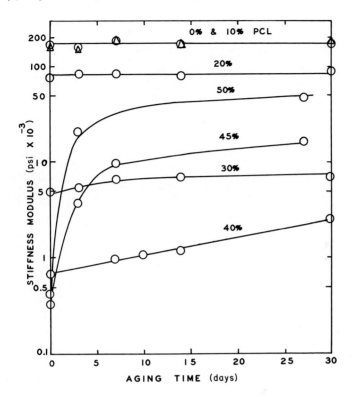

Fig. 1    Stiffness modulus as a function of aging time for poly(vinyl chloride) blended with various amounts of poly(ε-caprolactone). (From Koleske and Lundberg [17].)

obtained for the polymer quenched into liquid nitrogen from the molten state.

Robeson [23] studied the crystallization kinetics of PCL when it was blended with PVC. In agreement with the above data, he found that both crystallization rate and induction time for crystallization is critically affected by the concentration of components. Two stages of crystallization were apparent. By measuring change in modulus as a function of log time, it was found that both the nucleation-controlled stage and the diffusion-controlled stage have linear regions of response. The nucleation stage has a higher rate of change than the diffusion stage.

Investigations of the electron beam cross-linking of PCL–PVC compositions have been conducted [24]. High gel contents were obtained, and the compositions had improved mechanical properties.

Ethylene oxide–ε-caprolactone copolymers also form compatible blends

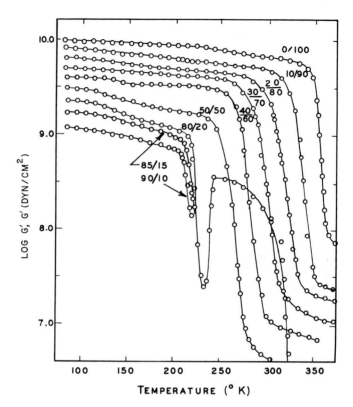

Fig. 2   Real component of the complex shear modulus for poly(vinyl chloride)–poly(ε-caprolactone) blends. For clarity all curves except that for PVC are vertically displaced about 0.1 log unit from each other. (From Koleske and Lundberg [17].)

when combined with PVC [25]. Mixtures of these materials have good low-temperature impact strength, excellent permanence and low volatility, flexibility at $<0°C$, and good color. Typical properties of such compositions are detailed in Table III.

## 2.   Poly(hydroxy ether) [1]

Blends of PCL with Phenoxy [Union Carbide Corporation poly(hydroxy ether) prepared from bisphenol A and epichlorohydrin] also form compatible systems. As shown in Fig. 3, the glass transition temperature of these blends is well fit by the Fox expression [19],

$$(1/T_{g1,2}) = (W_1/T_{g1}) + (W_2/T_{g2})$$                (5)

when 202°K is used for PCL and 376°K is used as the $T_g$ for phenoxy.

Table III

Properties of PVC Plasticized with an 86/14
ε-Caprolactone–Ethylene Oxide Copolymer[a,b]

| Properties | Copolymer (wt %) | | ASTM test |
|---|---|---|---|
| | 33 | 41 | |
| Hardness (durometer *D*) | 91 | 79 | D-1706 |
| Brittle temperature (°C) | −18 | −36 | D-746 |
| 1% Secant modulus, 25°C (psi) | 4000 | 900 | — |
| Spew (60°C, 100% RH, 2 weeks) | None | None | — |
| Water extraction, 24 hr, 70°C (%) | 1.2 | 1.3 | D-1203 |
| Oil extraction, 50°C | Nil | Nil | D-1203 |
| $T_4$, Temperature at which modulus = 10,000 psi (°C) | 20 | 2 | D-1043 |

[a] From Del Giudice and Dodd [26].
[b] Thermally stabilized with 1% Van Stay HT.

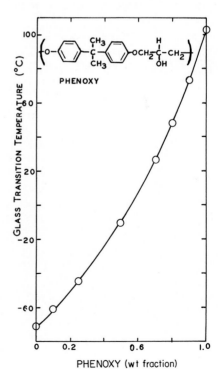

Fig. 3   Glass transition temperature from *G″* as a function of Phenoxy content. (Reprinted from G. L. Brode and J. V. Koleske, *J. Macromol. Sci.-Chem.* **A6**(6), 1109 (1972), copyright 1972, Marcel Dekker, Inc., by courtesy of Marcel Dekker, Inc.)

Thus, while it is readily apparent that these blends are truly compatible, it should be noted that blends containing more than 50% PCL will undergo a modulus increase and change in $T_g$ due to the PCL crystallizing after a few days at room temperature.

### 3. Cellulosic Polymers

Poly(ε-caprolactone) acts as a polymeric plasticizer for nitrocellulose [1]. Since the behavior of plasticized nitrocellulose is influenced by crystallinity [27], a rather complicated behavior was found for these blends. The plasticizing effect can be seen from the elongation–composition plot shown in Fig. 4. Values of $T_g$ taken from $G''$ and from mechanical loss both decreased as the PCL content in the blends increased. However, the difference in $T_g$ as measured by these two parameters was the expected few degrees only at very low PCL contents. At the 50% level, the difference in $T_g$ from these parameters had increased to 32 degrees. It was thought that the unusual behavior was attributable to the crystallinity of the nitrocellulose.

Elongation as a function of concentration is presented in Fig. 5 for blends of PCL and cellulose acetate butyrate [28]. The blends were transparent and tough in nature. Although no direct measurement of $T_g$ was made,

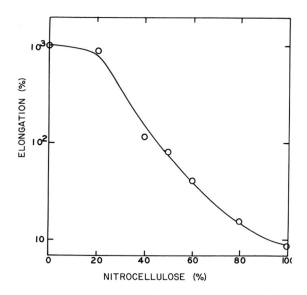

Fig. 4   Elongation of nitrocellulose–poly(ε-caprolactone) blends as a function of nitrocellulose content ($\frac{1}{2}$ sec nitrocellulose, 11% nitrogen, D.S. 2.0). (Reprinted from G. L. Brode and J. V. Koleske, *J. Macromol. Sci.-Chem.* **A6**(6), 1109 (1972), copyright 1972, Marcel Dekker, Inc., by courtesy of Marcel Dekker, Inc.)

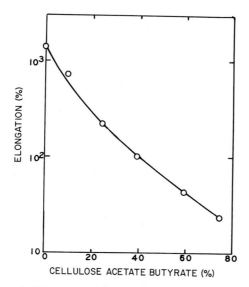

Fig. 5   Elongation of cellulose acetate butyrate–poly(ε-caprolactone) blends as a function of cellulose derivative content.

these data and blend appearance strongly suggest that the two polymers form compatible mixtures.

### 4.   Polyepichlorohydrin [1]

Polyepichlorohydrin is also compatible with PCL. The blends show phase separation at high levels of PCL when aged at room temperature; however, at $<30\%$ PCL, the blends are stable for extended periods of time. The effect of aging is described by the data presented in Table IV. Calculated values of $T_g$ agree well with the experimental values when the blends are studied immediately after preparation. However, when the blends are aged 10 days at room temperature, there is a marked difference in the value for the 50/50 blend and no change in $T_g$ for the 90/10 blend. The 50% PCL blend becomes translucent during the aging process. Since warming the blend to about 60°C returns the sample to a transparent state, the translucency may be attributed to PCL crystallization.

### 5.   Penton

Poly(ε-caprolactone) blended with Penton (Hercules' chlorinated polyether) forms compatible mixtures. Figure 6 is a comparison of mechanical loss for a quenched blend of these two polymers and Penton alone. The appearance of a single relaxation maximum for the blend meets the criteria

Table IV

$T_g$ for Polyepichlorohydrin–Polycaprolactone Blends[a]

| PCL (%) | $T_g$ | |
|---|---|---|
| | Experimental $G''$ (°C) | Calculated (°C)[b] |
| Not aged | | |
| 0 | −15 | — |
| 10 | −23 | −22 |
| 50 | −42 | −47 |
| 90 | −52 | −57 |
| | | |
| Aged ten days at room temperature | | |
| 0 | −15 | — |
| 10 | −23 | — |
| 50 | −30 | — |

[a] Reprinted from G. L. Brode and J. V. Koleske [1], *J. Macromol. Sci-Chem.* **A6**(6), 1109 (1972), copyright 1972 Marcel Dekker, Inc., by courtesy of Marcel Dekker, Inc.

[b] $1/T_{g_{1,2}} = W_1/T_{g_1} + W_2/T_{g_2}$; $T_{g_1} = 202°$K for PCL.

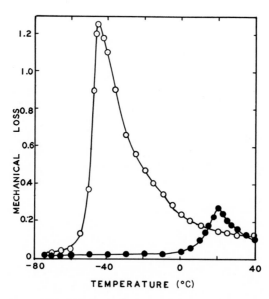

Fig. 6   Mechanical loss as a function of temperature: (○) 75/25 PCL/Penton; (●) Penton.

established for compatibility. A plot of the $T_g$ values determined from $G''$, as previously described [17], is given in Fig. 7. Since these data are well fit by Eq. (5), as shown by the solid line, compatibility for the system is demonstrated. Tensile properties of compression molded specimens of the blends are given in Table V. These data show that PCL is a plasticizer for Penton and increasing amounts of PCL produce a tougher product.

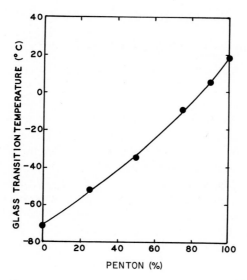

Fig. 7   Glass transition temperature as a function of blend composition: Penton and poly(ε-caprolactone) [16]. Values taken from $G''$ as a function of temperature.

Table V

Tensile Properties of Penton–PCL Blends[a, b]

| | PCL (%) | | | |
| --- | --- | --- | --- | --- |
| | 0 | 10 | 50 | 75 |
| 1% Secant modulus (psi) | 88,000 | 70,000 | 33,000 | 30,000 |
| Tensile strength (psi) | 3,100 | 3,700 | 3,700 | 5,200 |
| Elongation (%) | 105 | 120 | 1,000 | 1,800 |
| Rupture energy (lb/in³.) | 4,100 | 4,100 | 27,000 | 45,000 |

[a] From Koleske *et al.* [29].
[b] Strain rate for testing, 100% min⁻¹.

### 6. Styrene–Acrylonitrile Copolymers

Blends of PCL and a styrene–acrylonitrile copolymer have been studied [30]. Relaxation maxima from loss modulus data were well fit by Eq. (5) for all blends evaluated. However, in this particular case it was necessary to use a value of 213°K for the $T_g$ of PCL to fit the experimental data. No reason for this particular behavior was found.

## B. Mechanically Compatible Blends

### 1. Poly(vinyl acetate)

Dynamic mechanical properties of mechanically compatible blends of PCL and poly(vinyl acetate) have been described [1]. Glass transition temperatures that can be attributed to each of the homopolymers are readily apparent in these blends. However, the data presented in Table VI demonstrate that mechanical properties of the blends are not deleteriously altered. In fact, PCL seems to act as a plasticizer for the poly(vinyl acetate). Blends such as these vary from slightly hazy to opaque as the PCL concentration is increased. Since the characteristic $T_g$ maxima of the respective homopolymers do not blend into a single peak, it must be concluded that the domain size of each of the blended polymers is greater than a 10–30-Å size in mechanically compatible blends [31].

### 2. Polystyrene

The dynamic mechanical loss properties for PCL–polystyrene blends are presented in Fig. 8. Relaxation maxima in the vicinity of $-60°C$ and $100°C$ are apparent in these blends indicating that the polymers are incompatible. The

Table VI

Tensile Properties of PCL–Poly(vinyl acetate) Blends[a]

|  | PCL (%) | | | | |
|---|---|---|---|---|---|
|  | 0 | 10 | 30 | 50 | 90 |
| 1% Secant modulus (psi) | 200,000 | 180,000 | 70,000 | 22,000 | 22,000 |
| Yield stress (psi) | — | 5,300 | 3,300 | 1,500 | 1,600 |
| Elongation (%) | 8 | 9 | 680 | 1,270 | 1,200 |
| Tensile strength (psi) | 5,200 | 4,900 | 3,200 | 3,400 | 6,000 |

[a] Strain rate for testing $= 100\%$ min$^{-1}$.

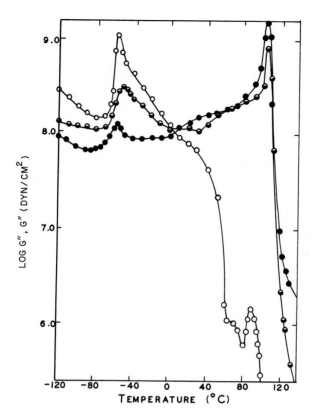

Fig. 8   Loss component of the complex shear modulus as a function of temperature for mechanically compatible blends of PCL and polystyrene (PS): (○) 90/10 PCL/PS; ( ◐ ) 50/50 PCL/PS; ( ● ) 10/90 PCL/PS.

shoulder apparent at 60°C in the 90% PCL blend is related to the melting point of polycaprolactone. This particular melting effect is not apparent in the other blends from these data. The small peak at 85°C in this blend is related to polystyrene. Since it is shifted to a lower temperature than is expected for this polymer, it appears that above its melting point, PCL has some degree of compatibility with polystyrene. Tensile properties of these blends are given in Table VII. From these data, it is apparent that the blends have good mechanical properties even though they are incompatible.

### 3.   Other Polymers

In the case of poly(methyl methacrylate), results similar to those found for blends with polystyrene and poly(vinyl acetate) were found. Dynamic

Table VII

Tensile Properties of PCL–Polystyrene Blends[a]

|  | PCL % | | | |
|---|---|---|---|---|
|  | 0 | 10 | 50 | 90 |
| 1% Secant modulus (psi) | 218,000 | 187,000 | 105,000 | 32,000 |
| Yield stress (psi) | — | 5,700 | 3,600 | 1,900 |
| Elongation (%) | 4 | 10 | 35 | 1,200 |
| Tensile strength (psi) | 5,800 | 5,400 | 3,700 | 6,100 |

[a] Strain rate for testing = 100% min$^{-1}$.

mechanical properties revealed the presence of two $T_g$'s, each of which could be related to the respective homopolymers. Mechanical properties of the blends are described by the data in Table VIII. Again, these blends vary, depending on the PCL concentration, from hazy to opaque in visual appearance.

Other polymers with which mechanical compatibility has been demonstrated include poly(vinyl butyral), poly(vinyl alkyl ethers), polysulfone, polycarbonates, and natural and synthetic rubbers. In the case of poly(vinyl butyral), the blends remain clear, tough, rigid composites at levels of up to 30% PCL. The $T_g$ for the blends is essentially unchanged at these levels from the 60°C value for poly(vinyl butyral).

With its broad compatibility with other polymers, it would be suspected that PCL would be compatible with poly[(ε-methyl)-co-(ε-caprolactone)].

Table VIII

Tensile Properties of PCL–Poly(methyl methacrylate) Blends[a]

| PCL (%) | 1% Secant modulus (psi) | Yield stress (psi) | Elongation (%) | Strength (psi) |
|---|---|---|---|---|
| 0 | 200,000 | — | 5 | 7600 |
| 10 | 173,000 | — | 7 | 7300 |
| 20 | 145,000 | 6800 | 14 | 5600 |
| 30 | 130,000 | 5000 | 32 | 4500 |
| 40 | 104,000 | 4300 | 125 | 4100 |
| 50 | 79,000 | 3500 | 510 | 4300 |
| 90 | 33,000 | 1600 | 1,300 | 4400 |

[a] Strain rate for testing = 100% min$^{-1}$.

However, a recent study [32] revealed that these two polymers were incompatible over the entire composition range. This points out that no obvious relationship exists between compatibility of two polymers and the chemical nature of their monomers. Apparently a careful balance of molecular interactions is required to achieve true compatibility.

## C. Crystalline Interaction

### 1. Polyethylene

Crystalline interactions have been found to exist when PCL is blended with polyethylene and polypropylene. In a 90/10 blend of PCL and low-density polyethylene, only the α relaxation of polyethylene was affected [1]. Since this relaxation maximum, which is related to vibrational or reorientation motion in polyethylene crystallites, was shifted to higher temperatures, it was hypothesized that the motions of the crystallites were restricted. Early in the investigations with these blends, x-ray evidence suggested that the blends might be cocrystalline in nature [33]. In general, very little reduction and in some cases an increase in the crystallinity level was observed when the polymers were well mixed. However, due to similarity in x-ray patterns for the two polymers, cocrystallinity could not be substantiated. Edwards and Phillips [34] found further evidence suggestive of cocrystallinity between PCL and polyethylene from hydrodynamically induced crystallization of the two polymers from solution. However, the matter has not been fully elucidated as yet. Another study [35] of these compositions indicated that the epitaxial crystallization of PCL on polyethylene resulted in an increase in the degree of crystallinity. In addition, the value of the Avrami exponent (~1.0) indicated that fibrillar lamellae of PCL grow perpendicular to the polyethylene surface.

### 2. Polypropylene

Well-mixed combinations of polypropylene and PCL have pronounced alterations in the relative intensities of certain x-ray reflections indicating an unusual association in the crystalline state [34]. This association is apparent from the x-ray diffraction patterns shown in Fig. 9 [36]. The intensity ratio of the 110–200 reflections in PCL is about 3/1. Polypropylene does not have a 200 reflection. While the 80/20 blend of the polymers has a reduction in intensity for the 110 reflection, the magnitude of the reduction is greater than would be predicted by the dilution effect. A surprisingly strong reflection appears for the blend at the 200 reflection of PCL. The ratio of intensities for the 110–200 reflections is essentially 1/4 in

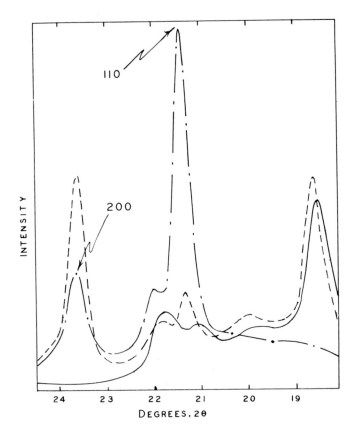

Fig. 9   X-ray diffraction patterns: (—·—·—·) PCL; (———) polypropylene (PP); (——) 80/20 PP/PCL. (From Walter and Koleske [36].)

the blend compared to the 3/1 found for PCL. This unexpected ratio for the blend was accompanied by a slight ( ~0.5%) contraction of the polypropylene along the $b$ axis. When blends such as these are highly oriented, as in a spun, drawn fiber, a strong first-layer line reflection corresponding to the 001 reflection of polypropylene is apparent. The 001 reflection usually is not evident in diffraction patterns of this polymer. While these effects were evident over the full composition range, the most pronounced effects were observed in the 5–30% PCL range.

### D.  Polyurethane Blends [37]

If a polyurethane is prepared with a low molecular weight polycapro-lactone as the soft segment, it may be thought of as an ε-caprolactone copolymer. The diisocyanate and short-chain diol extender would comprise the other comonomers. When such a polyurethane containing about 75% ε-caprolactone units was blended with polystyrene and with a styrene–acrylonitrile copolymer, mechanically compatible blends resulted. Blends with Phenoxy exhibited a single glass transition temperature, which was dependent on the relative concentration of the components. These latter mixtures were compatible in nature.

## III. UTILITY OF POLY(ε-CAPROLACTONE) BLENDS

In certain instances above, the plasticizing ability of PCL was pointed out. In the sense that a plasticizer lowers modulus and softens polymers, PCL appears to be a polymeric plasticizer for most other polymers (see Chapter 17). Its utility as a low-profile additive in glass-reinforced plastics is explained in Chapter 23.

When PCL is combined with polypropylene or polyethylene and spun into fibers, higher-tenacity fibers that are dyeable with disperse dyestuffs result [33, 38]. When ABA PCL–polyoxyethylene–PCL block copolymers are blended with these polyolefins and made into fibers and other shaped articles, the articles have dyeability with disperse, acid, premetallized and basic dyestuffs, improved stress crack resistance, low haze, and high gloss. When blended with nylon, these ABA copolymers yield fibers that have improved dyeability, static charge resistance (see Chapter 16), and improved soil resistance characteristics [39].

Poly(ε-caprolactone) alone [1] and in combination with poly(vinyl alkyl ethers) [40] form useful adhesive compositions. Such compositions are mechanically compatible and usually require temperatures of about 60°C to effect adhesion when 30% PCL is in the film. Blends containing about 80% PCL can be made into highly oriented films that will shrink and adhere when heated.

When PCL is combined with natural and synthetic rubbers and cross-linked, blends with improved processability, additive acceptance, and strength result. In addition, heat-shrinkable films and articles can be made. In the uncross-linked form, PCL improves creep resistance, handling, and forming characteristics of the gum stock [41].

Poly (ε-caprolactone) is an extremely efficient pigment dispersing aid [42]. It has been combined with a broad variety of pigments and fillers. Because of its unique ability to blend with many polymers, master batches such as these can be used in a wide variety of plastic masses. Combinations of PCL and nitrocellulose can be used as printing inks, wood lacquers, overprint lacquers, and the like [28].

PCL blended with phenolic resin compositions has enhanced mold-release properties [43]. Butadiene–ε-caprolactone–styrene block copolymers are useful impact modifiers for acrylonitrile–styrene copolymers [44, 45], and ozone resistance improves in styrene–butadiene rubbers [9]. Practice golf balls with good click, cutting resistance, and resilience are made from blends of PCL, rubber, and a filler [46]. Poly(vinyl chloride) blended with graft copolymers of ε-caprolactone onto a styrene–hydroxyethyl methacrylate backbone form wire and cable insulations with improved oil, abrasion, and solvent resistance [47]. Similar graft copolymers have been blended with diisocyanates and cured to form urethane coatings that have broad chemical resistance and good strength properties [48].

Potts *et al.* [49] have demonstrated that the excellent biodegradability characteristics of PCL make it and its blends useful for preparing containers to be used for growing and transplanting tree seedlings and other plants. The materials can be used for other packaging materials and containers. Cups of PCL will decompose in one year when buried in soil [50]. The rate of container biodegradation can be controlled by varying wall thickness and blend composition. Fields *et al.* [51] studied the degradation of PCL inoculated with the *P. pullulans* mold.

Thus, it is apparent that blends of PCL and other polymers have broad utility. Since PCL is a relatively new commercial polymer, further developments will increase the number of uses for this unique polymer.

## REFERENCES

1.  G. L. Brode and J. V. Koleske, *J. Macromol. Sci.-Chem.* **A6**(6), 1109 (1972).
2.  L. A. Pilato, J. V. Koleske, B. L. Josten, and L. M. Robeson, *Preparation and Behavior of ε-Caprolactone Graft Terpolymers in Miscible Polymer Blends.* 1976 San Francisco Amer. Chem. Soc. Meeting.
3.  F. E. Critchfield and J. V. Koleske, U.S. Patent 3,760,034 (September 18, 1973).
4.  R. Perret and A. Skoulios, *C. R. Acad. Sci. Paris* **268**, 230 (1969).
5.  J. V. Koleske and R. M. J. Roberts, U.S. Patent 3,670,045 (June 13, 1972).
6.  R. Perret and A. Skoulios, *Makromol. Chem.* **156**, 143 (1972); **162**, 147, 163 (1972).
7.  F. E. Bailey, Jr., and H. G. France, U.S. Patent 3,312,753 (1967).
8.  H. Imanaka and M. Sumoto, Japanese Patent 75/160,362 (1975) [*Chem. Abstr.* **84**, 151446k (1976)].
9.  R. F. Wright, U.S. Patent 3,652,720 (March 28, 1972) [*Chem. Abstr.* **77**, 21253q (1972)].
10. J. S. Fry, Canadian Patent 970,896 (July 8, 1975).
11. H. Bittiger, R. H. Marchessault, and W. D. Niegisch, *Acta Crystallogr.* **B26**, 1923 (1970).
12. C. G. Seefried, Jr., and J. V. Koleske, *J. Macromol. Sci.—Phys.* **B10**(4), 579 (1974).
13. J. V. Koleske and R. D. Lundberg, *J. Polym. Sci. Part A-2* **7**, 897 (1969).
14. M. R. Knecht and H. Elias, *Makromol Chem.* **157**, 1 (1972).
15. H. Sekiguchi and C. Clarisse, *Makromol. Chem.* **177**, 591 (1976).
16. J. V. Koleske, Union Carbide Corp., unpublished data (1976).
17. J. V. Koleske and R. D. Lundberg, *J. Polym. Sci. Part A-2* **7**, 795 (1969).

18. V. Crescenzi, G. Manzini, G. Calzolari, and C. Borri, *Eur. Polym. J.* **8**, 449 (1972).
19. T. G. Fox, *Bull. Am. Phys. Soc.* **2**, 123 (1956).
20. M. Gordon and J. S. Taylor, *J. Appl. Chem.* **2**, 493 (1952).
21. P. J. Flory, "Principles of Polymer Chemistry," p. 555. Cornell Univ. Press, Ithaca, New York, 1953.
22. Y. J. Shur and B. Rånby, *Compatibility studies of poly(vinyl chloride)/poly-ε-caprolactone blends based on gas permeation,* presented at the Int. Symp. Poly(vinyl chloride), 2nd. Lyons, France (July 5, 1976).
23. L. M. Robeson, *J. Appl. Polym. Sci.* **17**, 3607 (1973).
24. E. Saito, H. Fukazawa, T. Hayashi, T. Fukuda, and H. Nagai, Japanese Patent 76/17.958 (February 13, 1976) [*Chem Abstr.* **84**, 151636g (1976)]; Japanese Patent 76/23,556 (February 25, 1976) [*Chem. Abstr.* **85**, 6711j (1976)].
25. F. P. Del Giudice and R. D. Lundberg, U.S. Patent 3,629,374 (December 21, 1971).
26. F. P. Del Giudice and J. L. Dodd, Union Carbide Corp., unpublished data (1968).
27. K. Ninomiya and J. D. Ferry, *J. Polym. Sci. Part A-2* **5**, 195 (1967).
28. G. V. Olhoft, N. R. Eldred, and J. V. Koleske, U.S. Patent 3,642,507 (February 15, 1972).
29. J. V. Koleske, C. J. Whitworth, and R. D. Lundberg, U.S. Patent 3,925,504 (December 9, 1975).
30. C. G. Seefried, Jr., and J. V. Koleske, *J. Test. Eval.* **4**(3), 220 (1976).
31. C. F. Hammer, *Macromolecules* **4**(1), 69 (1971).
32. C. G. Seefried, Jr., and J. V. Koleske, *J. Polym. Sci. Polym. Phys. Ed.* **13**, 851 (1975).
33. E. R. Walter and J. V. Koleske, U.S. Patent 3,632,687 (January 4, 1972).
34. B. C. Edwards and P. J. Phillips, *J. Mater. Sci.* **9**, 1382 (1974).
35. T. Toshisada, N. Odani, and S. Nagase, *Kobunshi Ronbunshu* **32**(3), 173 (1975) [*Chem. Abstr.* **83**, 193867 (1975)].
36. E. R. Walter and J. V. Koleske, Union Carbide Corp., unpublished data (1969).
37. C. G. Seefried, Jr., J. V. Koleske, and F. E. Critchfield, *Polym. Eng. Sci.* **16**(11), 771 (1976).
38. J. V. Koleske and E. R. Walter, U.S. Patent 3,734,979 (May 22, 1973).
39. J. V. Koleske, C. J. Whitworth, and R. D. Lundberg, U.S. Patent 3,781,381 (December 25, 1973).
40. R. D. Lundberg, J. V. Koleske, D. F. Pollart, and W. H. Smarook, U.S. Patent 3,641,204 (February 8, 1972).
41. R. D. Lundberg, J. V. Koleske, and E. R. Walter, U.S. Patent 3,637,544 (January 25, 1972).
42. R. D. Lundberg, J. V. Koleske, and E. R. Walter, Canadian Patent 866,261 (March 16, 1971).
43. A. C. Soldatos, U.S. Patent 3,629,364 (December 21, 1971).
44. C. W. Childers and E. Clark, U.S. Patent 3,649,716 (March 14, 1972).
45. C. W. Childers and E. Clark, U.S. Patent 3,789,084 (January 29, 1974).
46. S. R. Harrison and A. Smith, British Patent 1,355,956 (June 12, 1974).
47. E. B. Harris and D. B. Braun, U.S. Patent 3,855,357 (December 17, 1974).
48. K. G. Sampson, V. F. Frederick, and A. J. Bunker, British Patent 1,426,062 (February 25, 1976).
49. J. E. Potts, R. A. Clendinning, and S. Cohen, *SPE Tech. Pap.* **21**, 567 (1975).
50. Anonymous, *Chem. Eng. News* **37** (September 11, 1972).
51. R. D. Fields, F. Rodriguez, and R. K. Finn, *J. Appl. Polym. Sci.* **18**(12), 3571 (1974).

# Chapter 23

# Low-Profile Behavior

## K. E. Atkins

*Chemicals and Plastics*
*Research and Development Department*
*Union Carbide Corporation*
*South Charleston, West Virginia*

## I. INTRODUCTION

Reinforced plastics, in general, and glass fiber-reinforced, thermosetting, unsaturated polyester resin, in particular, have shown remarkable growth over the past decade. Many products in the transportation, appliance, and electrical areas have adopted these versatile composites as true engineering materials. A major reason for this growth into new and sophisticated

391

applications has been the development of "low-profile" or "low-shrink" molding compounds based on fiber-reinforced, unsaturated polyester resins. The major disadvantages of materials based on such readily available, economical resins have been overcome with the development of these composites.

The "low-profile" or "low-shrink" character of these systems is achieved by using polymer blends. Selected thermoplastics of controlled structure and properties are blended into the unsaturated polyester resin composite, and upon peroxide curing under heat and pressure cause polymerization shrinkage to be essentially eliminated. This permits excellent reproduction of the mold surfaces and superior dimensional stability in the molded part.

It is the purpose of this chapter to trace the history and review the possible mechanisms of this "real-world" application of polymer blends.

## II. HISTORICAL USAGE

### A. Unsaturated Polyester Resins

Unsaturated polyester resins are a family of products based on the reaction (esterification) of unsaturated and saturated dicarboxylic acids (or their

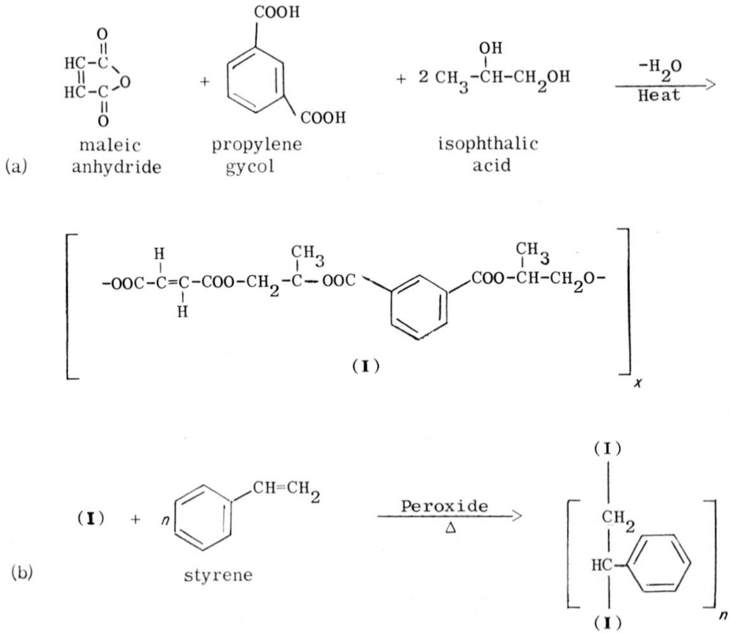

Fig. 1    (a) Polyesterification; (b) cross-linking (curing).

anhydrides) with various diols (glycols). These resins are then blended with monomers capable of cross-linking via peroxide catalysis with the unsaturation in the polyester. No volatile byproducts are created by this polymerization. Typical materials used in synthesizing these resins are unsaturated diacids such as maleic anhydride, saturated diacids such as ortho- and isophthalic acid, diols such as propylene glycol and cross-linking monomers such as styrene. A typical reaction of these materials is shown in Fig. 1.

A tremendous versatility in cured polymer properties is available based on selection of diacids, glycols, cross-linking monomers, catalysts, etc. A comprehensive review of this technology is beyond the scope of this chapter, but a number of excellent reviews are available [1, 2].

## B. Methods of Fabrication

### 1. General

Unsaturated polyester resins have been used in a variety of fiber reinforced fabrication techniques including hand lay-up, spray-up, filament winding, continuous pultrusion, continuous laminating and matched metal–die compression, transfer and injection molding. The lay-up and spray-up techniques have been utilized for many years and are still major volume users of resin for applications such as large marine parts, sanitary ware, and storage tanks.

However, the major growth in the fabrication of fiber-reinforced, unsaturated polyester composites is in matched metal–die molding applications. These techniques consist of curing a formulated compound at high temperatures (100–180°C) and pressure (100–2000 psi) in single- or multicavity hardened and chrome-plated molds. These methods provide the highest volume and highest part uniformity of any thermoset molding technique. Excellent design flexibility is available, including the use of inserts and attachments. It is in these techniques that "low-profile" and "low-shrink" compounds find their optimum usage. High temperatures are apparently necessary for the thermoplastic to provide compensation for the shrinkage of the thermosetting, unsaturated polyester. The discussion of low-profile behavior in this chapter is all in the context of match metal–die molding (e.g., cure under heat and pressure in a closed mold). This applies to compression, transfer, and injection molding applications.

General reviews of all types of fabrication and some formulation variables are given by Gaylord [3] and Penn [4].

## 2. Compression-Molding Techniques

Traditionally unsaturated polyester resins have been used in compression-molding applications when compounded with various fillers (clay, calcium carbonate, alumina trihydrate, etc.), peroxide catalysts, mold-release agents, and reinforcing fibers. The fibers are normally chopped glass, but others, such as sisal, are sometimes used. A wide variety of flow properties in the mold, part shapes that can be produced, and ultimate part strength are available, depending upon the resin structure selected, the cross-linking monomer, the fillers, the catalyst, and the reinforcing fiber. Also the compounding technique is crucial depending upon the application.

There are three major compounding techniques for match metal–die molding; these are briefly described below, and more detailed descriptions are given by Gaylord [3]. Examples of typical formulations and the wide strength properties available from these techniques are given in Tables I and II.

Table I

Typical Formulations for Compression Molding

|  | BMC | SMC | Preform |
|---|---|---|---|
| Resin (pbw)$^a$ | 100 | 100 | 100 |
| Filler (pbw) | 200–250 | 100–160 | 75 |
| Catalyst (pbw) | 1.0–1.5 | 1.0–1.5 | 0.75–1.5 |
| Mold release (pbw) | 2.0–4.0 | 2.0–4.0 | 0.75–4.0 |
| Thickening agent | 0.5–4.0 (optional) | 0.5–4.0 | — |
| Glass fiber (wt %) | 10–25 | 20–30 | 20–35 |
| Fiber length (in.) | 1/8–1/2 | 1 | 1–2 |

$^a$ Pbw = parts by weight.

Table II

Typical Strengths for Compression Molding SMC

|  | BMC | SMC | Preform |
|---|---|---|---|
| Izod impact, notched (ft-lb/in.) | 5 | 15 | 18 |
| Tensile strength (psi) | 5000 | 12000 | 15000 |
| Flexural strength (psi) | 15000 | 25000 | 30000 |
| Flexural modulus (psi × $10^6$) | 1.2 | 1.7 | 1.8 |
| Glass content (wt %) | 15 | 30 | 30 |

a.  *Sheet Molding Compound (SMC)*

Sheet molding compound (SMC) is produced on a machine onto which a mixture of resin, fillers, mold-release agent, catalyst, and thickening agent are doctored. Two layers of this mixture are used to sandwich chopped glass strand (about 1-in. lengths) between two polyethylene sheets. The thickening agent (e.g., MgO) reacts with carboxyl groups in the resin to increase viscosity to 10–80,000,000 cp. This high viscosity allows for good glass fiber distribution throughout a large part during mold flow. This thickening process is a complex phenomenon that has been the subject of several studies [5–9].

b.  *Premix or Bulk Molding Compound (BMC)*

These compounds comprise resin, filler, catalyst, mold-release agents, and short reinforcing fibers ($\frac{1}{8}$–$\frac{1}{2}$ in.) simply blended together in an appropriate mixer. Sometimes chemical thickening agents are added to these mixes. These compounds are usually employed to produce relatively small, complex parts.

c.  *Preform or "Wet" Compounds*

In this technique chopped glass fiber (1, 2-in. lengths) are preformed into the shape of the article to be molded and bound by cure of a small amount of polyester resin added as the glass is applied to the preforming mold. A mixture of resin system, filler, catalyst, and mold-release agent is poured over the glass preform at the molding press and the composite compression molded. This technique is used to produce large, relatively simple parts, in which uniform strength distribution is most critical.

## III.  COMPRESSION-MOLDING PROBLEMS

Even though standard polyester molding compounds showed considerable utility in compression molding applications, their growth into many potentially high-volume applications was inhibited by a number of problems. These problems included the following:

1.  Poor surface appearance, including fiber pattern show through, which (a) required costly sanding operations for painted, appearance applications, or (b) prohibited their use in high-appearance, internally pigmented applications.

2.  Warpage of molded parts and an inability to mold close tolerances.

3.  Internal cracks and voids, particularly in thick sections.

4.  A notable depression ("sink") on the surface opposite thick sections such as reinforcing ribs and bosses.

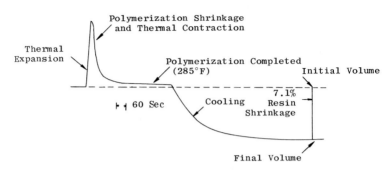

Fig. 2   Dialometric shrinkage measurement with conventional resin. (From Bartkus and Kroekel [16].)

## A.  Cause

The cause of these problems can be traced to the high polymerization shrinkage occurring during the copolymerization of the unsaturated polyester resin with the cross-linking monomer (Fig. 1). This effect has been widely observed and reported [1, 10–15]. The resin volume shrinkage causes the compound to pull away from the surface of the mold and shrink away from the fibrous reinforcement. This reduces the accuracy of mold surface reproduction and reveals fiber pattern at the surface. Stresses created by nonuniform shrinkage cause warpage and internal cracks, and prohibit molding to close tolerances.

Dilatometric experiments have been conducted by several workers [16, 17] to quantify the effect. Bartkus and Kroekel [16] have shown that the volumetric shrinkage is approximately 7% in the curing of typical unsaturated polyester resins (Fig. 2).

## B.  Overcoming the Problems

### 1.  Early Attempts

Over many years a wide variety of techniques have been employed in attempts to overcome the consequences of this shrinkage. These have included the use of large amounts of fillers, change in resin structure and comonomer, and even partial polymerization of the resin (B staging) before molding the composite [15]. For a variety of reasons, these techniques were unable to solve the problem.

### 2.  The Solution

#### a.  Addition of Thermoplastics

The ultimate solution to the problem has been the addition of certain thermoplastics during formulation of the molding composite. These thermo-

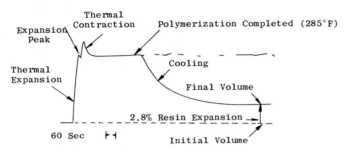

Fig. 3 Dialometric shrinkage measurement with low profile resins. (From Bartkus and Kroekel [16].)

plastics are usually present in only 2–5 wt % on the final molded part or 7–20 wt % of the organic portion of the formulation. This polymer mixture can essentially eliminate the unsaturated polyester polymerization shrinkage as shown by dilatometric measurement (Fig. 3) [16].

The degree of improvement in shrinkage control is dependent on the structure of the thermoplastic, its molecular weight, the amounts used, and the structure of the unsaturated polyester resin employed. However, with proper attention to these factors a "zero-shrinkage" composite can be assembled to yield moldings with excellent surface smoothness, no part warpage, and precise dimensional stability. The proper selection of materials can also allow excellent paint acceptance with minimum substrate preparation.

A simple premix molding formulation with and without a thermoplastic additive and the effects on molding shrinkage is shown in Table III. The effect of thermoplastic presence on physical properties is indicated in Table IV. The surface profiles as measured by a Bendix Surface Analyzer Microcorder are

Table III

Typical Formulation of Premix Molding Composites

|  | Conventional (wt %) | Low profile (wt %) |
|---|---|---|
| Unsaturated polyester resin[a] | 26.4 | 15.9 |
| Thermoplastic solution in styrene[b] | — | 7.9 |
| Styrene | — | 2.6 |
| Calcium carbonate | 53.6 | 53.6 |
| Zinc stearate | 3.0[c] | 3.0[c] |
| t-Butyl perbenzoate | 1.0[c] | 1.0[c] |
| Glass fibers ($\frac{1}{4}$ in. chopped) | 20 | 20 |

[a] Unsaturated polyester alkyd containing approximately 30 wt % styrene.
[b] A 40 wt % solution of thermoplastic in styrene.
[c] Weight percent based on resin.

Table IV

Typical Properties of Premix Formulations (See Table III)

|  | Conventional | Low profile |
|---|---|---|
| Molding conditions |  |  |
| Temperature (°C) | 150 | 150 |
| Time (min) | 2.0 | 2.0 |
| Pressure (psi) | 400 | 400 |
| Properties |  |  |
| Tensile strength (psi) | 6500 | 5700 |
| Flexural strength (psi) | 11,800 | 12,000 |
| Flexural modulus (psi × 10⁶) | 1.78 | 1.53 |
| Notched Izod impact (ft-lb/in.) | 4.4 | 3.8 |
| Water absorption, 24 hr (%) | 0.20 | 0.20 |
| Shrinkage (mils/in.) | 3.5 | 0.0 |
| Surface smoothness |  |  |
| ($\mu$ in waviness/in.) | 2000 | 150 |

shown in Figs. 4 and 5. A comparable measurement of automotive-grade steel is shown in Fig. 6. Note the "low profile" of the microcorder tracing of the surface on the molding containing thermoplastic. The name "low-profile additives" for the thermoplastics was derived from this measurement.

*b. Thermoplastic Structures*

A number of thermoplastics have been found to give varying levels of shrinkage control. Examples of the thermoplastics reported to perform to some degree are poly(vinyl acetates) [18–22], poly(methyl methacrylate)

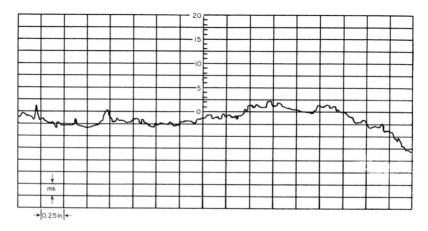

Fig. 4   Surface profile, standard molding *without* low profile.

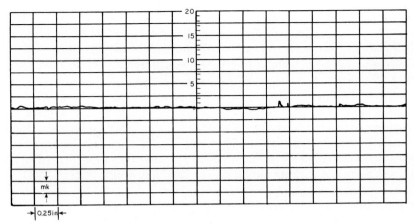

Fig. 5 Surface profile, standard molding *with* low profile.

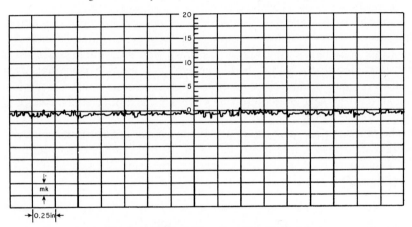

Fig. 6 Surface profile, automotive grade steel.

and copolymers with other acrylates [19, 23–27], vinyl chloride–vinyl acetate copolymers [18, 28, 29], polystyrene and various styrene–containing copolymers [18, 30–33], polyethylene [19], polycaprolactones [18, 34, 35], cellulose acetate 'butyrate [36, 37], and a variety of saturated polyesters and blends of saturated polyesters with poly(vinyl chloride) [38–41]. The addition of these thermoplastics has been effective in shrinkage control with all forms of compounding techniques for compression, transfer and injection molding. Some of these thermoplastics are also effective in pultrusion applications.

### c. Unsaturated Polyester Resin Structure

A wide variety of unsaturated polyester resin types has also been reported in these applications [20, 24, 25, 29, 42, 43]. These vary from those based on

maleic anhydride (1.0 mole) and propylene glycol (1.0 mole) [mol wt of
—C=C— is 156] to ones with (approximately) isophthalic or orthoph-
thalic acid (0.33 mole), maleic anhydride (0.67 mole), propylene glycol
(0.80 mole), and dipropylene glycol (0.20 mole) [mol wt of —C=C— is 275]. In
certain applications halogenated resins are also used.

Recently a different type of resin has been utilized in low-profile–low-
shrink applications. These are vinyl ester resins [32, 44], which are the
reaction products of epoxy resins with unsaturated carboxylic acids. These
resins then are mixed with an unsaturated monomer (e.g., styrene) and
peroxide cured similarly to standard unsaturated polyesters. These resins
can also be mixed with thermoplastics to yield low profile systems.

### d.  Relative Proportions Used

Until recently most of the exterior automotive applications had employed
the maleic anhydride–propylene glycol based resin. Its brittle resin matrix
and the advent of commercially available thermoplastic low-profile additives
that would work with less reactive (and tougher) unsaturated polyester
resin matrixes suggests that this type of resin will decrease in usage. Non-
transportation applications already generally employ an isophthalic acid
modified version. The selection of the polyester resin must be carefully
balanced with the thermoplastic additive for maximum control of polymerization shrinkage.

The amount of styrene monomer in the system is also important to its
overall performance and properties. The various applications govern the
exact amounts used. However, in general, the weight ratios of polyester
alkyd/thermoplastic/styrene are 30–50/10–20/40–50.

### e.  Commercially Utilized Thermoplastics

Obviously, a wide divergence in performance is obtained with these
various thermoplastics, that is, dependent upon their structure and that of the
remainder of the composite. In commercial practice, the poly(vinyl acetates)
and poly(methyl methacrylates) have emerged as the best additives for truly
"zero-shrinkage" low-profile applications. Somewhat lesser performances, but
still considerable improvements over conventional systems, have been found
in polycaprolactones, polystyrenes, and polyethylene, in decreasing order.
Many applications have also been found for these materials.

## IV.  COMMERCIAL APPLICATIONS

Examples of applications in which thermoplastic containing unsaturated
polyesters have been found to be true engineering materials are automotive

parts (e.g., front ends) [45, 46], business machine components [47, 48], power tool housings [49], air conditioner parts [50], appliance parts [51], truck components, farm implements [52], and a variety of other applications [52, 53].

## V. POSSIBLE MECHANISMS OF SHRINKAGE CONTROL

A number of different mechanisms possibly governing shrinkage control in these complex thermoset–thermoplastic systems have been proposed. Some of these have been very specific as to the polyester resin structure employed, which, of course, has an important effect on the total systems performance.

### A. Optical Heterogeneity and Boiling Monomer

One theory [16, 24] indicates that within polyester structures having a molecular weight-to-double-bond ratio of 150–186 microscopic examination

1 cm = 2 μm

Fig. 7　Scanning electron microscope examination of acrylic polymer in a MA–PG polyester (magnification × 500).

of the thermoplastic-cured thermoset system must reveal an "optically heterogeneous" structure (Fig. 7) where distinct thermoplastic domains are dispersed in a cured thermoset matrix. This proposal suggests that during cure unreacted styrene monomer (dissolved in these globules of thermoplastic) boils from these domains, creating an internal pressure, which compensates for the polymerization shrinkage and leaves voids that can be seen within the thermoplastic domains. This work was done with a polyester based on maleic anhydride (1.0 mole) and propylene glycol (1.0 mole) [mol/wt —C=C— is 156] and an acrylic-based thermoplastic. In this system the thermoplastic is incompatible with the polyester resin even before cure.

## B.   Strain Relief through Stress Cracking

Other studies of this system have resulted in suggested modifications of the theory. Pattison et al. [54] have proposed a mechanism by which yet unreacted liquid styrene monomer thermally expands as the polyester–styrene cross-linking occurs to compensate for polymerization shrinkage. As the cross-linking proceeds and very little unreacted styrene remains, these workers propose a second pathway. Strain, due to shrinkage, develops in the system, particularly at the thermoplastic domain–thermoset matrix interface. This strain increases to the point that stress cracking propagates through the weak thermoplastic phase, relieving this strain, forming voids, and compensating for the overall shrinkage. A similar explanation has been offered by Rabenold [55], except his work includes the proposal that the large excess of styrene monomer (compared to conventional molding resins) can result in styrene homopolymerization inside the thermoplastic domains, thus accounting for some void formation and compensation for shrinkage. Walker [56] also suggests that void formation is critical to polymerization shrinkage control.

Mechanism studies have also been expanded to other unsaturated polyester–thermoplastic systems. Pattison et al. [57] have studied a one-component resin system (thermoplastic totally soluble in polyester before molding) based on a polyester resin from phthalic anhydride (0.1 mole), maleic anhydride (0.9 mole), and propylene glycol (1.0 mole), employing poly(vinyl acetate) as the thermoplastic portion. A similar mechanism to that previously proposed for the two-phase (acrylic thermoplastic) system is suggested. One major difference is that the system dramatically differed from the theory of "optical heterogeneity" proposed as necessary for true low profile performance [16, 17]. Microscopic examination [48] clearly showed no distinct thermoplastic domains such as those in Fig. 7.

## C. Thermoplastic Expansion

While the above studies appear reasonable in explaining shrinkage control in a given unsaturated polyester–thermoplastic matrix, they are not addressed to the performance differences observed in even the same polyester resin with different thermoplastic additives. These differences are well known in commercial practice and should be understood.

Atkins and co-workers [58] have studied this problem and have concluded that these performance differences can be explained by careful examination of certain thermoplastic properties. These properties are glass transition temperature ($T_g$), thermal coefficient of expansion, and polarity. Three thermoplastic additives in two unsaturated polyesters were studied: poly(vinyl acetate), a methyl methacrylate–ethyl acrylate copolymer, and polystyrene. The polyesters were based on (1) maleic anhydride (1.0 mole) and propylene glycol (1.0 mole), mol wt —C=C— of 156, and (2) isophthalic acid (0.25 mole), maleic anhydride (0.75 mole) and propylene glycol (1.0 mole), mol wt —C=C of 224. A critical molding formulation was used to evaluate these materials to simulate differences seen in complex commercial molds. In every case the polyester alkyd/thermoplastic/styrene weight ratio was held constant at 42/10/48.

In both unsaturated polyester resins the shrinkage control was best with the poly(vinyl acetate), next best with the acrylic based thermoplastic, and considerably worse with polystyrene. Still shrinkage with the polystyrene was much less than that with no thermoplastic present (Tables V and VI).

Table V

Critical Molding Formulation[a]

|  | Parts by weight |
|---|---|
| Polyester alkyd | 42 |
| Thermoplastic additive | 10 |
| Styrene | 48 |
| Calcium carbonate (filler) | 175 |
| *t*-Butyl perbenzoate (catalyst) | 1.2 |
| Zinc stearate (mold release) | 4.0 |
| Glass fiber (wt %) | 10.0 |

[a] Molding conditions: 150°C/2.0 min/500 psi.

Scanning electron microscope examination of each of the cured resin matrixes in the absence of fillers and reinforcements was conducted. Only the acrylic additive with Polyester 1 showed the "optically heterogeneous" structure previously proposed as necessary for low-profile behavior [16, 24]

(Fig. 7). With Polyester (1) the poly(vinyl acetate) gave an "optically homogeneous" appearance (Fig. 8). Higher-magnification examination revealed the presence of $< 1$-$\mu$m diameter voids. Torsion pendulum studies (Fig. 9) were required to demonstrate that the thermoplastic–thermoset components were

Table VI

Mold Shrinkage with Formulation of Table V

| Polyester (mol wt —C=C) | Thermoplastic | Mold shrinkage (mils/inch) |
|---|---|---|
| 156 | Poly(vinyl acetate) | 0.3 |
| 156 | Poly(methyl methacrylate) | 0.9 |
| 224 | Poly(vinyl acetate) | 0.7 |
| 224 | Poly(methyl methacrylate) | 1.4 |
| 224 | Polystyrene | 2.9 |
| 156 | None | 6.0 |
| 224 | None | 7.0 |

1 cm = 2 $\mu$m

Fig. 8 Scanning electron microscope examination of poly(vinyl acetate) in a MA–PG polyester (magnification × 500).

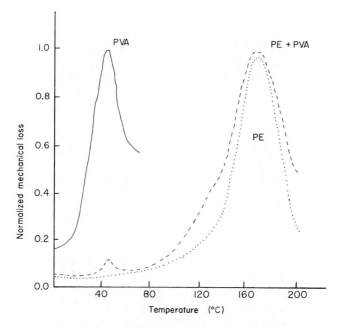

Fig. 9  Mechanical loss spectrum of poly(vinyl acetate) in a MA–PG polyester.

incompatible (see Volume 1, Chapter 5). Initial interpretation could be that voids localized in distinct domains were less effective than generally dispersed voids in controlling shrinkage. Studies with Polyester (2) showed that this was untrue.

The acrylic additive in Polyester (2) gave the same cured resin optical appearance as the poly(vinyl acetate) polymer did in Polyester (1) (Fig. 10). The poly(vinyl acetate) system also retained the same appearance as in Polyester (1) (Fig. 11) with a general distribution of voids. Still the shrinkage control was superior with the poly(vinyl acetate).

These differences in ability to control shrinkage appear to be best explained by the thermal coefficients of expansion of these thermoplastics [59]. A graph of the specific volume of the thermoplastic versus temperature is given in Fig. 12. The inflection point represents the $T_g$ (measured on the torsion pendulum) of each material. At every temperature, it is evident that the poly(vinyl acetate) occupies more volume than the acrylic polymer because of its higher thermal coefficient of expansion. This evidence suggests that differences in the ability of the thermoplastic to expand with the application of heat is important to its ability to control shrinkage. Obviously, the thermoplastic must become chemically incompatible with the thermosetting system before this effect can be realized.

⊢⎯⎯⎯⎯⊣
1 cm = 2 $\mu$m

Fig. 10   Scanning electron microscope examination of the acrylic polymer with isophthalic polyester (magnification × 500).

Polystyrene as the thermoplastic additive is an illustration of another principle of the importance of thermoplastic properties. Figure 12 shows that the specific volume of polystyrene at any temperature is greater than either that of poly(vinyl acetate) or the acrylic material. Yet it is not nearly as effective as either of these materials for controlling shrinkage. Some explanation for this can be seen in the microscopic examination of this material in both Polyesters (1) and (2) (Figs. 13 and 14). Particularly with Polyester (2) the matrix appears visibly compatible and no voids are evident even at a magnification of 5000. Yet some incompatibility was noted by torsion pendulum studies.

The factor of greatest influence here is suggested to be polarity. Before cross-linking, the unsaturated polyester is reasonably polar (dipole moment 2.0–2.5), while after cross-linking the introduction of styrene segments lowers the polarity (e.g., dipole moment approximately 0.8). Both poly(vinyl acetate) and the acrylic polymer are polar (dipole moments 1.6 and 1.3, respectively), while polystyrene is relatively nonpolar (dipole moment 0.3). Therefore, as the polyester cross-links, the polar thermoplastics have a greater driving force to become incompatible and thus to be available for expansion to compensate for polymerization shrinkage. Conversely, the polystyrene, being nonpolar, has less of a tendency to become incompatible with the thermosetting system; therefore, lesser amounts of it are available for expansion and compensation of shrinkage.

These workers feel that void formation is due to styrene homopolymerization in the thermoplastic phase and shrinkage of the thermoplastic phase away from the thermoset phase during cooling of the matrix from curing temperature. The greater "compatibility" of the polystyrene systems is also believed to be the reason that these composites can accept internal pigmentation and yield uniform colors, while use of the more polar additives normally will not allow this approach.

1 cm = 2 $\mu$m

Fig. 11   Scanning electron microscope examination of poly(vinyl acetate) polymer with isophthalic polyester (magnification × 500).

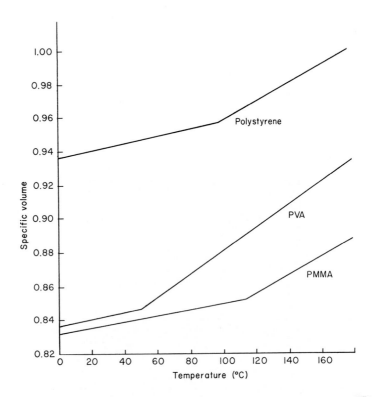

Fig. 12   Volume-temperature relationship for poly(methyl methacrylate) (PMMA), poly(vinyl acetate) (PVA), and polystyrene.

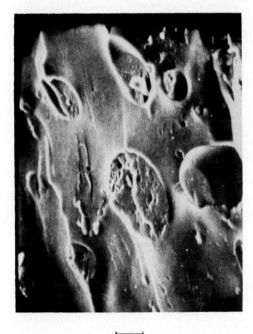

$\vdash\!\!\!-\!\!\!\dashv$
1 cm = 2 $\mu$m

Fig. 13  Scanning electron microscope examination of polystyrene with MA–PG polyester (magnification × 500).

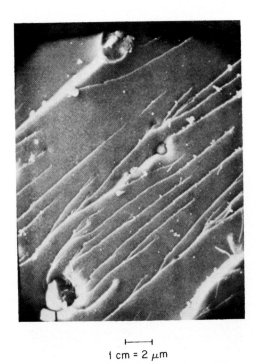

├────┤
1 cm = 2 μm

Fig. 14    Scanning electron microscope examination of polystyrene with isophthalic polyester (magnification × 500).


## VI.    PERFORMANCE IN CHEMICALLY THICKENED SYSTEMS

The same principles of mechanism apply to molding compounds that are chemically thickened such as sheet molding and bulk molding compounds (SMC and BMC). In this chemical thickening process the residual carboxyl groups are reacted with inorganic materials such as magnesium oxide, magnesium hydroxide, calcium oxide, and calcium hydroxide, and matured at 70–100°F to raise the viscosity of the compound before molding. The usefulness of this process was described in Section II.B.2. In many cases, the thermoplastic, low-profile additive can separate from the thickening unsaturated polyester during this process and exude to the surface of the composite. The shrinkage control of such a composite is generally unchanged, but potential molding problems can be incurred.

To obviate these possible problems the thermoplastics are generally modified with carboxyl groups [20, 60–62] to allow participation in the thickening reaction and, hence, avoid exudation.

A subtle danger lies in this approach and has been overlooked until recently. If the structure of the polyester resin, the thermoplastic additive, and the thickening agent are not carefully balanced, a reduction in shrinkage control, sink resistance, increased long-term waviness, etc. can result [63]. The reason that the traditional "low-profile" characteristics can be reduced as this; if the carboxyl groups on the thermoplastic are bound too strongly with the carboxyl groups on the polyester through the thickening agent, less of the thermoplastic will become incompatible upon cure. Obviously, the type of polyester and thermoplastic structure is very important here. Some workers [64] have suggested that thickening has an effect on cure rate.

Other factors are also very important in making this "polymer blend" approach commercially successful. All components of the composite contribute to its utility and each must be carefully selected to assemble the total system. Molding parameters (pressure, closure time, temperature, etc.) and mold design [63, 65] are also critical, and these must be carefully considered.

## VII. NEW DEVELOPMENTS

Many new developments are currently underway with low-profile–low-shrink polyester composites. These include energy absorbing systems utilizing resilient polyester resins [66], low-profile composites to yield even lower-density parts [67], ultra-high-strength materials of high glass content [68], conductive parts containing carbon fiber directly from the press for electrostatic spraying and radiofrequency interference (RFI) suppression [69] and granular solid polyester molding compounds directly usable in typical thermoset injection molding equipment [70].

## VIII. SUMMARY

Polyblends of specially selected thermoplastics and thermosetting, unsaturated polyesters are combined to yield fiber-reinforced composites with true engineering properties after compression, transfer, or injection molding.

The thermoplastic present, about 10–20% of the amount of the thermoset, apparently becomes chemically incompatible with the thermoset system as it cures. The thermoplastic expands under the curing conditions to contribute to compensation for the polymerization shrinkage of the unsaturated polyester resin. Thermoplastic properties that are important are thermal coefficient of expansion, $T_g$, polarity, and molecular weight. Generally unsaturated polyesters with a potentially high glass transition temperature (or a high cross-link density) perform best. Complex moldings with excellent dimensional stability, over a wide temperature range, esthetic surfaces, high heat distortion resistance, good insert holding properties, and high strengths (including impact resistance) are being produced commercially with these composites.

## REFERENCES

1.  H. U. Boenig, "Unsaturated Polyesters." Elsevier, Amsterdam, 1964.
2.  B. Parkyn, F. Lamb, and B. V. Clifton, "Polyesters," Vol. 2. London Iliffe Books, New York and American Elsevier, New York, 1967.
3.  M. W. Gaylord, "Reinforced Plastics—Theory and Practice," 2nd ed. Cahners Publ., 1974.
4.  W. S. Penn, "GRP Technology," Maclaren, London, 1967.
5.  K. N. Warner, *Annu. Tech. Conf., Reinforced Plast. Composites Inst., 28th, Plast. Ind.* Reprint Sect. 19E, pp. 1–12 (1973).
6.  F. Fekete, *Annu. Tech. Conf., Reinforced Plast. Composites Inst., 27th, Soc. Plast. Ind.* Preprint Sect. 12-D, pp. 1–24 (1972).
7.  PPG Industries U.S. Patent 3,631,217 (December 28, 1971).
8.  F. B. Alvey, *J. Polym. Sci. Part A-1* **9**, 2233–2245 (1971).
9.  I. Vancsco-Szmercsanyi, *Kunststoffe* **60**(12), 1066–1070 (1970).
10. W. Schönthaler, *Lethmathei. W Kunststoffe* **60**, 951–953 (1970).
11. E. J. Bartkus and C. H. Kroekel, *Appl. Polym. Symp. No. 15* pp. 113–135 (1970).
12. H. J. Boos and K. R. Hauschildt, *Erlangen Kunststoffe* **63**, 181–184 (1973).
13. British Industrial Plastics, British Patent 936,351 (September 11, 1963).
14. Deutsche Gold Und Silber-Scheideanstalt, British Patent 937,703 (September 25, 1963).
15. N. A. Cutler, J. E. Ferriday, and F. J. Parker, *Annu. Meeting Reinforced Plast. Divi. 17th, Soc. Plast. Ind.* Sect. 5-F, pp. 1–16 (1962).
16. E. J. Bartkus and C. H. Kroekel, *Appl. Polym. Symp. No. 15*, pp. 113–115 (1970).
17. H. J. Von Boob and K. R. Hauschildt, *Kunststoffe* **63**, 181–184 (1973).
18. K. E. Atkins, M. A. Harpold, L. R. Comstock, and P. L. Smith, *Annu. Tech. Conf., 28th, Reinforced Plast. Composite Inst.* Paper (1973).
19. British Industrial Plastics, British Patent 936,351 (September 11, 1963).
20. Union Carbide Corp., U.S. Patent 3,718,714 (February 27, 1973).
21. Freeman Chemical, Ltd., British Patent 1,321,683 (June 27, 1973).
22. Chemische Werke Huls, German Patent 2,163,089 (June 20, 1973).
23. Chemische Werke Albert, Netherlands Patent 72-03034 (September 12, 1972).
24. Rohm and Haas Co., U.S. Patent 3,701,748 (October 31, 1972).
25. Rohm and Haas Co., U.S. Patent 3,772,241 (November 3, 1973).
26. Toyo Spinning Co., Japanese Patent 4,851,090 (July 18, 1973).
27. Toyo Spinning Co., Japanese Patent 4,849,884 (July 13, 1973).
28. Union Carbide Corp., German Patent 2,104,575 (August 12, 1971).
29. Diamond Shamrock Corp., U.S. Patent 3,721,642 (March 20, 1973).
30. W. R. Grace and Co., U.S. Patent 3,503,921 (March 1, 1970).

31. Koppers Co., Netherlands Patent 70-15386 (April 26, 1971).
32. Dow Chemical Co., U.S. Patent 3,674,893 (July 4, 1972).
33. Takeda Chemical Ind., German Patent 2,252,972 (May 24, 1973).
34. Union Carbide Corp., U.S. Patent 3,549,586 (December 22, 1970).
35. Union Carbide Corp., U.S. Patent 3,688,178 (June 6, 1972).
36. W. W. Blount, R. H. Calendine, and J. H. Davis, *Annu. Tech. Conf., 27th, Reinforced Plast./ Composite Inst.* Paper 12-C (1972).
37. Rohm and Haas Co., U.S. Patent 3,642,672 (February 15, 1972).
38. Distillers Co. Ltd., U.S. Patent 3,489,707 (January 13, 1970).
39. Reichold, Japanese Patent 4,601,783 (October 10, 1971).
40. Koppers Co., Netherlands Patent 70-14568 (June 17, 1971).
41. Toyo Shibaura Electric Co. Ltd., U.S. Patent 3,736,728 (May 29, 1973).
42. Chemische Werke Albert, Netherlands Patent 72-03034 (September 12, 1972).
43. Rohm and Haas Company, German Patent 2,203,199.
44. J. E. Cutshall and D. W. Pennington, *Annu. Tech. Conf., 27th, Reinforced Plast./Composite Inst.* Paper 15-A (1972).
45. *Mod. Plast.* pp. 46–47 (November 1971).
46. *Plast. World* p. 18 (December 1971).
47. F. W. Tortolano, *Plast. World* pp. 40–42 (May 19, 1975).
48. R. R. Thomas, *Annu. Tech. Conf., 31st, Reinforced Plast./Composite Inst.* Paper 1-C (1976).
49. A. Hall, *Plast. World* pp. 38–41 (December 16, 1974).
50. *Plast. World* p. 10 (January 21, 1975).
51. J. Cross and C. G. Nenadal, *Annu. Tech. Conf., 31st, Reinforced Plast./Composite Inst.* Paper 1-E (1976).
52. *Mod. Plast.* pp. 58–61 (February 1973).
53. *Plast. Eng.* pp. 49–50 (August 1975).
54. V. A. Pattison, R. R. Hindersinn, and W. T. Schwartz, *SPE J.* **19**, 553 (May 7, 1973).
55. R. Rabenold, *Annu. Tech. Conf., 27th, Reinforced Plast./Composite Inst.* Paper 15-E (1972).
56. A. C. Walker, *SPE Tech. Paper* **17**, 454 (1971).
57. V. A. Pattison, Raymond R. Hindersinn, and Willis T. Schwartz, *J. Appl. Polym. Sci.* **19**, 3045–3050 (1975).
58. K. E. Atkins, J. V. Koleske, P. L. Smith, E. R. Walker, and V. E. Matthews, *Annu. Tech. Conf., 31st, Reinforced Plast./Composite Inst.* Paper 2-E (1976).
59a. J. Brandrup and E. H. Immergut, "Polymer Handbook." Wiley (Interscience), New York, 1966.
59b. R. H. Boundy and R. F. Boyer, "Styrene, Its Polymers, Copolymers and Derivatives." Van Nostrand–Reinhold, Princeton, New Jersey, 1952.
59c. R. C. Spencer and G. D. Gilmore, *J. Appl. Phys.* **20**, 502 (1949).
59d. R. J. Clash, Jr., and L. M. Rynkiewicz, *Ind. Eng. Chem.* **36**(3), 279–282 (1944).
60. Diamond Shamrock, Netherlands Patent 72-08269 (December 19, 1972).
61. Rohm and Haas Co., Belgian Patent 740,581 (April 21, 1970).
62. Kuraray Co. Ltd., Japanese Patent 73-21,788 (March 19, 1973).
63. H. Jeff Boyd, *Annu. Tech. Conf., 31st, Reinforced Plast./Composite Inst.* Paper 2-C (1976).
64. H. Kubota, *J. Appl. Polym. Sci.* **19**, 2279–2297 (1975).
65. R. B. Jutte, *Soc. Automotive Eng. Meeting* (January 10–12, 1973).
66. J. T. Shreve and F. E. Tropp, *Annu. Tech. Conf., 31st, Reinforced Plast./Composite Inst.* Paper 2-A (1976).
67. G. A. Sundstrom, J. Collister, and C. W. Hays, *Anniversary Tech. Conf., 30th, Reinforced Plast./Composite Inst.* Paper 1-B (1975).
68. J. Maaghul and E. J. Potkanowicz, *Annu. Tech. Conf., 31st, Reinforced Plast./Composite Inst.* Paper 7-C (1976).

69. K. E. Atkins, R. R. Gentry, and D. C. Hiler, *SPE National Technical Conf. High Performance Plastics* p. 148 (October 1976).
70. R. D. Lake, H. T. Shreve, and R. L. Lovell, *Annu. Tech. Conf., 31st, Reinforced Plast./Composite Inst.* Paper 13-C (1976).

# Appendix

# Conversion Factors to SI Units

FORCE (newton, N)
$$1 \text{ N} = 1 \text{ m·kg/sec}^2$$
$$= 10^5 \text{ dyn}$$
$$= 0.1020 \text{ kgf}$$
$$= 0.2248 \text{ lbf}$$

PRESSURE OR STRESS (pascal, Pa)
$$1 \text{ Pa} = 1 \text{ N/m}^2$$
$$= 10.0 \text{ dyn/cm}^2$$
$$1 \text{ kPa} = 7.50 \text{ mmHg} = 7.50 \text{ Torr}$$
$$1 \text{ MPa} = 10.00 \text{ bar}$$
$$= 10.20 \text{ kg/cm}^2$$
$$= 145.0 \text{ psi}$$

ENERGY (joule, J)
$$1 \text{ J} = \text{N·m}$$
$$= 10^7 \text{ erg}$$
$$= 0.2387 \text{ cal}$$
$$1 \text{ kJ} = 0.9471 \text{ Btu}$$
$$1 \text{ MJ} = 0.2778 \text{ kWh}$$

FORCE PER UNIT LENGTH
$$1 \text{ mN/m} = 1 \text{ dyn/cm}$$
$$1 \text{ kN/m} = 5.710 \text{ lbf/in}$$

VISCOSITY
$$1 \text{ Pa·sec} = 10.00 \text{ P}$$
$$1 \text{ mPa·sec} = 1.000 \text{ cP}$$
$$1 \text{ m}^2\text{/sec} = 10^4 \text{ stokes}$$

# Index to Volume 1

# Index to Volume 2